T0329538

Developments in Porous, Biological and Geopolymer Ceramics

Developments in Porous, Biological and Geopolymer Ceramics

A Collection of Papers Presented at the 31st International Conference on Advanced Ceramics and Composites January 21–26, 2007 Daytona Beach, Florida

Editors

Manuel Brito
Eldon Case
Waltraud M. Kriven

Volume Editors

Jonathan Salem
Dongming Zhu

The American Ceramic Society

BICENTENNIAL
1807
WILEY
2007
BICENTENNIAL

WILEY-INTERSCIENCE
A John Wiley & Sons, Inc., Publication

Published by John Wiley & Sons, Inc., Hoboken, New Jersey.
Published simultaneously in Canada.

For general information on our other products and services or for technical support, please contact our Customer Care Department within the United States at (800) 762-2974, outside the United States at (317) 572-3993 or fax (317) 572-4002.

Wiley also publishes its books in a variety of electronic formats. Some content that appears in print may not be available in electronic format. For information about Wiley products, visit our web site at www.wiley.com.

Wiley Bicentennial Logo: Richard J. Pacifico

Library of Congress Cataloging-in-Publication Data is available.

ISBN 978-0-470-19640-3

10 9 8 7 6 5 4 3 2 1

Contents

Preface   ix

Introduction   xi

POROUS CERAMICS

Hierarchical Porosity Ceramic Components from Preceramic   3
Polymers
    P. Colombo, L. Biasetto, E. Bernardo, S. Costacurta, C. Vakifahmetoglu, R. Peña-
    Alonso, G.D. Sorarù, E. Pippel, and J. Woltersdorf

Electrophoretic Deposition of Particle-Stabilized Emulsions   13
    Bram Neirinck, Jan Fransaer, Omer Van der Biest, and Jef Vleugels

Application of Porous Acicular Mullite for Filtration of Diesel Nano 27
Particulates
    Cheng G. Li and Aleksander J. Pyzik

Microstructural Development of Porous β-Si$_3$N$_4$ Ceramics Prepared   41
by Pressureless-Sintering Compositions in the Si-Re-O-N
Quaternary Systems (Re=La, Nd, Sm, Y, Yb)
    Mervin Quinlan, D. Heard, Kevin P. Plucknett, Liliana Garrido, and
    Luis Genova

Compositional Design of Porous β-Si$_3$N$_4$ Prepared by   49
Pressureless-Sintering Compositions in the Si-Y-Mg-(Ca)-O-N
System
    Mervin Quinlan, Kevin P. Plucknett, Liliana Garrido, and Luis Genova

Aligned Pore Channels in 8mol%Yttria Stabilized Zirconia by   57
Freeze Casting
    Srinivasa Rao Boddapati and Rajendra K. Bordia

Preparation of a Pore Self-Forming Macro-/Mesoporous Gehlenite   67
Ceramic by the Organic Steric Entrapment (PVA) Technique
    Dechang Jia, DongKyu Kim, and Waltraud M. Kriven

Nonlinear Stress-Strain Behavior Evaluation of Porous Ceramics 77
with Distributed-Micro-Crack Model
Hidenari Baba and Akihiko Suzuki

Thermal Shock Behavior of NITE-Porous SiC Ceramics 89
Yi-Hyun Park, Joon-Soo Park, Tatsuya Hinoki, and Akira Kohyama

Optimization of the Geometry of Porous SiC Ceramic Filtering 95
Modules Using Numerical Methods
S. Alexopoulos, G. Breitbach, B. Hoffschmidt, and P. Stobbe

Weibull Statistics and Scaling Laws for Reticulated Ceramic 105
Foam Filters
Mark A. Janney

Effect of Test Span on Flexural Strength Testing of Cordierite 113
Honeycomb Ceramic
Randall J. Stafford and Rentong Wang

BIOCERAMICS

Biomaterials Made with the Aid of Enzymes 127
Hidero Unuma

Use of Vaterite and Calcite in Forming Calcium Phosphate 135
Cement Scaffolds
A. Cuneyt Tas

Deposition of Bone-Like Apatite on Polyglutamic Acid Gels in 151
Biomimetic Solution
Toshiki Miyazaki, Atsushi Sugino, and Chikara Ohtsuki

Surface Structure and Apatite-Forming Ability of Polyethylene 159
Substrates Irradiated by the Simultaneous Use of Oxygen Cluster
and Monomer Ion Beams
M. Kawashita, S. Itoh, R. Araki, K. Miyamoto, and G.H. Takaoka

Conversion of Borate Glass to Biocompatible Phosphates in 171
Aqueous Phosphate Solution
Mohamed N. Rahaman, Delbert E. Day, and Wenhai Huang

Fabrication of Nano-Macro Porous Soda-Lime Phosphosilicate 183
Bioactive Glass by the Melt-Quench Method
Hassan M.M. Moawad and Himanshu Jain

Pores Needed for Biological Function Could Paradoxically Boost 197
Fracture Energy in Bioceramic Bone Tissue Scaffolds
Melissa J. Baumann, I.O. Smith, and Eldon D. Case

Novel Routes to Stable Bio-Templated Oxide Replicas 209
Y. Cheng, L. Courtney, and P.A. Sermon

GEOPOLYMERS

Role of Alkali in Formation of Structure and Properties of a 221
Ceramic Matrix
Pavel Krivenko

Will Geopolymers Stand the Test of Time? 235
John L. Provis, Yolandi Muntingh, Redmond R. Lloyd, Hua Xu,
Louise M. Keyte, Leon Lorenzen, Pavel V. Krivenko, and
Jannie S.J. van Deventer

New Trends in the Chemistry of Inorganic Polymers for Advanced 249
Applications
Kenneth J.D. MacKenzie, Suwitcha Komphanchai, and Ross A. Fletcher

Influence of Geopolymer Binder Composition on Conversion 257
Reactions at Thermal Treatment
A. Buchwald, K. Dombrowski, and M. Weil

Laser Scanning Confocal Microscopic Analysis of Metakaolin- 273
Based Geopolymers
Jonathan L. Bell, Waltraud M. Kriven, Angus P.R. Johnson, and Frank Caruso

Synthesis and Characterization of Inorganic Polymer Cement from 283
Metakaolin and Slag
L. Cristina Soare and A. Garcia-Luna

Recent Development of Geopolymer Technique in Relevance to 293
Carbon Dioxide and Waste Management Issues
Ko Ikeda

Aqueous Leachability of Geopolymers Containing Hazardous 309
Species
D.S. Perera, S. Kiyama, J. Davis, P. Yee, and E.R. Vance

Practical Applications of Geopolymers 321
Chett Boxley, Balky Nair, and P. Balaguru

Geopolymer-Jute Composite: A Novel Environmentally Friendly 337
Composite with Fire Resistant Properties
Amandio Teixeira-Pinto, Benjamin Varela, Kunal Shrotri, Raj S. Pai Panandiker,
and Joseph Lawson

Design, Properties, and Applications of Low-Calcium Fly 347
Ash-Based Geopolymer Concrete
B. Vijaya Rangan

Rate of Sulphuric Acid Penetration in a Geopolymer Concrete 363
Xiujiang Song, Marton Marosszeky, and Zhen-Tian Chang

Corrosion Protection Assessment of Concrete Reinforcing Bars with 373
a Geopolymer Coating
W.M. Kriven, M. Gordon, B.L. Ervin, and H. Reis

Author Index 383

Preface

This issue contains a collection of 32 papers on the topics of porous ceramics, bioceramics and geopolymers that were presented during the 31st International Conference on Advanced Ceramics and Composites, Daytona Beach, FL, January 21–26, 2007. Specifically, papers were submitted from the following 3 Symposia/Focused Sessions:

Symposium on Porous Ceramics: Novel Developments and Applications

This volume of the Ceramic Engineering and Science Proceedings incorporates papers presented at the Symposium on Porous Ceramics: Novel Developments and Applications within the 31st International Conference and Exposition on Advanced Ceramics and Composites, held in Daytona Beach, Florida, January 21–26, 2007. The previous symposia or focused sessions in this series were held in Cocoa Beach, Florida since 2000, which demonstrate a continuous interest in this matter. The objective of this symposia series is to provide an international forum for discussion of research results and developments related to a wide variety of ceramic materials with different applications from the perspective of porosity -its structure, control and functionality- as the important parameter.

The papers in this issue cover several aspects of porosity in ceramics, including development of systems, materials for different applications and their processing and performance. These papers taken together provide up-to-date comprehensive information on the status of the technology and related research activities for the community.

The editor thanks Professor Paolo Colombo, University of Padova and Dr. Gary Crosbie, Ford Motor Company, for their assistance in organizing and managing the symposium, to the authors and reviewers for their hard-work, and to The American Ceramic Society staff for their assistance in publishing this issue.

MANUEL BRITO
National Institute of Advanced Industrial Science and Technology (AIST)

Bioceramics and Biocomposites

The bioceramics papers included in this CESP issue represent a cross-section of current bioceramics research. For example, the use of enzymes to control the synthesis and properties of biomaterials is included along with a discussion of techniques for the rapid formation of apatitic calcium phosphate from calcium carbonate precursors. Also, two papers deal with bioceramic/polymeric composites, including organic/inorganic hybrid bioactive materials and techniques to aid in the attachment of polymeric ligament replacement materials to bone. Another two papers address research in bioactive glasses including the conversion of borate glass to biocompatible phosphates and the engineering of the porosity in bulk phosphosilicate glasses. The links between pores size distributions and biomedical/mechanical properties are also explored as well as the formation of oxide overcoats on geometrically complex bioforms. Thus, this issue highlights some of the most recent advances in the critical areas of bioceramic synthesis, ceramic/polymer hybrids, bioactive glasses, the function of porosity in engineered bone tissue, and biotemplated replicas.

ELDON CASE
Michigan State University

Geopolymers

The 31st International Conference on Advanced Ceramics and Composites is the fourth occasion in which a focused session on geopolymers was hosted by The American Ceramic Society (ACerS). The first took place in 2003 during the ACerS 105th Annual Meeting in Nashville, Tennessee; the second session was part of ACerS 106th Annual Meeting in Indianapolis, and the third was part of the ACerS 107th Annual meeting held in Baltimore, MD, 2005. Papers from these geopolymer focused sessions were published in the ACerS Ceramic Transactions Series, Volumes 153, 165 and 175.

In 2007 the Geopolymer Focused Session was moved to the 31st International Conference and Exposition on Advanced Ceramics and Composites, leading to the papers published in this issue of the Ceramic Engineering and Science Proceedings. In this focused session there were 22 speakers during the one and a half day focused session and the resulting 13 papers are published in this issue. This relatively large, international gathering of geopolymer researchers was directly due to the generous financial support for speakers, provided by the US Air Force Office of Scientific Research through Dr. Joan Fuller, Program Director of Ceramic and Non-Metallic Materials, Directorate of Aerospace and Materials Science.

WALTRAUD M. KRIVEN
University of Illinois at Urbana-Champaign

Introduction

2007 represented another year of growth for the International Conference on Advanced Ceramics and Composites, held in Daytona Beach, Florida on January 21-26, 2007 and organized by the Engineering Ceramics Division (ECD) in conjunction with the Electronics Division (ED) of The American Ceramic Society (ACerS). This continued growth clearly demonstrates the meetings leadership role as a forum for dissemination and collaboration regarding ceramic materials. 2007 was also the first year that the meeting venue changed from Cocoa Beach, where it was originally held in 1977, to Daytona Beach so that more attendees and exhibitors could be accommodated. Although the thought of changing the venue created considerable angst for many regular attendees, the change was a great success with 1252 attendees from 42 countries. The leadership role in the venue change was played by Edgar Lara-Curzio and the ECD's Executive Committee, and the membership is indebted for their effort in establishing an excellent venue.

The 31st International Conference on Advanced Ceramics and Composites meeting hosted 740 presentations on topics ranging from ceramic nanomaterials to structural reliability of ceramic components, demonstrating the linkage between materials science developments at the atomic level and macro level structural applications. The conference was organized into the following symposia and focused sessions:

- Processing, Properties and Performance of Engineering Ceramics and Composites
- Advanced Ceramic Coatings for Structural, Environmental and Functional Applications
- Solid Oxide Fuel Cells (SOFC): Materials, Science and Technology
- Ceramic Armor
- Bioceramics and Biocomposites
- Thermoelectric Materials for Power Conversion Applications
- Nanostructured Materials and Nanotechnology: Development and Applications
- Advanced Processing and Manufacturing Technologies for Structural and Multifunctional Materials and Systems (APMT)

- Porous Ceramics: Novel Developments and Applications
- Advanced Dielectric, Piezoelectric and Ferroelectric Materials
- Transparent Electronic Ceramics
- Electroceramic Materials for Sensors
- Geopolymers

The papers that were submitted and accepted from the meeting after a peer review process were organized into 8 issues of the 2007 Ceramic Engineering & Science Proceedings (CESP); Volume 28, Issues 2-9, 2007 as outlined below:

- Mechanical Properties and Performance of Engineering Ceramics and Composites III, CESP Volume 28, Issue 2
- Advanced Ceramic Coatings and Interfaces II, CESP, Volume 28, Issue 3
- Advances in Solid Oxide Fuel Cells III, CESP, Volume 28, Issue 4
- Advances in Ceramic Armor III, CESP, Volume 28, Issue 5
- Nanostructured Materials and Nanotechnology, CESP, Volume 28, Issue 6
- Advanced Processing and Manufacturing Technologies for Structural and Multifunctional Materials, CESP, Volume 28, Issue 7
- Advances in Electronic Ceramics, CESP, Volume 28, Issue 8
- Developments in Porous, Biological and Geopolymer Ceramics, CESP, Volume 28, Issue 9

The organization of the Daytona Beach meeting and the publication of these proceedings were possible thanks to the professional staff of The American Ceramic Society and the tireless dedication of many Engineering Ceramics Division and Electronics Division members. We would especially like to express our sincere thanks to the symposia organizers, session chairs, presenters and conference attendees, for their efforts and enthusiastic participation in the vibrant and cutting-edge conference.

ACerS and the ECD invite you to attend the 32nd International Conference on Advanced Ceramics and Composites (http://www.ceramics.org/meetings/daytona2008) January 27 - February 1, 2008 in Daytona Beach, Florida.

JONATHAN SALEM AND DONGMING ZHU, Volume Editors
NASA Glenn Research Center
Cleveland, Ohio

Porous Ceramics

HIERARCHICAL POROSITY CERAMIC COMPONENTS FROM PRECERAMIC POLYMERS

P. Colombo[1]*, L. Biasetto, E. Bernardo, S. Costacurta, C. Vakifahmetoglu
University of Padova, Dipartimento di Ingegneria Meccanica – Settore Materiali
via Marzolo, 9
35131 Padova, Italy
[1]Department of Materials Science and Engineering, The Pennsylvania State University
University Park
PA 16802, USA

R. Peña-Alonso, G.D. Sorarù
University of Trento, Dipartimento di Ingegneria dei Materiali e Tecnologie Industriali
via Mesiano 77
38050 Trento, Italy

E. Pippel, J. Woltersdorf
Max Planck Institut für Mikrostrukturphysik
Weinberg 2
06120 Halle, Germany

ABSTRACT

Cellular SiOC ceramic components with a porosity > 70 vol% and cell size dimension ranging from 10 µm to a few mm approximately have been developed using a silicone resin. These cellular materials have been further functionalized by developing a high specific surface area (SSA), which is useful for applications such as absorbers or catalyst supports. Several approaches were followed to achieve this goal, including controlling the heating process to retain the transient porosity, depositing a SiO_2-based meso-structured coating on the surface of the SiOC cellular material, or etching the foams in HF. The samples were characterized using various techniques, including SEM, high resolution and analytical TEM methods and N_2 adsorption-desorption. The SSA value produced varied according to the processing strategy followed.

INTRODUCTION

Porous ceramic components, in particular cellular ceramics, find use in a very broad range of engineering applications, for instance as filters for molten metal or particulate in gas streams, as catalyst support, as load-bearing lightweight structures or as scaffolds for biomedical applications. Most of these applications have rather strict requirements in terms both of the amount of porosity and of its morphology (cell size, cell window size, interconnectivity between the cells) that can be met by choosing the appropriate fabrication method [1,2].

The development of components with hierarchical porosity (micro-, meso- and macro-porosity) is of particular interest because they possess at the same time high permeability and tortuosity of the flow paths (improving mixing and heat transfer), provided by the macro-pores (with a dimension d > 50 nm), and a high specific surface area (SSA), given by the micro- (d < 2 nm) and meso-pores (2 < d < 50 nm). Components with hierarchical porosity can thus be employed in separation, gas storage, removal of pollutants and catalysis applications.

---

* Member, The American Ceramic Society. Corresponding author; paolo.colombo@unipd.it

Preceramic polymers allow the fabrication of cellular ceramics (micro- and macro-cellular foams [3,4]) as well as of components with high SSA [5,6], taking advantage both of the shaping possibilities given by the use of a polymeric material and of the transient porosity that is generated upon pyrolysis. Moreover, functionalization of the porous components is possible, by simply adding suitable fillers to the preceramic polymers, in order to enhance/modify some physical properties (for instance electrical conductivity or magnetic properties) [7].

In this paper, we briefly discuss the use of several possible approaches for developing a high specific surface area in open-cell macro-porous ceramic components obtained from preceramic polymers.

EXPERIMENTAL

SiOC microcellular foams were prepared using a preceramic polymer (MK Wacker-Chemie GmbH, Germany) and poly-methylmetacrylate (PMMA) microbeads (Altuglas BS, Altuglas International, Arkema Group, Rho (MI), Italy) of nominal size ranging from 10 to 185 μm acting as sacrificial filler [4]. The powders were mixed at a constant weight ratio (20 wt% MK, 80 wt% PMMA) by ball milling for 1 hour and then warm pressed (130 to 180°C, 20 MPa). The warm pressing temperature was adjusted as a function of the PMMA microbeads size, in order to optimize the viscous flow of the molten polymer through the beads and the degree of crosslinking of the preceramic polymer. The green samples were then treated in air at 300°C for 2h (heating rate = 0.5 °C·min^{-1}) in order to burn out the PMMA microbeads and allow for the crosslinking of the preceramic polymer. Macro-cellular foams were produced by direct foaming according to the procedure reported in reference 3. All the foams, in their polymeric stage, were pyrolyzed under nitrogen flow at 1200°C for 2 hours (heating rate = 2 °C·min^{-1}). During pyrolysis, the preceramic polymer samples were subjected to a large volume shrinkage (~50%; about 23% linear shrinkage) due to the polymer-to-ceramic transformation, occurring with the elimination of organic moieties and leading to the formation of an amorphous Si-O-C ceramic.

A mesoporous silica precursor solution was prepared containing tetraethyl orthosilicate (TEOS) as the silica source, and the block copolymer Pluronic F127 (EO_{106}-PO_{70}-EO_{106}, $EO=CH_2CH_2O$, $PO=CH(CH_3)CH_2O$) as the structure-directing agent. A starting solution was prepared by mixing 4 g TEOS and 2.5 g EtOH, and successively adding 0.355 g of a 0.768 M HCl solution. The solution was then stirred for 45 minutes in order to pre-hydrolize the silane units, and added to a solution containing 12 g EtOH, 1.3 g Pluronic F127 and a 0.057 M HCl solution. This was stirred for further 15 minutes before impregnation into the SiOC foams. The SiOC foams were immersed into the silica solution and withdrawn after 1 minute; excess solution was then removed from the samples with absorbing paper. The impregnated samples were dried in air at ambient temperature for a few days. The impregnated samples were heat treated in order to remove the surfactant from the mesopores and promote further condensation of the silica framework. Calcination was carried out in air at 350°C for 1 hour (heating rate = 1°C·min^{-1}).

SiOC foams were etched using an HF solution (20 vol% in H_2O). Samples were placed in a polypropylene container with the hydrofluoric acid (HF) solution. The weight ratio between the SiOC samples and the HF solution was kept constant to 1:50. The solution was gently stirred at room temperature for 9h, and then the samples were extracted and rinsed with distilled water to remove any residual HF. They were then heated up to 110°C to eliminate the remaining water. Some samples were heat treated in air at 650°C for 2h after pyrolysis (before etching), to remove completely any traces of residual carbon possibly deriving from the incomplete burn out of the PMMA microbeads.

MK silicone resin was mixed (50 wt%) with a liquid poly-dimethyl-siloxane (Rhodorsil RTV 141, Rhodia, France), cast into a mold and, after crosslinking, the sample was pyrolyzed at 1200°C for 2 hours (heating rate = 2 °C·min^{-1}).

The samples were characterized using Scanning Electron Microscopy (SEM, JEOL JSM-6490, Japan), Transmission Electron Microscopy (Philips CM20 FEG microscope, operated at 200 keV), and nitrogen adsorption-desorption (ASAP 2020, Micromeritics; analysis performed at 77 K). Specific Surface Area (SSA) was determined from a BET (Brunauer, Emmet, and Teller) analysis [8]. Pore size distribution was calculated from the adsorption branch of the isotherm through the BJH analysis [8].

RESULTS AND DISCUSSION

Several approaches can be followed in order to produce porous components using preceramic polymers. Macro-porous cellular components with a large cell size (macro-cellular foams, average cell size from ~250 μm to ~3 mm) or a small cell size (micro-cellular foams, average cell size from ~1 μm to ~200 μm) can be fabricated using different procedures: macro-cellular foams can be obtained from direct foaming or by using polyurethane precursors as expanding agents [3,9], while micro-cellular foams are produced with the aid of sacrificial fillers [4,10]. Additionally, micro- and/or meso-porosity can be developed within cellular ceramics, thus obtaining ceramic components with hierarchical porosity.

Several routes that can be pursued to accomplish this goal are listed in Table I and are addressed in this paper.

Table I. Possible strategies for producing hierarchical porosity ceramic components with high SSA from preceramic polymers, with typical SSA values obtained

Strategy	SSA (m^2/g)	Reference
Controlled thermal treatment of preceramic polymers	400-700*	[5,6]
Addition of high SSA fillers	400-650*	[5,6]
Deposition of zeolites	150-300	[11]
Deposition of meso-porous coatings	60	[12]
Infiltration of aerogels into foams	150-220	[13]
Etching	150-600	[14]
Mixing of preceramic polymers with different characteristics	20-30	[15]

* at 600°C

The first approach listed in Table 1 is the controlled thermal treatment of preceramic polymers. This approach takes advantage of the transient porosity generated upon heat treating a preceramic polymer in the temperature range at which the polymer-to-ceramic conversion occurs (generally 400-800°C) [16,17]. The build-up of internal pressure in the component, provoked mainly by the decomposition of the organic moieties in the preceramic polymer (with generation typically of CH_4 and H_2 gas) leads to porosity (with pore size typically below 50-100 nm). This (micro)porosity is transient, in that it is eliminated when the pyrolysis temperature leading to the completion of the ceramization process is increased. Figure 1a shows the gas evolution of MK polysiloxane, with its associated weight loss, which is maximum in the polymer-to-ceramic transformation temperature interval. Figure 1b reports the data for the pore volume present in a macro-cellular foam pyrolyzed at various temperatures. It can be seen that with increasing pyrolysis temperature, the micro- and meso-pores are closed, with consequent drastic reduction of the Specific Surface Area values. A significant influence of the pyrolysis rate on the amount of the generated SSA at each temperature has also been reported (the slower the heating, the higher the SSA values) [5,6].

The amount of produced specific surface area is stable up to the pyrolysis temperature reached, even after prolonged heating at the same temperature. In one experiment, a macro-cellular foam was heated at 600°C for times ranging from 1 to 48h, and the SSA value decreased only of about 10% (to 380 m²/g) after the longest heating time, confirming that the closure of the small pores is driven by a densification mechanism based on surface reaction/pyrolysis accommodated by viscous flow (and thus is temperature dependant) [17].

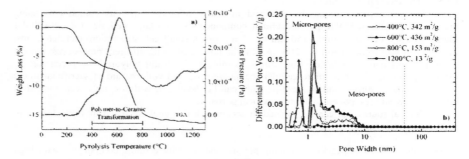

Fig. 1. a) Gas generation and weight loss upon pyrolysis; b) Evolution of micro- and meso-pore volume and of SSA with pyrolysis temperature (macro-cellular foam).

Despite the fact that high values of SSA can be achieved by controlling the heating schedule and limiting the pyrolysis temperature [5,6], it is very difficult to retain the micro- and meso-porosity when heating at temperatures above 600-800°C, which can be desirable for some applications. A possible solution is the second approach listed in Table 1: the mixing of micro-/meso-porous fillers with the preceramic polymer, followed by shaping into a cellular ceramic component. For instance, the addition of carbon black (activated carbon) powders to a polysiloxane, led to SSA values of about 100-130 m²/g up to a pyrolysis temperature of 1200°C [5].

The third and fourth entries in Table 1 refer to the deposition of high SSA materials on the cell walls, namely zeolites and meso-porous coatings. Macro-porous ceramic foams, in fact, are an ideal candidate as substrates for the development of porous bodies with hierarchical (micro-, meso- and macro-) porosity. Their very large geometric surface area (for micro-cellular foams, in particular, of the order of a few m·cm⁻³) provides a wide region for the deposition of coatings of suitable materials possessing micro-/meso-porosity. Zeolites have been deposited both on macro-cellular foams [18] and on micro-cellular foams [19]. In the latter case, a high SSA value for the ceramic components with hierarchical porosity was achieved (221 m²/g for ZSM-5 coatings, 293 m²/g for silicalite-1 coatings, corresponding to a zeolite loading of about 80 wt%). We therefore decided to further develop this strategy by depositing a SiO₂ mesoporous layer obtained by block copolymer-templated self-assembly from a sol-gel solution [12].

Figure 2a and 2b show some TEM micrographs of such coating deposited on the walls of a microcellular SiOC foam (average cell size 185 μm). TEM investigations revealed the mesoporous nature of the layer, with an average pore size of ~4 nm. The pores are assembled in an 3D ordered fashion, according to a BCC unit cell with a lattice parameter of about 13 nm. It can be observed that closer to the foam there was a lower degree of order in the pores, which became regularly arranged further away from the substrate. From SEM and TEM investigations, the coating thickness was assessed to be in the range 1 to 5 μm, depending on the region of the foam observed. In some areas the coating appeared to be partially detached form the substrate, either because of stress caused when sectioning the sample or because of uneven drying shrinkage. The issue will be further investigated in forthcoming experiments.

The nitrogen adsorption-desorptiom isotherms indicated that the coated microcellular SiOC foams followed the typical behavior of mesoporous materials, showing also the presence of micropores (deriving from the interconnection among the mesopores). The analysis of the data gave a SSA value of 60 m^2/g for the entire sample. Considering that the uncoated SiOC foam has a SSA value of only ~3 m^2/g, and taking into account the amount of meso-porous coating deposited on the foam (from weight change measurements before and after impregnation and calcination), this corresponds to a SSA for the meso-porous silica layer of about 560 m^2/g, which is typical for mesoporous materials. Thus, we can state that the nature and morphology of the foam substrate did not affect the self-assembling process of the molecular inorganic nano-building blocks, which is guided by the presence of supramolecular structure-directing agents formed by the self-assembly of amphiphilic species.

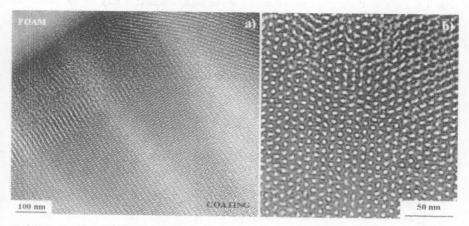

Fig. 2. a) Cross-sectional TEM of SiOC foam coated by a meso-porous SiO_2 layer; b) particular of the meso-porous coating, showing a 4-fould symmetry axes for the pores.

Besides meso-porous coatings, other infiltrants could be used for the generation of high SSA and hierarchical porosity in macro-porous ceramics; sol-gel derived aerogels, for instance, which have a SSA as high as 1000 m^2/g, can be infiltrated into open cell foams with or without the aid of vacuum [13]. If completely infiltrated, the resulting component has a reduced permeability. but the ceramic foam structure provides strength and support to the otherwise mechanically weak aerogel material.

Another strategy that has been used to increase the specific surface area in cordierite honeycombs is etching [20], which has also been applied with success to SiOC materials in powder form [14,21]. Because of the nano-structured microstructure that SiOC ceramics possess, which is comprised of nano-domains (typical size 1-5 nm) based on clusters of silica tetrahedra encased within an interdomain wall constituted from mixed bonds of SiOC and from a network of sp^2 carbon [22], HF attack provokes the preferential removal of silica from the material. This leads to the development of micro- and meso-porous structures with a high SSA [14]. The etching experiments performed on an as-pyrolyzed foam gave only a limited increase in the specific surface area (30 m^2/g). The reasons for this discrepancy with the results published by other authors concerning SiOC glass powders could be several. We can exclude a non complete infiltration of the foam by the HF solution, because we performed etching tests on crushed foams as well, but we did not observe a significant difference in the SSA data. Therefore, a possibility could be the presence of some residual carbon at the surface of the cells (deriving from an incomplete burn-out of the PMMA microbeads), which could act as a barrier towards chemical attack [23]. We therefore proceeded to heat treat a SiOC foam in air at 650°C, in order to remove any residual C present. After etching of the oxidized foam, an increase of SSA (46 m^2/g) was indeed observed, but not as significant as one could have supposed. Therefore, the reason for the limited development of specific surface area with etching in our present experiments must lie elsewhere. At this stage of the investigation, we believe that a possible motivation could be related to the size of the silica nano-domains in our SiOC ceramic; if they are too small, this would hinder the access of the etchant to the silica tetrahedra [14]. Small Angle X-ray Scattering experiment will have to be performed to clarify this issue. In Figure 3a and 3b are shown the absorption-desorption isotherms, which indicate the presence of both micro-pores and meso-pores, as well as the pore size distribution for both the as-pyrolyzed and oxidized SiOC etched foams. We can observe that most of the pores have a dimension below 10 nm, but the distribution is very broad (slightly more so for the oxidized SiOC foam).

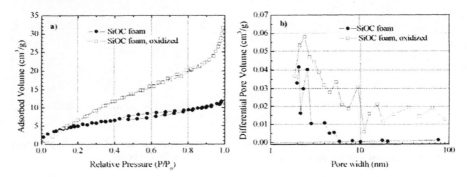

Fig. 3. a) Absorption-desorption isotherms for etched SiOC foam (as received and after oxidation at 650°C); b) Pore size distribution.

The last strategy listed in Table 1 is the mixing of precursors with different architecture (poly-silsesquioxanes typically possessing a cage-like structure or linear poly-siloxanes). This is another promising strategy for the generation of (macro-)porous ceramics from preceramic polymers [15]. Some preliminary work conducted in our laboratory has shown that porous structures of various morphology can be produced depending on the composition of the starting mixture. As a typical example, in Figure 4a and 4b are shown SEM micrographs of a porous ceramic obtained by mixing a silsesquioxane (MK resin) with a linear silicone (RTV 141). The generation of porosity is due to the different behavior that the two polymers have upon pyrolysis, in terms of weight loss, shrinkage and amount of gas generated (all much higher for the siloxane with a linear structure than for the silicone resin, with a more complex structure). It can be seen that a large amount of open porosity was generated in the sample, and that the specimen had quite a complex morphology. Besides large pores of a few hundred microns in size, a large amount of smaller, micron-sized pores and cavities is present. At this stage of the investigation we do not have any information relative to the amount of specific surface area produced by this processing method, but literature data indicate that relatively high SSA values could be achieved [15].

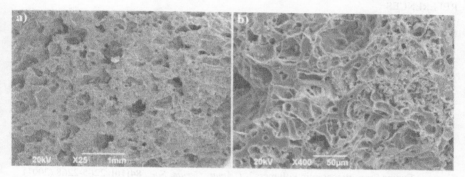

Fig. 4. SEM micrograph of the fracture surface of a macro-porous ceramic obtained by mixing preceramic polymers with different characteristics. a) general view; b) higher magnification detail.

Finally, it is worth noting that in some current applications (e.g. Diesel particulate filters with a wash-coat [24]), the amount of SSA present in the component is typically well below 100 m^2/g, so the required target value for high SSA in hierarchical porosity ceramics should be assessed depending on the application considered.

CONCLUSIONS

Using preceramic polymers, in particular polysiloxanes, it is possible to fabricate macro-porous (micro- and macro-cellular) ceramics. Several strategies are available for adding to these structures a high specific surface area (SSA), thus creating ceramic components with hierarchical porosity. Interrupting the pyrolysis process in the polymer-to-ceramic temperature range allows to retain the transient porosity in the material, but increasing pyrolysis temperature leads to a drastic reduction of the SSA. The addition of (micro- and meso-)porous fillers seems a good way of retaining high SSA values at high pyrolysis temperature. The deposition of coatings possessing (micro- and meso-)pores is also a viable route to produce ceramic components with hierarchical porosity. Infiltration of the macro-porous ceramic components with aerogels is another possibility worth further exploring. Etching can also be used, but in the present experiments only a limited increase in SSA was observed, possibly due to the intrinsic nature of the SiOC ceramic tested. The mixing of preceramic polymers with different characteristics upon pyrolysis allows to fabricate (macro-)porous ceramics with a varied pore morphology.

ACKNOWLEDGMENTS

The authors wish to thank Dr. S. Carturan of Laboratori Nazionali di Legnaro (LNL) Padova, Italy, for BET measurements and the helpful discussions. P. C., G.D. S. and C. V. acknowledge the support of the European Community's Sixth Framework Programme through a Marie-Curie Research Training Network (PolyCerNet MRTN-CT-019601). R. P.-A. gratefully acknowledges the Spanish Ministerio de Educación y Ciencia for the financial support for a post-doctoral fellowship.

REFERENCES

[1]M. Scheffler and P. Colombo. Eds. *Cellular Ceramics: Structure, Manufacturing, Properties and Applications*. Wiley-VCH Verlag GmbH, Weinheim, Germany, 2005.

[2]P. Colombo, "Conventional and novel processing methods for cellular ceramics," *Phil. Trans. R. Soc. A*, **364[1838]**, 109-124 (2006).

[3]P. Colombo and J. R. Hellmann, "Ceramic foams from preceramic polymers," *Mat. Res. Innovat.*, **6** , 260-272 (2002).

[4]P. Colombo, E. Bernardo and L. Biasetto, "Novel Microcellular Ceramics from a Silicone Resin," *J. Am. Ceram. Soc.*, **87 [1]**, 152–154 (2004).

[5]H. Schmidt, D. Koch , G. Grathwohl and P. Colombo, "Micro-Macro Porous Ceramics From Preceramic Precursors", *J. Amer. Ceram. Soc.*, **84[10]**, 2252-2255 (2001).

[6]M. Wilhelm, C. Soltmann, D. Koch and G. Grathwohl, "Ceramers – functional materials for adsorption techniques," *J. Europ. Ceram. Soc.*, **25**, 271-276 (2005).

[7]P. Colombo, T. Gambaryan-Roisman, M. Scheffler, P. Buhler, P. Greil, "Conductive Ceramic Foams From Preceramic Polymers", *J. Amer. Ceram. Soc.*, **84[10]**, 2265-2268 (2001).

[8](a) S. J. Gregg and K.S.W. Sing, *Adsorption, surface area and porosity*, Academic Press, London, 1982, pp. 285-286. (b) E. P. Barret, L. G. Joyner, and P. H. Halenda, "The determination of pore volume and area distributions in porous substances. I. Computations from nitrogen isotherms," *J. Am. Chem. Soc.*, **53**, 373-380 (1951).

[9]P. Colombo and M. Modesti, "Silicon oxycarbide ceramic foams from a preceramic polymer", *J. Am. Ceram. Soc.*, **82** [3], 573-578 (1999).

[10]P. Colombo, E. Bernardo, "Macro- and Micro-cellular Porous Ceramics from Preceramic Polymers", *Comp. Sci. Tech.*, **63**, 2353–2359 (2003).

[11]A. Zampieri, P. Colombo, G.T.P. Mabande, T. Selvam, W. Schwieger and F. Scheffler, "Zeolite coatings on microcellular ceramic foams: a novel route to microreactor and microseparator devices," *Adv. Mater.*, **16 [9-10]**, 819-823 (2004).

[12]S. Costacurta, L. Biasetto, P. Colombo, E. Pippel and J. Woltersdorf, "Hierarchical Porosity Components via Infiltration of a Ceramic Foam," submitted to *J. Am. Ceram. Soc.*

[13]L. Biasetto, P. Colombo, L. VanGinneken, J. Luyten, unpublished results.

[14]R. Peña-Alonso, G.D. Sorarù, R. Raj, "Preparation of Ultrathin-Walled Carbon-Based Nanoporous Structures by Etching Pseudo-Amorphous Silicon Oxycarbide Ceramics," *J. Am. Ceram. Soc.*, **89[8]**, 2473-2480 (2006).

[15]F. Berndt, P. Jahn, A. Rendtel, G. Motz and G. Ziegler, "Monolithic SiCO Ceramics with tailored Porosity," *Key Eng. Mater.*, **206-213**, 1927-1930 (2002).

[16]E. Kroke, Y-L. Li, C. Konetschny, E. Lecomte, C. Fasel and R. Riedel, "Silazane derived ceramics and related materials," *Mater. Sci. Eng. R*, **26**, 97-199 (2000).

[17]J. Wan, M.J. Gasch and A.K. Mukherjee, "In Situ Densification Behavior in the Pyrolysis Consolidation of Amorphous Si-N-C Bulk Ceramics from Polymer Precursors," *J. Am. Ceram. Soc.*, **84[10]**, 2165–69(2001).

[18]F.C. Patcas, "The methanol-to-olefins conversion over zeolite-coated ceramic foams", *J. Catal.*, **231[1]**, 194-200, (2006).

[19]A. Zampieri, P. Colombo, G.T.P. Mabande, T. Selvam, W. Schwieger and F. Scheffler, "Zeolite coatings on microcellular ceramic foams: a novel route to microreactor and microseparator devices," *Adv. Mater.*, **16 [9-10]**, 819-823, (2004).

[20]Q. Liu, Z. Liu, and Z. Huang, "CuO Supported on Al_2O_3-coated cordierite-honeycomb for SO_2 and NO removal from flue gas: Effect of acid treatment of the cordierite", *Ind. Eng. Chem. Res.*, **44**, 3497-3502 (2005).

[21]A.M. Wilson, G. Zank, K. Eguchi, W. Xing, B. Yates, and J.R. Dahn, "Pore Creation in Silicon Oxycarbides by Rinsing in Dilute Hydrofluoric Acid," *Chem. Mater.*, **9[10]**, 2139-2144 (1997).

[22]A. Saha, R. Raj and D.L. Williamson, "A Model for the Nanodomains in Polymer Derived SiCO," *J. Am. Ceram. Soc.*, **89[7]**, 2188-2195 (2006).

[23]G. D. Soraru, S. Modena, E. Guadagnino, P. Colombo, J. Egan and C. Pantano, "Chemical Durability of Silicon Oxycarbide Glasses," *J. Am. Ceram. Soc.*, **85[6]**, 1529-1536, (2002).

[24]C. Agrafiotis and A. Tsetsekou, "The effect of powder characteristics on washcoat quality. Part I: Alumina washcoats", *J. Euro. Ceram. Soc.*, **20**, 815-824 (2000).

ELECTROPHORETIC DEPOSITION OF PARTICLE-STABILIZED EMULSIONS

Bram Neirinck, Jan Fransaer, Omer Van der Biest and Jef Vleugels[*]
Department of Metallurgy and Materials Engineering (MTM), K.U.Leuven,
Kasteelpark Arenberg 44, B-3001, Heverlee, Belgium

ABSTRACT

A new route for the production of freestanding porous structures based on the electrophoretic deposition (EPD) of particle-stabilized emulsions is described. Emulsions of paraffin oil in ethanol, stabilized with α-alumina particles were prepared using a magnetic stirrer. These emulsions were consolidated using EPD under a constant electric field into disk shaped specimens. The pore forming liquids are extracted prior to sintering by means of evaporation. The pores in the sintered disks are spherical and their size can be tuned within the 2 to 200 μm range. Both open and closed porosity can be obtained by varying the emulsion composition. Since no pore forming fugitive solid organic binder is used, the very delicate and time-consuming debinding step during conventional processing is eliminated.

Keywords:
pores/porosity; porous materials; alumina; debonding; foams

INTRODUCTION

In the last decades, porous ceramics have received ever-increasing attention. The interest in these materials has grown from several application areas, which fuel the demand for cheap but high performance porous ceramics, resulting in more rigorous demands on the material properties. Stricter particle emission norms for instance have lead to widespread implementation of ceramic sooth filters in diesel engines. Each of these applications stipulates it's own set of optimal porosity properties and minimal structural strength. Thus for specific implementations tailored materials are useful. A final factor affecting the porous ceramic market is an additional concern on material safety. A ban of fiber based porous materials, such as those used for refractories, is being considered due to the hazardous effect of free fibers on the respiratory track when inhaled.

Spherical pores are considered ideal when mechanical strength or thermal shock resistance is of primary importance. One of the most widespread methods for producing such spherical cavities is the use of sacrificial template materials such as polymer beads or starch particles. The main problem with this method is the removal of large quantities of sacrificial material, which is usually achieved by debinding or burnout at elevated temperatures. This is an arduous and time consuming step, especially when thick pieces are to be produced.

In this paper, the use of particle stabilized emulsions for obtaining spherical pores with a controlled pore size distribution is proposed. Upon consolidation of the emulsion and extraction of the internal liquid, hollow spherical pores with tailored diameter can be formed in a solid particle powder compact. This type of emulsion was first described by Pickering in 1907[1], and are referred to as Pickering emulsions ever since. Whereas most commonly known emulsions are stabilized by surfactants, Pickering emulsions are stabilized by solid particles located at the interface. While both types of emulsions behave similarly at first sight, there are some significant differences. First of all, the surfactants lower the surface tension of the liquid-liquid interface. Particles located at this interface do not alter its surface tension, but merely reduce the available area. Furthermore, surfactant molecules desorb and resorb uninterruptedly at the fluid-fluid plane. The particle location at the interface on the other hand is a thermodynamically driven process. The energy needed to remove a particle from the interface formed by liquid phases a and b, $-\Delta G_{ab}$, can be calculated as[2]:

$$-\Delta G_{ab} = r^2 \pi \gamma_{ab} \left(1 \pm \cos\theta_{ab}\right)^2 \tag{1}$$

This energy depends on the particle radius r, the surface tension of the liquid-liquid interface γ_{ab} and the liquid-liquid-solid contact angle θ_{ab}, measured in phase b (Fig. 1a). The sign of the cosine is chosen so that the term in brackets is smaller than 1. In other words if θ_{ab} is smaller than 90° the sign is negative and vice versa. When θ_{ab} is close or equal to 90°, the energy will be maximal. The liquid that preferentially wets the particles is favored as the continuous phase as shown in Fig. 1. For most systems, the energy needed to remove the particle is substantial, up to a few 1000 times kT. The particles are irreversibly adsorbed at the interface.

Finally, the stabilization of the droplets themselves with surfactants is based on steric or electrosteric repulsion while particles adsorbed at the interface create a mechanical barrier against coagulation. When a sufficient amount of particles is available, the interface will exhibit a hexagonal close packed (HCP) layer, as schematically represented in Fig. 1b, completely shielding the droplet from its surroundings. The stability of the emulsion itself will thus depend on the continuous phase much like the stability of a suspension depends on the suspending medium.

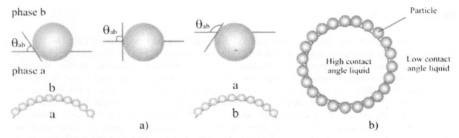

Fig. 1: Schematic representation of a) contact angle influence on particle positioning at the interface and b) a Pickering emulsion droplet

In this work electrophoretic deposition (EPD) is used for the consolidation of the particle-stabilized emulsions. Hamaker and Verwey described the electrophoretic deposition process as early as 1940[3]. However, it was not until the beginning of the 90's that EPD was suggested as a commercial viable method for the production of ceramic objects and coatings[4]. Electrophoretic deposition itself is fundamentally a two-step process. In the first step, charged particles in an electrical field move towards an electrode (electrophoresis). In the second step, the particles collect at the electrode and form a coating or solid object on this electrode. The composition of the formed deposit depends directly on the composition of the suspension at the time of deposition. Hence controlling the suspension properties during processing can be used to create tailored structures such as gradient materials, laminates or porous ceramics. Since the droplets are covered with particles, it is expected that the droplets will behave similarly during electrophoretic deposition. Hence a co-deposition of particles and droplets yields a porous structure[5].

This paper reports on the initial results obtained on the production of free-standing porous structures from alumina powder stabilized ethanol/paraffin oil emulsions, consolidated by electrophoretic deposition (EPD).

EXPERIMENTAL PROCEDURE

In most cases, solid particle stabilized emulsions are studied for water and oil systems. Water is however not desirable for EPD experiments since the electrolytic decomposition limits the maximum voltage that can be used to 3-4 V. For this reason, electrophoretic deposition is commonly carried out in organic media such as ethanol. In order to form emulsions, a second liquid has to be added. A small test series with common non-polar solvents revealed that liquid paraffin, which is a mixture of several long chain alkanes, is sufficiently insoluble in ethanol.

Pickering emulsions were made using submicrometer α-Al_2O_3 powder (Baikowski grade SM8) with an average particle size of approximately 200 nm using absolute ethanol (VWR Prolabo) and liquid paraffin (Nujol mull grade for IR-spectroscopy, Acros). N-butylamine (pro analysis., Merck) was used as charging agent and surface modifier. A controlled amount of demineralized water was added to the absolute ethanol in order to facilitate electrophoretic deposition. The α-Al_2O_3 particle size distribution, depicted in Fig. 2, was measured using an acoustic particle sizer (Matec APS 100).

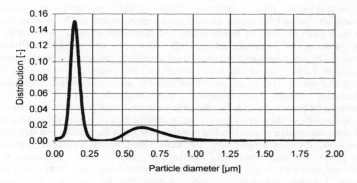

Fig. 2: Particle size distribution of SM8 alumina powder

The optimum amount of n-butylamine was determined by titration of an ethanol-based 5 vol% alumina suspension while measuring the zeta potential (Matec ESA 9800) (Fig. 3). A maximum zeta potential was reached upon the addition of 1 to 5 vol% of butylamine. Above 5 vol%, the zeta potential gradually decreased. This is primarily attributed to a decrease the electrophoretic mobility due to an increase in ionic strength, which is illustrated by an increase in conductivity Fig. 3 [6].

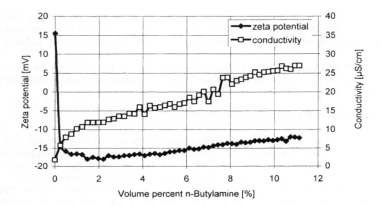

Fig. 3: Zeta potential of the ethanol-based alumina suspension as a function of the n-butylamine content.

For each experiment, two suspensions are prepared, as summarized in Table I. To one suspension paraffin oil is added to prepare an emulsion. The other suspension is gently mixed with this emulsion prior to electrophoretic deposition. The emulsion composition is used to control the pore size. The particles in the second suspension remain in the continuos phase when added to the previously prepared emulsion. These particles are incorporated in-between the stabilized droplets during EPD, hence forming the interpore structure. The alumina is suspended in ethanol with equal amounts of n-butylamine and demineralized water addition by means of magnetically stirring for 15 minutes, subsequently followed by ultrasonification (Branson 2510) for 15 minutes. The emulsion part is prepared from a similarly prepared suspension with adjusted composition (see Table I) by adding liquid paraffin oil and vigorously stirring for 10 minutes using a magnetic stirrer. This emulsion is then added to the corresponding suspension and the obtained Pickering emulsion is kept in motion by gentle magnetic stirring.

The emulsions were prepared with a specific targeted porosity and pore size, as included in Table I. The macro porosity of the green deposits was estimated by calculating the volume fraction of paraffin to the total volume of paraffin and alumina powder in the system. For this estimate it is assumed that the droplets deposit with the same speed as the individual particles in the continuous phase. The pore sizes were estimated by calculating the maximum surface the alumina powder used to stabilize the droplets could cover when they form a hexagonal close packed layer. For this calculation it was assumed that the particles are spherical and monodisperse with a diameter of 0.2µm. In addition the assumption was made that the droplet curvature is low enough compared to the particle so that the total interface surface can be considered as a flat surface with respect to the particle. This results in a surface coverage of 90%. From the calculated surface and the paraffin volume, the droplet diameter and hence the green pore size, can be estimated.

The electrophoretic deposition experiments are carried out in an experimental set-up as schematically shown in Fig. 4. This set-up is comprised of a flow-through horizontal deposition cell and an emulsion circulation system driven by a peristaltic pump (Watson-Marlow, 505Du). The pump speed is restricted to 60 RPM in order not to exert too much stress on the emulsion droplets. The total volume of emulsion used in this system is 200 ml. The distance between the two circular electrodes is 35 mm and their diameter is 37.5 mm. The deposition was performed at a constant voltage of 350 or 175 V for 600 or 1200 s using a DC power supply (F.U.G., MCN 1400-50). For both voltage settings a reference deposit is made (A and J in Table I). The deposition process was monitored by recording the cell current (Onron, K3NX).

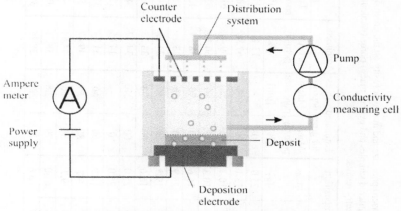

Fig. 4: Schematic representation of the EPD set-up

The resulting deposits were subsequently dried for 2 hours in a vacuum chamber. A dilatometer (Netzsch, DIL 402 C) experiment was performed in order to determine the appropriate sintering temperature. The dried deposits were subsequently sintered in air for 30 minutes at 1400°C with a heating and cooling rate of 10 °C/min.

The porosity of the sintered deposits was investigated by means of scanning electron microscopy (SEM, XL30-FEG, FEI). The density of the sintered pieces was measured by the Archimedes method in ethanol. A lacquer with known density was used to seal the pores before submerging the pieces. Mercury porosimetry (Micrometrics Autopore IV 9500) was used to check the open or closed porous character of the sintered samples and to measure the porosity. In the case of an open pore structure, the window size distribution in-between the pores is quantified.

Table I: Summary of experimental parameters, estimated values for macro pore size and macro porosity on the green deposit, porosity values measured by Archimedes and mercury porosimetry for the sintered samples and overall macro porosity calculated from the mercury porosimetry results.

Experiment number	Emulsion composition				Suspension composition			Estimated values for green deposit		Archimedes	Mercury porosimetry	
	V_{EtOH} [ml]	V_{BA} V_{H2O} [ml]	m_{SM8} [g]	V_{Para} [ml]	V_{EtOH} [ml]	V_{BA} V_{H2O} [ml]	m_{SM8} [g]	Macro porosity [%]	Macro pore size [μm]	Total porosity [%]	Total porosity [%]	Macro porosity [%]
A	-	-	-	-	180	10.0	25.0	0	-	13.11	12.1	0
B	13.5	0.75	1.75	35.0	135	7.50	15.1	90	56	67.88	70.18	65.72
C	13.5	0.75	0.75	15.0	153	8.50	10.3	85	56	59.47	61.31	55.52
D	13.5	0.75	0.75	15.0	153	8.50	10.3	85	56	60.10	63.21	57.71
E	18.0	1.00	0.50	10.0	153	8.50	16.6	70	56	38.54	40.36	31.45
F	18.0	1.00	0.50	10.0	153	8.50	25.9	60	56	33.43	35.34	25.68
G	22.5	1.25	0.25	5.00	153	8.50	19.5	50	56	24.17	24.91	13.69
H	18.0	1.00	10.0	10.0	153	8.50	16.6	60	2.8	33.25	33.52	23.58
I	13.5	0.75	0.175	15.0	153	8.50	25.0	60	240	34.17	38.69	29.52
J	-	-	-	-	190	5.00	25.0	0	-	14.05	13.73	0
K	19.0	0.50	0.50	10.0	161.5	4.25	15.9	60	56	31.74	31.62	21.41
L	19.0	0.50	0.50	10.0	161.5	4.25	9.7	80	56	49.47	51.8	44.60
M	19.0	0.50	0.50	10.0	161.5	4.25	9.7	80	56	44.84	49.9	42.41
N	19.0	0.50	10.0	10.0	161.5	4.25	6.60	70	2.8	38.71	38.77	29.63
O	19.0	0.50	0.10	10.0	161.5	4.25	16.6	70	280	38.44	39.28	30.21

RESULTS AND DISCUSSION

Emulsion preparation

The particle surface chemistry determines which of the liquids used to prepare an emulsion preferentially wets the particle. As schematically represented in Fig. 1 this liquid is favored as the continuous phase. However, this is not valid for the entire volume concentration range. Above a certain volume fraction of "preferred" emulsified phase, it is no longer possible to keep this phase in its emulsified form because there is simply not enough continuous phase to surround it. The emulsified phase then becomes the continuous phase and vice versa. Due to the usually sudden appearance of this phenomenon, it is described as the critical phase inversion[2, 7]. In order to determine the concentration range for emulsion preparation in the ethanol/paraffin oil system, the critical inversion point was experimentally determined by means of electrical conductivity measurements and visual observation.

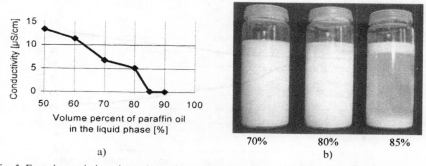

Fig. 5: Experimental phase inversion point determination by a) conductivity measurements and b) visual inspection.

The inversion point is located around 85 vol% paraffin oil, as shown in Fig. 5. The viscosity however increased significantly upon addition of > 70 vol% paraffin oil, rendering handling of the suspensions unnecessary difficult. Therefore 70 vol% was chosen as maximal volume percent of paraffin used in the emulsion preparation. Sample B was prepared using this concentration. For all other samples concentrations between 16 and 50 vol% were used.

The acidic hydroxyl groups on the alumina surface react with the amine group of n-butylamine. As a result, a layer of butylamine is adsorbed on the alumina surface with the amine group directed towards the surface, forming an organic C-H chain adlayer. It has been reported that a 4 carbon chain results in an optimum adlayer thickness and zeta potential[8].

The adlayer of n-butylamine renders the surface of the alumina particles slightly more hydrophobic. This minimal increase in hydrophobic behavior is sufficient to ensure efficient incorporation of the particles at the liquid-liquid interface without causing phase inversion.

The effect of n-butylamine adsorption is not restricted to the fixation of the particles at the interface. It has also been reported that the addition of butylamine inhibits the formation of a potential drop over the deposit during EPD, theoretically allowing to obtain deposits of unlimited thickness[9].

Electrophoretic deposition

A first series of emulsion systems was tested in EPD experiments with an electric field strength of 100 V/cm. Smooth deposits were obtained for the systems with an estimated low (≤ 60 %) and high porosity (≥ 80 %).. Irregular green deposits however were obtain for the systems with intermediate porosity, what is attributed to the partial coverage of the deposition electrode with non-conductive paraffin oil. This effective reduction of electrode surface results in an inhomogeneous electric field which gives rise to areas with a high current density and spikes in the EPD cell current profile during deposition, as shown in Fig. 6. This resulted in damaged inhomogeneous deposits, as shown in Fig. 7a. In order to reduce current density problems, further experiments were performed with an electric field strength of 50 V/cm and the conductivity of the emulsion systems was lowered by reducing the amount of water and butylamine from 5 to 2.5 vol% each. Although the reduction of the effective electrode surface still occurred, the altered experimental conditions resulted in smooth and homogeneous deposits as shown in Fig. 7b.

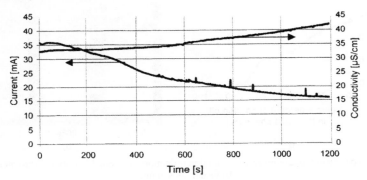

Fig. 6: Evolution of current and conductivity as a function of time during deposition of sample E

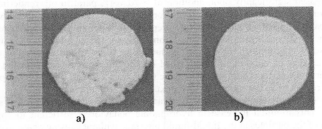

a)   b)

Fig. 7: Photographs of sintered deposits made with an electric field strength of 100 V/cm (a) (experiment I) and 50 V/cm (b) (experiment N).

The deposits with higher paraffin content have a higher electrical resistance. The electrical resistance of a non-conducting deposit is directly related to the resistance against ion transport through the deposit. Paraffin oil does not allow this ion transport. As more paraffin is incorporated in the deposit, the resistance increases and the current through the deposit decreases, as shown in Fig. 6.

The conductivity of the suspension however was measured to gradually increases, as shown in Fig. 6. Since the conductivity of the reference suspensions rose as well, be it less pronounced, the explanation must be sought after in the suspension system. Three possibilities come to mind. The first is that some of the adsorbed butylamine is released when the particle is deposited. The second is that the electrical field causes decomposition of either water or butylamine, thus increasing the ion content. A third possibility is the formation of reaction products in the organic medium. Previous experiments have shown that reaction occurs in amine ketone systems[6], resulting in a similar conductivity increase. This reaction, which already takes place in the absence of an electric field, is apparently accelerated when a voltage is applied. During the reaction water is formed as a byproduct causing the conductivity to increase. Further examination of this subject is needed.

Characterization of sintered samples

The sintered density of the deposit was determined using the Archimedes method in ethanol. A lacquer with know dry density was used to seal the open pores. The total porosity was calculated from the measured values assuming a theoretical density of 3.89 g/cm^3 for alumina. The total open pore volume was measured with mercury porosimetry. From this pore volume a second porosity value was calculated. Results for both measurements are summarized in Table I. If we compare the measured values to the estimated target values a shift towards lower porosity is apparent. The estimated green macro porosity values are obtained by calculating the volume fraction of paraffin to the total volume of paraffin and powder as described in the experimental section. However in this estimate the global shrinkage of the green structures during sintering has not yet been taken into account. The green density of the reference samples A and J is 47.9% and 49.3% respectively. This means that 50% of porosity needs to be removed to reach full density. This typically results in 20% of linear shrinkage for fully dense ceramics. In our case, the two reference deposits reached a final density of 86 and 88 %. Assuming that the solid structure in the porous samples displays a similar shrinkage, the total pore volume per gram solid material will decrease likewise. The extent of pore shrinkage however will not be equal to the total shrinkage of the blanks. The interparticle pore decrease will partly be translated in a macropore volume increase. Thus some global shrinkage will occur, but to a lower extent than for the "pore free" samples, reducing the global porosity of the samples. Measured data show that this extent of shrinkage is

Overall the porosity of the samples shows high reproducibility. Samples with the same estimated green macroporosity, such as E, M and N for 70%, yield deposits with similar sintered porosities. This holds both for identical experiments such as C and D, L and M, as well as for samples prepared with different pore sizes and deposition parameters. As a result we can conclude that the deposition speed of the droplets shows little dependence on their size and that deposition is homogeneous for both filed strengths.

The fact that both reference grades have not been sintered to full density implies residual interparticle porosity. Therefore, not all porosity measured with both the Archimedes method and mercury porosimetry can be attributed to the macropores created by the droplets. The actual amount of macropores and interparticle pores can be calculated. Since the density of alumina,

3.89 g/cm^3, is know the volume of solid material in 1 g of samples can be calculated. The total strut volume can hence be obtained by assuming the strut density and the total sample volume is calculated using the total porosity. The volume of macropores is the difference between these two numbers. Macro porosity values calculated using the porosity data obtained by mercury porosimetry and assuming a strut density equal to the reference sample density, 87%, are summarized in Table I.

The interparticle pore network is an open porous network. Densification must be higher than 95% before closed interparticle pores are obtained. This also means that no real closed macroporous samples were obtained. Even when a pore is not in contact with the surface through windows to other pores, it will be filled during mercury porosimetry through the interparticle pores. Hence in mercury porosimetry no real pore sizes are measured. Instead, the obtained values reflect the minimal hydrodynamic diameter of the channel used to fill a pore with mercury. Windowless pores will thus contribute to what otherwise seams to be the interparticle pore volume. Fig. 8 show an example of this phenomenon. Sample J contains only interparticle pores, and is used to determine the nominal interparticle pore range. Sample L has a porosity of 51.8%. At this porosity only a fraction of the pores are interconnected. These pores result in the broad peak around 10µm. The peak near 0.1µm represents both the interparticle pores in the struts, and an amount of "closed" pores. If the samples are sintered to sufficient density to obtain fully closed porous samples, than the closed pores can be quantified by comparing mercury porosimetry results with density measurements according to Archimedes.

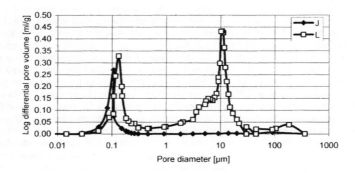

Fig. 8: Differential pore volume as measured with mercury porosimetry for reference sample J and sample K

The pore sizes were estimated using the calculation scheme given in the experimental section. Since mercury porosimetry does not provide an accurate value for the pore size, scanning electron micrographs were used to visually check the pore size and morphology of fracture surfaces. The SEM images shown in Fig. 9 reveal that in some cases the pores have been deformed, especially for the samples with pores larger than 200 µm. The spherical cavities were in the direction nominal to the electrode, resulting in an ellipsoidal cross section. This deformation of the pores can be attributed to two factors. The first one is the force applied on the droplets during EPD. Since the droplets themselves are not rigid, the electrophoretic force might be high enough to deform them. The second is the way in which the deposits were dried. In order not to damage the deposits, they were kept on the electrode during drying. Due to this most shrinkage will occur in the direction perpendicular to the electrode surface.

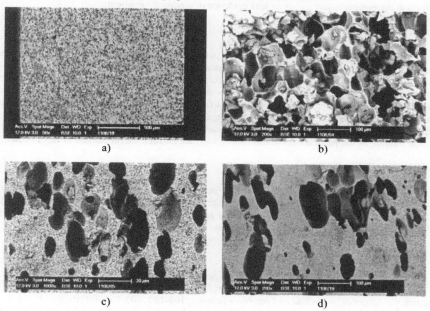

a)   b)

c)   d)

Fig. 9: SEM micrographs of fracture surfaces of a) sample K b) sample B c) sample C and d) sample O

The pore size observed by SEM qualitatively corresponds to the estimated values, especially for the 56 µm sized pores (see Fig. 9a & b). The narrow minimal pore size of 2.8 µm distribution however was not obtained. It is suspected that the mixing method did not provide enough energy to prepare an emulsion consisting of small droplets. For example for sample C, a pore size range between approximately 2 and 20 µm was obtained (Fig. 9c). The larger targeted pore size of 240 or 280 µm turned out to be smaller than predicted. This is shown for sample O in Fig. 9d. There are several possible reasons for this pore size reduction. The most obvious is that the pores shrink during drying and sintering. A second possibility is that the droplets brake up during recirculation of the emulsion trough the EPD-cell, limiting the maximum attainable pore size using the suggested method. Using a higher viscosity emulsified phase might extend the ability of the droplets to withstand shear. However, for all samples with large pores, deformation and merging of pores renders it difficult to provide a correct number for the average pore size. In all cases image analysis on polished samples or tomography is needed in order to obtain quantitative information on the actual pore size distribution.

In theory, larger particles or droplets should deposit slower than smaller particles of the same material, resulting in a gradient in particle size and/or porosity throughout the deposit, i.e., a pore free layer at the electrode side and an increased porosity with increasing distance from the electrode surface. In concentrated systems however, the faster particles drag slower moving specimens along, resulting in a homogeneous deposit. SEM observation indeed shows no edge effects at the electrode surface, or a clear evidence of a porosity gradient.

CONCLUSION

Experimental results showed that solid particle stabilized emulsions can be consolidated through electrophoretic deposition. The obtained deposits were sintered without a preceding debinding step in order to obtain porous ceramics. Judicious control over the emulsion composition and used mixing methods offers widespread control over the droplet size and final pore diameter. Controlling the volume of emulsified liquid relative to the mass of powder in the continuous phase assures that a broad range of porosity can be produced with good reproducibility. Both open and closed porosity can be obtained through this porosity control. The strong pinning of the particles at the fluid-fluid interface ensures the robustness of this method.

ACKNOWLEDGEMENTS

This work was supported by the Flemish Institute for the Promotion of Scientific and Technological Research in Industry (IWT) under grant SB/51092 and the Research Fund K.U. Leuven under project GOA/2005/08-TBA. The authors thank the Civil Engineering Department of K.U.Leuven for using the mercury porosimeter.

REFERENCES

[1]S. U. Pickering, "Emulsions", *J. Chem Soc.*, **91**, 2001-21 (1907).
[2]R. Aveyard, B. P. Binks and J. H. Clint, "Emulsions stabilised solely by colloidal particles", *Advances in Colloid and Interface Science*, **100-102**, 503-46 (2003).
[3]H. Hamaker and E. Verwey, "Formation of a deposit by electrophoresis", *Trans. Farad. Soc.*, **36**, 180-5 (1940).
[4]O. VanderBiest and L. J. Vandeperre, "Electrophoretic deposition of materials", *Annual reviews of material science*, **29**, 327-52 (1999).
[5]B. Neirinck, J. Vleugels, J. Fransaer and O. Van der Biest, "A novel process for producing sintered porous materials from solid particle stabilized emulsions without debinding", (2005), UK patent GB 0512904.4
[6]G. Anné, K. Vanmeensel, B. Neirinck, O. Van der Biest and J. Vleugels, "Ketone-amine based suspensions for electrophoretic deposition of Al2O3 and ZrO2", *Journal of the European Ceramic Society*, **26**, 3531-7 (2006).
[7]P. M. Kruglyakov and A. V. Nushtayeva, "Phase inversion in emulsions stabilised by solid particles", *Advances in Colloid and Interface Science*, **108-109**, 151-8 (2004).
[8]B. Siffert, A. Jada and J. E. Letsango, "Location of the Shear Plane in the Electric Double Layer in an Organic Medium", *Journal of Colloid and Interface Science*, **163**, 327-33 (1994).
[9]G. Anné, B. Neirinek, K. Vanmeensel, O. Van der Blest and J. Vleugels, "Origin of the potential drop over the deposit during electrophoretic deposition", *Journal of the American Ceramic Society*, **89**, 823-8 (2006).

* Corresponding author: Tel: +32 16 321244 Fax: +32 16 321992 E-mail address: jozef.vleugels@mtm.kuleuven.be

APPLICATION OF POROUS ACICULAR MULLITE FOR FILTRATION OF DIESEL NANO PARTICULATES

Cheng G. Li,
Dow Automotive, 3900 Automation Ave. Auburn Hills, MI 48326
Aleksander J. Pyzik,
Core R&D, New Products, The Dow Chemical Company, Midland, MI 48764

ABSTRACT

Diesel particulate filters (DPF) made from an acicular mullite (ACM) have demonstrated high filtration efficiency, low pressure drop, high temperature handling capability and good mechanical integrity at a porosity of 60% or higher. Due to the ability to control microstructure, total porosity and particle size distribution, DPF's made from acicular mullite can be tailored to meet the requirements of deep bed filtration and fine particle emission control. In addition, the ACM DPF can be used with a broad range of catalysts and over a wide range of catalyst loadings while retaining its filtration performance. This study has shown that acicular mullite selected for a DPF design can meet the current diesel particulate emission control requirements as well as the requirements of future nano-size particle emission control.

INTRODUCTION

Diesel engines have been widely used in automotive, transportation, construction equipment, stand-by power generators, marine and off-road applications because of their high performance and durability. Today, diesel engines are found on almost 50% of European passenger cars and all heavy duty trucks worldwide. Besides enjoying the diesel engine's efficiency, the diesel engine also brings lower exhaust gaseous emissions such as hydrocarbons (HC), carbon monoxide (CO) and nitrous oxides (NO_x) as well as lower carbon dioxide than conventional spark ignition engines due to its lean combustion nature.

By 2007 and beyond, new legislation requires up to a 90% diesel particulate reduction from engine tailpipe emissions. Therefore, effectively removing diesel particulates from engine exhaust without adversely affecting the diesel engine operation has become a challenge to the automotive industry. Since the late 1990's, emission control technology development in Europe has been focused on diesel particulate removal from exhaust, mainly diesel particulate filter technology development [1 - 3]. It has been understood that future Euro VI legislation, projected to be in place by 2011, will include nano-sized diesel particle reduction on top of today's particulate mass reduction. This brings the challenge of developing new materials for diesel particulate emission control.

In the early 1990's, Dow developed a technology that enabled the synthesis of porous ceramics in the presence of Fluorine-containing gas. The development has focused on chemistries that could produce highly porous materials with controllable grain and pore size distribution. Mullite was selected as the base material due to its ability to nucleate and grow high aspect ratio needles that have high stability at elevated temperatures, high corrosion resistance and high strength [4]

The objective of this paper is to review the filtration and regeneration characteristics of the mullite filters that have acicular microstructure and mean pore diameter from 12-15 microns. The special emphasis is on designing diesel particulate filters (DPF's) that not only satisfy today's particulate emission control requirements, but also future nanoparticle filtration needs.

MATERIAL DESIGN AND FILTER FABRICATION

The Dow's approach is based on a gas-solid catalyzed reaction to form mullite by low temperature desorption of silicon tetrafluoride (SiF_4) from fluorotopaz ($Al_2SiO_4F_2$) [5]. In the first stage of this process, fluorotopaz is formed from clay and alumina (Al_2O_3) in the presence of silicon tetrafluoride (SiF_4) gas. On subsequent heating at temperatures between 1000 and 1200°C, fluorotopaz melts to form mullite and free SiF_4 gas [6]. The unique feature of this process is the resulting highly elongated (acicular) grain structure as illustrated in Figure 1. The formation and growth of acicular mullite grains "interlocks" the microstructure and results in retained high porosity. This porosity remains virtually unchanged up to 1400-1450°C. In addition, the size of the mullite needles can be controlled depending on starting raw materials and processing conditions. Typically, the aspect ratio remains about 20, but needle diameters can be altered from about 3 to 50 microns. This allows for addressing the requirements of the DPF application, where an average pore size distribution between 15 and 20 microns in the honeycomb wall is preferred. Since the micro pore and pore size in the filter wall can be controlled through the reaction process, adjustments can be made to produce filters for filtration applications on the nanoparticle scale [7-9].

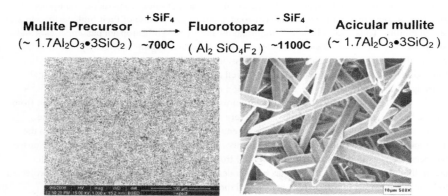

Mullite Precursor $\xrightarrow{+SiF_4}$ **Fluorotopaz** $\xrightarrow{-SiF_4}$ **Acicular mullite**
(~ $1.7Al_2O_3 \bullet 3SiO_2$) ~700C ($Al_2 SiO_4F_2$) ~1100C (~ $1.7Al_2O_3 \bullet 3SiO_2$)

Figure 1. Low temperature desorption of silicon tetrafluoride (SiF_4) from fluorotopaz ($Al_2SiO_4F_2$) leads to formation of mullite with high aspect ratio grains.

Filters used in this study had a diameter of 5.66 inch and a 6 inch length. After extrusion, honeycombs were dried and calcined. Calcined honeycombs were reacted with SiF_4 gas under controlled pressure and time conditions to achieve the desired mullite grain morphology and an average pore size distribution between 12 and 15 microns.

During diesel engine operation, the particulate matter, (referred to as soot), is generated from the incomplete combustion of the diesel fuel. To separate soot from a diesel exhaust stream, wall flow filters are required. There are several criteria in filter design. The most critical are:

- Filter medium should have a proper pore size distribution that enables achievement of 95% filtration efficiency for all regulated particulate matter sizes.

- Filter should have a high porosity so that high soot loadings can be sustained without excessive back pressure generation.

- During uncontrolled regeneration to burn off trapped soot, the filter material should be able to tolerate temperatures of up to 1400°C and potentially large temperature gradients.

- Filter should have sufficient mechanical strength to tolerate the canning process.

- Filter material should have sufficient chemical resistance at high temperatures to tolerate contact with the ash that accumulates in a DPF during usage and good adhesion compatibility with catalyst coatings typically applied to DPF's.

Based upon these application requirements, an ACM DPF was designed to have 60% porosity (or higher) with an average pore size distribution of 12-15 micrometers, a cell density of 200 cells per square inch (cpsi) with a wall thickness of 0.3 mm (14 mil).

Figure 2 shows a Diesel Particulate Filter made from acicular mullite. Physical and geometric characteristics of these filters are summarized in Table 1.

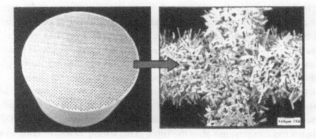

Figure 2. Acicular mullite wall flow filter and its typical microstructure.

Table 1. Summary of ACM filter physical and geometrical properties

Filter diameter	5.66 inch
Filter length	6 inch
Cell density	~ 200 cpsi
Wall thickness	~14 mil
Porosity	~ 60 %
Mean pore size	~12 -15 ☐m
Substrate density	0.5 kg/L

DPF PERFORMANCE VALIDATION

To validate the wall flow type diesel particulate filter (DPF) performance, a series of experiments were performed on both lab reactor and engine bench with real diesel exhaust. A 5.66 x 6" DPF made from ACM was selected as the standard size of DPF for the entire study. The validation included three major areas:

- ACM DPF filtration for both particulate matter (PM) mass based and PM size distribution
- Typical soot loading and regeneration
- Catalyst coating and pressure drop

ENGINE TEST BENCH

 A full-scale engine performance evaluation was designed and conducted to address the light duty diesel engine application. The test engine used in this study was a modified Volkswagen 1.9-liter four-cylinder direct injection (DI) diesel engine with four valves and a common rail injection system. The engine's control system was modified in order to gain free access to all relevant engine components. This engine, with an engine gas recirculation system (EGR), was calibrated to meet Euro III emission standards using standard European diesel fuel. The test bench was equipped with an automatic data acquisition system for the relevant engine-operating parameters, including engine rpm, torque, turbocharger temperature, and pressure. Two online, real-time exhaust analyzers were attached to the engine bench. One analyzer was located "upstream", just after the turbocharger, to directly measure the gas emission composition coming out of the engine. The second analyzer was located "downstream", just after the DPF, to measure the exhaust gas composition after the filter. This configuration of emission gas analyzers provided the capability to monitor gas emission changes during particulate matter loading and regeneration. Two pressure sensors were installed; one upstream and the other downstream of the DPF for measurement of the pressure drop across the DPF. Nine type K thermocouples were installed inside the DPF to monitor filter performance and temperature distribution, especially during the filter regeneration process. The thermocouples were inserted to the locations shown on the left side of Figure 3 by insertion into the appropriate exit side channel. A commercial diesel oxidation catalyst with precious metal loading was inserted upstream of the DPF for part of the study. The canned ACM DPF sample mounted on the engine bench used in this study is shown on the right hand side of Figure 3.

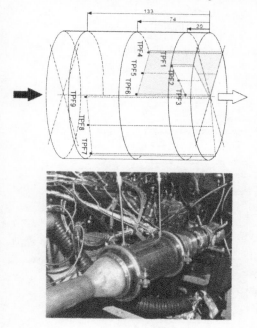

Figure 3. Nine thermocouple locations inside of DPF (left) and the canned ACM DPF on the engine bench (right)

For the nanoparticle filtration study, a modified engine bench was needed in order to accommodate the ACM DPF together with the necessary pressure, temperature, gas and particulate sampling sections upstream and downstream of the filter. The exhaust scheme is shown in Figure 4.

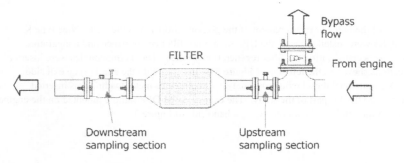

Figure 4. The engine test set up for particle size study

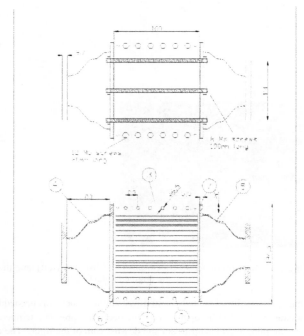

Figure 5. Filter canister assembly

In order to easily remove the DPF, a split-canning system was designed and used. With this design, the tested DPF could be quickly removed or assembled via the bolts shown in Figure 5.

MASS BASED FILTRATION EFFICIENCY

Mass based filtration efficiency was determined by using the industry standard particle sampling procedure [10], which employs two sample ports, one located upstream and the other downstream of the DPF. The exhaust sample passes through a mini-dilution tunnel and enters a Teflon-coated paper filter trap. Particulate matter from the diluted exhaust streams was collected on standard paper filters. This test protocol included total mass collection followed by wet chemical extraction and analysis to determine the total filtration efficiency, the soluble organic fraction (SOF), the non-soluble organic fraction (NSOF), and the sulfate content. After particle sampling and conditioning, the paper filter is weighed. The difference of mass on these paper filters gave the overall particulate matter reduction amount across the DPF and allowed calculation of the overall particulate matter filtration efficiency.

The filtration efficiency of the ACM DPF was determined under engine operating conditions of 8 bar Brake Mean Effective Pressure (BMEP) at 2000 rpm when the filter was fresh. Figure 6 summarizes the results of the filtration efficiency measurements. The overall particulate filtration efficiency by mass was found to be at least 96% when the DPF was fresh. The filtration efficiency increased when the filter was loaded. The non-soluble organic fraction filtration efficiency was 98%. These results demonstrate that the ACM DPF has high filtration efficiency and will satisfy the particulate matter filtration requirements set by current legislation standards. It was also noticed that the overall filtration efficiency for an ACM DPF did not change under the various engine operating conditions in this study.

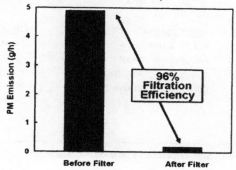

Figure 6. Mass based filtration efficiency of diesel particulate filter made from acicular mullite

MECHANISM OF FILTRATION FOR NANO-SIZE SOOT PARTICLES

In general, diesel particulate filtration can be considered as a process of gas-solid separation. When a gas-solid mixture flows through the porous material medium, the relatively large solid particles will directly impact the porous material surface and form a particle layer while relatively small particles continuously flow through the porous medium. This is typically called collision interruption. Depending on the filter medium characteristics, most of small particles are trapped inside of medium pores (interception filtration). Regarding nano-size particles, both the particle size and their terminal velocity are so small, that they are not impacted or intercepted by the filter medium; however, they can interact with the surface of the filter medium material. It has been found that filters made out of fibers can be significantly more effective in interactions with nanoparticles which move in a typical random Brownian motion. Figure 7 provides a brief description of the different types of filtration mechanisms, depending on the size of the particles.

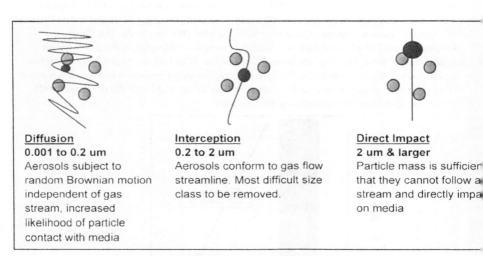

Diffusion
0.001 to 0.2 um
Aerosols subject to random Brownian motion independent of gas stream, increased likelihood of particle contact with media

Interception
0.2 to 2 um
Aerosols conform to gas flow streamline. Most difficult size class to be removed.

Direct Impact
2 um & larger
Particle mass is sufficient that they cannot follow a stream and directly impact on media

Figure 7. Size-Dependent Filtration Mechanisms

Real time on-line measurement to detect small particle sizes and size distribution in hot exhaust streams is a great challenge today. Considering the need for both accurate and fast response, a Scanning Mobility Particle Sizer (SMPS) was used in this study. SMPS is an electrical mobility method based instrument. It is connected to a 3-stage mini diluter system and consists of a Differential Mobility Analyzer (DMA) combined with an Ultrafine Condensation Particle Counter (U-CPC). It provides the concentration amount in each particle size group from 10 to 430 nm. Integration of the concentration over the entire particle size range provides the total particle concentration. SMPS can also be set to measure continuously (real time) number particle concentration for a specific particle size (e.g. 80 nm).

In this study, particle sizes between 10 to 430 nm were examined at three levels of exhaust gas. Soot loadings varied from clean filter status to 0.8 g/m².

Figure 8 presents the upstream and downstream particle size distributions obtained by SMPS at an average linear flow velocity of 1.5 cm/s. Here average linear flow velocity is defined as the exhaust gas across the filter medium surface area, as calculated from the exhaust flow rate and total filter surface areas. The particle size distribution and concentration of particles in the engine exhaust are measured by SMPS. It can be seen from Figure 8 that about 80% filtration efficiency for particle sizes under 100 μm is observed when the ACM DPF is clean i.e. without soot loaded. As soon as small amounts of soot are loaded on the filter, for example, 0.2 g/m², the filtration efficiency reaches above 90-95% for all nanoparticle sizes in this study.

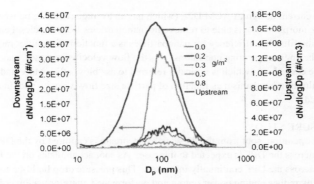

Figure 8. Particle size distributions across DPF during soot loading at flow velocity (FV) of 1.5 cm/s.

Figure 9 shows the same measurement of particle size distribution and particle concentration at a flow velocity of 6.2 cm/s. The general trend observed is the same as in Figure 8; however the filtration efficiency on both of the clean filter and the soot loaded filter is lower than experienced with a flow velocity of 1.5 cm/s. This indicates that an increase of linear velocity removes some of the adsorbed nanoparticles from the filter surface. However, it has been noticed that even when a flow velocity used to generate data presented in Figure 9 is four times higher than in Figure 8, the 60% porous ACM DPF still has over 50% filtration efficiency.

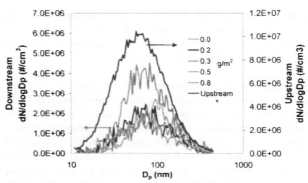

Figure 9. Particle size distributions across DPF during soot loading at FV of 6.2 cm/s.

In general, the current design of ACM DPF (which has an average pore size between 12 and 15 ⊏m and needle morphology), is able to remove the nanoparticles from the diesel engine exhaust. It is clear that the filtration efficiency for nanoparticles is a function of the filter medium structure and the engine operating conditions. As the flow velocity increases, nano-size particles can be carried out (which is typically referred to as "blow off"). More understanding of the diffusion filtration mechanism as function of particle size flow velocity and microstructural design is needed.

BACK PRESSURE STUDY
During engine operation, soot from the diesel exhaust stream is captured by the DPF and the pressure drop across the DPF is expected to increase. As soot accumulates on the DPF, the pressure drop across the DPF continually increases. This pressure drop build-up to certain level will significantly reduce engine power output and performance, therefore, a lower DPF pressure drop is desired. Figure 10 shows the increase of pressure drop across the DPF during engine soot loading. A commercial silicon carbide (SiC) DPF was used as a reference for comparison with the ACM DPF. Both the ACM and the SiC DPFs have the same size, i.e. 5.66"x6" and 200 cpsi and each of the DPFs was loaded under identical engine operating conditions and loading times. The pressure drop across the DPF's was monitored during engine operation. At the end of the soot loading, the filters were removed from the bench and the actual soot loading was determined by the weight difference of the DPF samples at the beginning and end of the soot-loading test. For this testing condition, the soot loading level was about 8 g/L.

As shown in Figure 10, the ACM DPF exhibited a lower pressure drop increase during soot loading for the same size filter and same soot loading. Figure 11 is a comparison of back pressure drop performance as a function of exhaust flow rate after each DPF had been loaded to 8 g/l. The superior low back pressure performance of the ACM filter is most likely due to a combination of ACM's unique surface structure, porosity, pore size distribution, and the deep bed filtration potential inherent in the ACM structure.

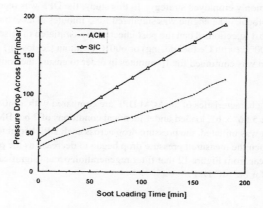

Figure 10. Comparisons of the pressure drop increase across the filter during soot loading at 50Nm @ 3000 RPM for ACM and SiC DPF, Soot loading =8g/l.

As a DPF accumulates particulate matter, the pressure drop across the filter rises. Thus, periodic regeneration of the DPF to remove the accumulated soot is necessary to maintain engine performance

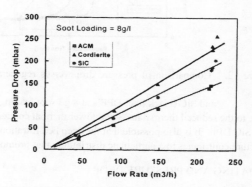

Figure 11. Comparison of the pressure drop across ACM, SiC, and cordierite DPF's.

during normal operation. An effective method is to thermally oxidize the soot by raising the exhaust temperature. The exhaust temperature during this study was increased by varying the engine control parameters. The use of post injection (PI) in conjunction with a diesel oxidation catalyst is a commonly employed strategy. In this study, the DPF was regenerated according to the following protocol. The engine was stabilized for 5 minutes at an exhaust temperature of 180°C without post injection. Then the post injection was initiated to raise the exhaust temperature to 500°C (with Ce fuel dosing) or 600°C (without Ce dosing) in less than 2 minutes. The post injection was continued for 10 minutes in order to ensure the completion of the regeneration.

The regeneration characteristics of an ACM DPF are compared with a SiC DPF in Figure 12. Both DPF's were 5.66" x 6", loaded under identical conditions of 2 bar BMEP to 8 g/L. When the post injection was initiated, the pressure drop across each DPF began to rise as the exhaust gas got hotter. Then the measured pressure drop began to decrease as the particulate matter was oxidized. It is clear from Figure 12 that filter regeneration occurs significantly faster for the ACM DPF than for the SiC DPF.

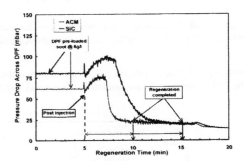

Figure 12.　Comparison of pressure drop versus regeneration time for ACM (red line)

and SiC (black line) DPF's at 8 g/l soot loading.

This is likely due to the reduced thermal mass and lower thermal conductivity of the ACM DPF compared to the SiC DPF. It is also possible that the deep bed filtration that can occur with the ACM microstructure contributes to a particulate distribution that promotes faster soot oxidation.

CATALYST COATING AND PRESSURE DROP

Typically, to effectively regenerate soot and eliminate gaseous emission, a catalyst is coated directly onto the DPF substrate. In general, a catalyst coated on a DPF will result in a back pressure increase. Therefore, a DPF that has a high catalyst loading capacity without increasing back pressure is desired. A catalyst coating process has been studied on diesel particulate filters made from acicular mullite. Figure 13 shows the plot of pressure drop comparison between a diesel particulate filter made from acicular mullite with a 31 g/l (870 g/ft^3) catalyst coating, and an uncoated commercial SiC DPF (as reference B). It has been found that at a 2.2 m^3/min (80 CFM) flow rate, the clean SiC DPF has a pressure drop that is 30% higher than the catalyst coated ACM DPF.

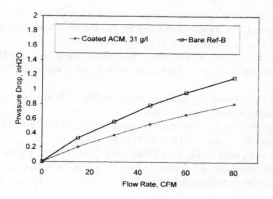

Figure 13. Pressure drop of catalyst coated ACM DPF (red line)
vs. uncatalyzed SiC DPF (blue line)

CONCLUSIONS

Technology to produce high porosity ceramic structures characterized by a controlled pore size distribution and interconnected acicular mullite grain structure has been developed by The Dow Chemical Company. The ability to control microstructure allows acicular mullite diesel particulate filters to be tailored to meet the requirements of deep bed filtration and nanoparticulate emission control. Diesel particulate filters made from acicular mullite with well-controlled process conditions have illustrated high filtration efficiency, low pressure drop, and fast regeneration. The acicular mullite DPF tested in this work showed more than 96% filtration efficiency in mass based soot and higher than 80% filtration efficiency or high for particle sizes between 10-430 nm when the flow velocity is at 1.5 cm/s. This already meets the specification range of the current Euro IV and Euro V regulations. However, the diesel particulate filter made from acicular mullite with variations in the controlled process should be able to provide even higher nanoparticle size filtration efficiency that is expected in future Euro VI legislation. This is due to the flexibility of microstructure design in acicular mullite system.

REFERENCES
1. Konstandopoulos A.G., Kostoglou M., Skaperdas E., Papaioannou E., Zarvalis D. and Kladopoulou E. (2000) "Fundamental Studies of Diesel Particulate Filters: Transient Loading, Regeneration and Aging", SAE Tech.Paper 2000-01-1016 (SP-1497).
2. Konstandopoulos, A.G. & Johnson, J.H., (1989) "Wall-Flow Diesel Particulate Filters-Their Pressure Drop and Collection Efficiency", SAE Trans. 98 sec. 3 (J. Engines) Paper No. 890405, pp. 625-647.
3. A. Mayer et al. "VERT particulate trap verification", SAE 2002-01-0435
4. J.R. Moyer and N.N. Houghes, A Catalytic Process for Mullite Whiskers, Journal of American Ceramic Society, 77,4, 1083-86, 1994
5. J.R. Moyer- Phase Diagram for Al2O3-SiF4, unpublished work, (1993).
6. J. R. Moyer – Phase Diagram for Mullite-SiF4, J. Am. Ceram. Soc.,78(12), 3253-58, (1995)
7. Cheng G. Li etc., Properties and Performance of Diesel Particulate Filters of An Advanced Ceramic Material, SAE 2004-01-0955
8. F. Mao and Cheng G. Li, Validation of Advanced Diesel Particulate Filter With High Catalyst Loading Capacity, SAE 2005 FFL-40, 2005
9. A. J. Pyzik and Cheng G. Li, New Design of a Ceramic Filter for Diesel Emission Control Application, International Journal of Applied Ceramic Technology, 2,6, 440-51, 2005
10. PM collection methodology definition, U. S. Environmental Protection Agency (EPA) 1990, www.epa.gov.

MICROSTRUCTURAL DEVELOPMENT OF POROUS β-Si₃N₄ CERAMICS PREPARED BY PRESSURELESS-SINTERING COMPOSITIONS IN THE Si-Re-O-N QUATERNARY SYSTEMS (Re=La, Nd, Sm, Y, Yb)

Mervin Quinlan, D. Heard and Kevin P. Plucknett[1]
Materials Engineering Program, Department of Process Engineering and Applied Science, Dalhousie University, 1360 Barrington Street, Halifax, Nova Scotia, B3J 1Z1, CANADA

Liliana Garrido
CONICET, Centro de Tecnologia de Recursos Minerales y Cerámica (CETMIC, CIC-CONICET-UNLP), Cam. Centenario y 506, C.C.49 (B 1897 ZCA) M.B. Gonnet. Pcia. De Buenos Aires, Argentina

Luis Genova
IPEN Instituto de Pesquisas Energéticas e Nucleares, CCTM Centro de Ciência e Tecnologia de Materiais, Cidade Universitária, Travessa R 400, 05508-900 São Paulo, Brazil

ABSTRACT

It has recently been demonstrated that porous β-Si₃N₄ can be engineered to provide similar mechanical properties to dense β-Si₃N₄, with the additional benefits of reduced mass. Porous β-Si₃N₄ applications could potentially include filtration media, thermal shock resistant components and high-strength, lightweight structural components. Prior work in Japan has shown that flexural strengths exceeding 1 GPa can be achieved by tailoring the microstructure of the β-Si₃N₄ grains, such that moderately high aspect ratios are obtained after sintering (up to 10:1) with considerable anisotropic grain alignment, while maintaining porosity levels of ~10-15 %. In the present study, β-Si₃N₄ ceramics containing higher porosity levels are generated via pressureless sintering in a nitrogen atmosphere, with the addition of single rare earth oxide sintering aids. The oxide additives have been chosen from both the lanthanide (La, Nd, Sm, Yb) and Group IIIB (Y) elements. In each case, the compositions were milled for 24 hours, cold isostatically pressed and then sintered in nitrogen (0.1 MPa) at temperatures ranging from 1500 to 1750°C. Post sinter microstructural characterization was performed using scanning electron microscopy and x-ray diffraction. It is noted that densification and phase transformation kinetics are both strongly influenced by the oxide chosen.

INTRODUCTION

In recent years there has been an increasing drive to develop high performance porous ceramics. These materials have a wide range of potential applications, including gas and molten-metal filters, bio-prosthetic implants and bioreactor supports. In addition, they may be considered for use as lightweight structural materials, with some porous ceramics exhibiting properties comparable to their dense equivalents. Porous silicon nitride (Si₃N₄) ceramics have been shown to exhibit an excellent combination of mechanical and thermal properties, and microstructurally tailored β-Si₃N₄ can possess strengths above 1 GPa while retaining porosity levels of ~15 vol. %.[1,2] This response is achieved through the growth of highly anisotropic and interlocking β-Si₃N₄ grains. The use of single sintering aids that do not promote significant

---

[1] Address all correspondence to this author (kevin.plucknett@dal.ca)

densification, but do promote the α- to β-Si₃N₄ transformation and hence the formation of an anisotropic grain structure, are preferred. Sintering additives that can achieve this include many of the lanthanide or Group IIIB elements (e.g. Y_2O_3, Yb_2O_3, Sm_2O_3, La_2O_3, Lu_2O_3, etc.).[3-5] This approach results in sintered ceramics with extremely fine scale porosity, and the materials can be expected to retain good mechanical properties. Texture development can be applied to further strengthen these materials, using either aligned β-Si₃N₄ seed crystals or sinter-forging.[1,2]

The present work reports upon the formation of porous β-Si₃N₄ ceramics using a variety of single metal oxide additions. The additives selected, notably La_2O_3, Nd_2O_3, Sm_2O_3, Y_2O_3 and Yb_2O_3, belong to either the lanthanide or Group IIIB elements, and are typical sintering additives for the preparation of α- to β-Si₃N₄ ceramics. The influence of additive upon densification behavior, α- to β-Si₃N₄ transformation and the final sintered microstructure will be discussed.

EXPERIMENTAL PROCEDURES

All samples in the current work were prepared using Ube SN E-10 α-Si₃N₄ powder, together with the rare earth oxides summarized in Table I. The compositions that will be discussed in the present work are all based on single oxide sintering additions. The baseline composition was prepared with 5 wt. % Y_2O_3, following prior work,[3] while the remaining compositions were prepared with molar percentages equivalent to the Y_2O_3 example (La_2O_3 was pre-calcined at 900°C for 2 hours prior to mixing to decompose any stable hydroxide component). In each case, 50 g batches of the appropriate composition were prepared by ball milling for 24 h in isopropyl alcohol using TZP milling media.

Milled powders were then dried and sieved through a 75 μm mesh stainless steel sieve. The powders were then uniaxially pressed (~31 MPa) into 31 mm diameter x 4 mm thick discs, vacuum bagged and cold-isostatically pressed (~175 MPa). Pressureless sintering was performed in a graphite resistance furnace with the samples sited within a powder bed comprised of 50 wt. % BN/50 wt. % Si₃N₄, held in a high purity graphite crucible. Sintering was conducted under a static nitrogen atmosphere (0.1 MPa) to minimize Si₃N₄ dissociation, with a heating rate of 10°C/min used up to the sintering temperature, which was held for 2 hours, followed by cooling at 30°C/min until the furnace cooled naturally at slower rates.

Sintering aid	Purity (wt. %)	Supplier
La_2O_3	99.99	Metall Rare Earth Ltd., Shenzhen, China
Nd_2O_3	99.99	Treibacher Industries, Toronto, Canada
Sm_2O_3	99.99	Treibacher Industries, Toronto, Canada
Y_2O_3	99.99	Metall Rare Earth Ltd., Shenzhen, China
Yb_2O_3	99.99	Treibacher Industries, Toronto, Canada

Table I. Summary of the oxide sintering additives used in the present work, together with their purity and supplier information.

Due to the high volume of retained porosity, the densities of sintered samples were determined by immersion in mercury. Post-sinter microstructure characterisation was performed using x-ray diffraction (XRD) and field emission scanning electron microscopy (FE-SEM) of carbon coated fracture surfaces. The volume fraction of β-Si₃N₄ was determined from XRD using the relationship:

$$R_\beta = 1.4434 \times \frac{I_\beta(101)}{I_\beta(101) + I_\alpha(201)} - 0.4434 \left[\frac{I_\beta(101)}{I_\beta(101) + I_\alpha(201)} \right]^2 \times 100 \qquad \text{Eqn. 1}$$

where R_β is the fraction of β-Si₃N₄ formed and I_α and I_β are the measured intensities of the respective XRD peaks.[6]

RESULTS AND DISCUSSION

The densification behavior of Si₃N₄ samples prepared with selected Group IIIB and lanthanide metal oxides is shown in Figure 1. Data is presented as a percentage of theoretical density, based upon a theoretical value determined from a rule of mixtures taking into account the densities of the individual components. It is apparent that the use of single oxides alone does not promote significant densification for any of the examined compositions. This behaviour can be generally anticipated from prior studies.[3,7] In particular, it is apparent from Figure 1 that samples prepared with Y₂O₃ additions show minimal densification above 1500°C.

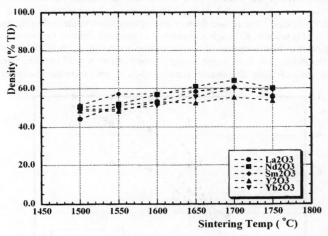

Figure 1. Densification behavior for porous β-Si₃N₄ ceramics prepared with single oxide additions of various lanthanide and Group IIB elements.

One of the major issues with preparing β-Si₃N₄ ceramics by pressureless-sintering is the tendency towards decomposition at the higher sintering temperatures. Potential decomposition mechanisms have been reviewed and discussed in a comprehensive study by Lange.[8] It can be anticipated that such decomposition mechanisms will be exacerbated when preparing porous β-Si₃N₄ ceramics, as the effective surface area will be dramatically higher throughout the sintering cycle. For the current materials there is an approximately linear increase in weight loss with increasing temperature for each of the additives examined. Unsurprisingly, the highest weight losses were observed at 1750°C, while minimal weight losses occurred at 1500°C. At 1750°C, the highest losses were observed with Sm_2O_3 and Yb_2O_3 additions (~7.5 wt. %), while the lowest were observed with Y_2O_3 additions (~4 wt. %).

An even more pronounced difference in the behavior of the individual oxide additives is apparent when assessing the α- to β-Si₃N₄ transformation kinetics (Figure 2). The use of pure Sm_2O_3, Y_2O_3 or Yb_2O_3 as a sintering aid promotes the transformation of α- to β-Si₃N₄, particularly above 1600°C, with complete transformation observed at 1700°C for all three additives (Figure 3) and nearly complete transformation at 1650°C for both Sm_2O_3 and Yb_2O_3. In comparison, when using either La_2O_3 or Nd_2O_3 additions, there is a significant reduction in the extent of α- to β-Si₃N₄ transformation for the higher sintering temperatures, although at 1500°C both exhibit higher transformation than the other three additives. For the case of La_2O_3, even after heat-treatment at 1750°C for 2 hours, some residual α-Si₃N₄ is retained.

These general observations are in accordance with the recent work of Dai et al, who examined the α- to β-Si₃N₄ transformation kinetics for a wider selection of trivalent lanthanide oxide additions (Ce, Dy, Er, Eu, Gd, Nd, Sm, Yb).[9] It is notable that the work of Dai et al also involved examination of nominally pressureless sintered materials, prepared at temperatures up to 1800°C. Conversely, in the work of Kitayama et al, the reported levels of α- to β-Si₃N₄ transformation were severely retarded during heat-treatment at temperatures up to 1850°C for Gd, La, Nd and Yb oxides.[10] The reason for this discrepancy when compared to the present study and that of Dai et al is unclear, although Kitayama et al assessed samples that were pre-densified by hot pressing prior to high temperature heat-treatment, such that any potential contributions to α- to β-Si₃N₄ transformation from vapor transport mechanisms would be eliminated. Work is presently ongoing to more fully characterize the α- to β-Si₃N₄ transformation behavior.

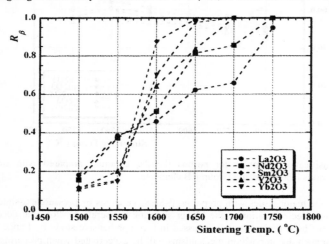

Figure 2. The fraction of β-Si₃N₄ formed during sintering, determined from XRD using Eqn. 1.

Figure 3 demonstrates the early stage microstructural changes that occur when sintering with Y_2O_3 additions. In this instance, at 1500 and 1550°C, there is only limited formation of β-Si_3N_4 (Figure 2) and the microstructure retains the typical globular appearance due to a predominance of equiaxial α-Si_3N_4 particles. Additionally, at 1550°C there is increased evidence of liquid wetting and the formation of a more continuous structure than at the lower temperature. Conversely, by 1650°C, there is clear evidence of β-Si_3N_4 formation, with a high percentage of high aspect ratio grains; at this stage the microstructure is comprised of approximately 20 % α- and 80 % β-Si_3N_4, as noted from XRD (Figure 4a). After sintering at 1750°C for 2 hours, β-Si_3N_4 grain growth is ready apparent.

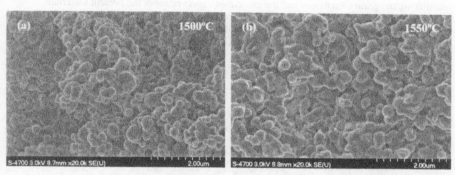

Figure 3. Microstructure development in Si_3N_4 prepared with 5 wt. % Y_2O_3 and pressureless sintered at (a) 1500 and (b) 1550°C. Note the apparent increase in liquid wetting and particle connectivity when sintering at 1550°C.

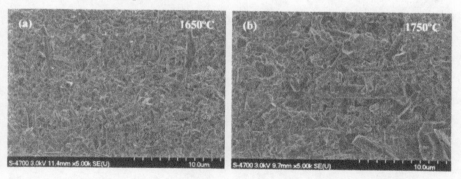

Figure 4. Microstructure development in Si_3N_4 prepared with 5 wt. % Y_2O_3 and pressureless sintered at (a) 1650 and (b) 1750°C. Note the β-Si_3N_4 coarsening that has occurred after sintering at 1750°C.

The microstructures of samples sintered at 1750°C with the remaining oxide additives are presented in Figure 5. The use of either La_2O_3 or Nd_2O_3 results in some degree of grain size refinement when compared to the other additives (including Y_2O_3). In particular, there is significant coarsening for Yb_2O_3 additions relative to La_2O_3. Qualitatively, it is also apparent that the La_2O_3 containing samples exhibit the highest aspect ratios. Becher *et al* have assessed the effects of different rare earth oxides on aspect ratio development in high-density β-Si₃N₄ ceramics.[11] It was demonstrated that La_2O_3 additions typically promote the highest grain aspect ratios. Related to the present study, they noted that the aspect ratio increased in the order Yb < Y < La; similar observations were also made by Satet and Hoffmann.[12] Future work is aimed at quantifying the grain dimensions and the associated aspect ratios of the present materials.

High magnification imaging of samples prepared with Sm_2O_3 additions highlights several microstructural features typical of these materials (Figure 6). The fracture surfaces demonstrate glass 'residues' on the surface of β-Si₃N₄ grains (it is likely that these are actually crystalline or partially crystalline oxynitride mixtures), which include recessed regions where other grains had previously been bonded. Similar residues are also noted bonding together β-Si₃N₄ grain clusters.

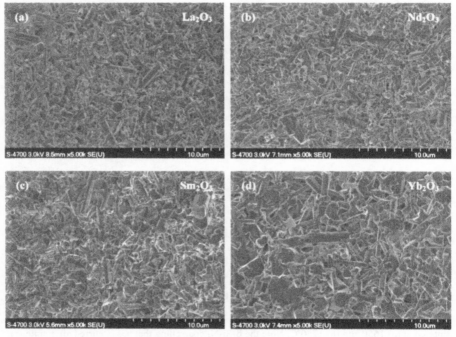

Figure 5. Microstructure development of Si₃N₄ ceramics pressureless sintered at 1750C for 2 hours, prepared with sintering additions of (a) La_2O_3, (b) Nd_2O_3, (c) Sm_2O_3, (d) Yb_2O_3 (c.f. Figure 4(b) showing the microstructure formed with Y_2O_3 addition).

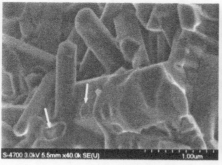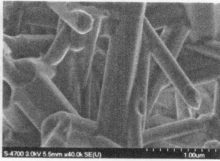

Figure 6. High magnification FE-SEM images of the typical structure development obtained after sintering at 1750°C with Sm_2O_3 sintering additions. (a) Glassy/crystalline oxynitride regions are apparent, together with broken contact points from grains removed during fracture (arrowed). (b) Similar regions are seen bonding individual β-Si₃N₄ grains.

CONCLUSIONS

An initial investigation of the pressureless sintering behavior of porous Si_3N_4 ceramics has been undertaken, utilizing single oxide additions of either lanthanide (La_2O_3, Nd_2O_3, Sm_2O_3, Yb_2O_3) or Group IIIB (Y_2O_3) elements. It was apparent that the oxide additive selected has a significant effect upon the α- to β-Si₃N₄ transformation kinetics and, to a certain extent, the densification behavior. Higher levels of densification were noted with La_2O_3, Nd_2O_3 and Yb_2O_3. Retained porosity levels varied between 35 and 45 vol. % when sintering at 1700 and 1750°C. Additions of Sm_2O_3, Y_2O_3 and Yb_2O_3 were observed to promote the α- to β-Si₃N₄ phase transformation. In particular, essentially complete transformation was achieved with Sm_2O_3 additions at temperatures as low as 1650°C, after a hold at temperature of 2 hours. Conversely, when La_2O_3 additions were used, residual α-Si₃N₄ was still present after sintering at 1750°C for 2 hours.

FE-SEM examination was used to demonstrate the microstructural changes that occurred during sintering. After sintering at 1750°C for 2 hours, it was observed that Y_2O_3 and Yb_2O_3 additions result in the most coarse grained β-Si₃N₄ structure, while La_2O_3 and Nd_2O_3 produce the finest microstructures. Qualitatively, it was also apparent that La_2O_3 produces grains with the highest aspects ratios, in accordance with previous studies on multi-additive systems; future work is targeted at quantitative evaluation of aspect ratio as a function of oxide additive. In addition, as part of a broader study, the influence of additive content will be examined, together with additional trivalent oxides within both the lanthanide and Group IIIB elements.

ACKNOWLEDGEMENTS

The authors would like to acknowledge NSERC for provision of funding through the program Inter-American Research in Materials (CIAM).

REFERENCES

1. Y. Inagaki, Y. Shigegaki, M. Ando and T. Ohji, *J. Eur. Ceram. Soc.*, **24** 197-200 (2004).
2. Z.-Y. Deng, Y. Inagaki, J. She, Y. Tanaka, Y.-F. Liu, M. Sakamoto and T. Ohji, *J. Am. Ceram. Soc.*, **88** 462-465 (2005).

3. K.P. Plucknett and M.H. Lewis, *J. Mater. Sci. Lett.*, **17** 1987-1990 (1998).
4. J.-F. Yang, Z.-Y. Deng and T. Ohji, *J. Eur. Ceram. Soc.*, **23** 371-378 (2003).
5. J. Yang, J.-F. Yang, S.-Y. Shan, J.-Q. Gao and T. Ohji, *J. Am. Ceram. Soc.*, **89** 3843-3845 (2006).
6. T. Hirata, K. Akiyama and T. Morimoto, *J. Eur. Ceram. Soc.*, **20** 1191-1195 (2000).
7. S. Hampshire and K.H. Jack, pp. 225-230 in *Progress in Nitrogen Ceramics*, Ed. F.L. Riley, Martinus Nijhoff Publishers, The Hague, 1983.
8. F.F. Lange, *J. Am. Ceram. Soc.*, **65** C120-121 (1982).
9. J. Dai, J. Li and Y. Chen, *Mater. Chem. Phys.*, **80** 356-359 (2003).
10. M. Kitayama, K. Hirao and S. Kanzaki, *J. Am. Ceram. Soc.*, **89** 2612-2618 (2006).
11. P.F. Becher, G.S. Painter, N. Shibata, R.L. Satet, M.J. Hoffmann and S.J. Pennycook, *Mater. Sci. Eng.*, **A422** 85-91 (2006).

COMPOSITIONAL DESIGN OF POROUS β-Si₃N₄ PREPARED BY PRESSURELESS-SINTERING COMPOSITIONS IN THE Si-Y-Mg-(Ca)-O-N SYSTEM

Mervin Quinlan and Kevin P. Plucknett[1]
Materials Engineering Program, Department of Process Engineering and Applied Science, Dalhousie University, 1360 Barrington Street, Halifax, Nova Scotia, B3J 1Z1, CANADA

Liliana Garrido
CONICET, Centro de Technologia de Recursos Minerales y Cerámica (CETMIC, CIC-CONICET-UNLP), Cam. Centenario y 506, C.C.49 (B 1897 ZCA) M.B. Gonnet. Pcia. De Buenos Aires, ARGENTINA

Luis Genova
IPEN Instituto de Pesquisas Energéticas e Nucleares, CCTM Centro de Ciência e Technologia de Materiais, Cidade Universitária, Travessa R 400, 05508-900 São Paulo, BRAZIL

ABSTRACT

Porous β-Si₃N₄ can be microstructurally engineered to provide similar mechanical properties to dense β-Si₃N₄, with the additional benefits of reduced mass. However, while flexural strengths exceeding 1 GPa can be achieved by tailoring the microstructure of the material, through alignment of anisotropic β-Si₃N₄ grains, the porosity levels of such materials are still relatively low (< 15 vol. %). This also requires moderately complex processing techniques to be employed, such as tape-casting and sinter-forging. In the present study, an alternate approach is taken where compositional design is advocated, with the aim of producing a more porous β-Si₃N₄ structure (ideally 20-40 vol. % porosity) with high grain aspect ratios (i.e. >10:1). This approach makes use of a low volume fraction of multiple sintering aids, where each additive plays one or more roles in the sintering behavior of Si₃N₄ (i.e. densification aid, transformation aid and/or whisker growth aid). Compositions are based on various ratios of Y₂O₃:MgO, both with and without CaO additions. Sintering has been conducted in a nitrogen atmosphere (0.1 MPa) between 1400 and 1700°C. The effects of processing parameters (e.g. composition, sintering temperature and time) will be discussed, paying particular attention to microstructure development, including both the densification behavior and α- to β-Si₃N₄ transformation kinetics.

INTRODUCTION

There is increasing interest in the development of porous ceramics for a number of advanced applications, including use as filters, bio-prosthetic implants, bioreactor supports, and lightweight structural materials. Among the materials that are used in such applications, silicon nitride (Si₃N₄) ceramics possess arguably the most favorable combination of mechanical and thermal properties. Several approaches have been developed to prepare β-Si₃N₄ based ceramics with either nano-, micro- or macro-porosity. For the preparation of materials with nano-/micro-porosity, the primary approach involves the selection of a single sintering aid that promotes the α- to β-Si₃N₄ transformation, but has little beneficial effect on densification.[1-5] Suitable sintering aids include many of the lanthanide or Group IIIB elements (e.g. La₂O₃, Lu₂O₃, Sc₂O₃, Sm₂O₃, Y₂O₃, Yb₂O₃, etc). The result of this approach is to produce materials with extremely fine scale

---

[1] Author to whom correspondence should be addressed (kevin.plucknett@dal.ca)

porosity, between the individual β-Si₃N₄ grains. With this approach, the materials can be expected to retain good mechanical properties, through the attainment of an anisotropic, interlocking β-Si₃N₄ grain structure. Extending this approach, through the use of controlled seeding and texture development, strengths in excess of 1 GPa can be achieved for materials with a porosity content of less than 15 vol. %, in combination with high toughness.[6,7]

Alternatively, macro-porosity can be controlled through the addition of fugitive pore formers, which can be either particulate (e.g. starch) or continuous in nature (e.g. open cell foams).[8-10] Recently, novel processes based on suspension freezing have also been developed, where displacement of the ceramic powder occurs at the solidification front, resulting in interconnected porosity after sublimation of the suspension medium (e.g. water, camphene).[11]

The present work reports upon a new approach to develop nano-/micro-porous β-Si₃N₄ ceramics through the use of a low volume fraction of multiple sintering additives, where each additive component is chosen to promote one or more of the following functionalities: (1) α- to β-Si₃N₄ phase transformation, (2) anisotropic β-Si₃N₄ grain growth, or (3) densification. This method is partly adapted from the prior work of Pyzik *et al* in developing tough, *in-situ* reinforced β-Si₃N₄ ceramics.[12] It is possible to present each potential oxide additive on a "functionality map" that highlights the role(s) of that specific additive in sintering β-Si₃N₄ ceramics, as shown in Figure 1. In this figure, the position of each additive is dictated by its role(s) in sintering; for example an additive that contributes to each of functions 1-3 will sit in the middle of the triangle, while one that only contributes to one role will sit in that respective corner, etc.

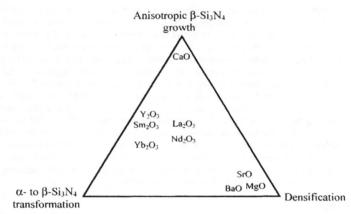

Figure 1. Schematic Si₃N₄ sintering aid "functionality map" (adapted in part from ref. 12).

EXPERIMENTAL PROCEDURES

All samples in the current work were prepared using Ube SN E-10 α-Si₃N₄ powder, together with the oxides summarized in Table I. The compositions that were prepared in this study are summarized in Table II. In each case, 50 g batches of the appropriate composition were prepared by ball-milling for 24 h in isopropyl alcohol using TZP media, followed by drying and sieving through a 75 μm mesh stainless steel sieve. The powders were then uniaxially pressed (at ~31 MPa) into 31 mm diameter x 4 mm thick discs, followed by cold-isostatic pressing at ~175 MPa. Sintering of MgO containing samples was conducted within a powder bed comprised of 50 wt. % BN/49 wt. % Si_3N_4/1 wt. % MgO, while the pure Y_2O_3 additives samples were sintered in a mixture of 50 wt. % Si_3N_4/50 wt. % BN. All samples were sintered within a graphite crucible, under a static nitrogen environment (0.1 MPa), at temperatures between 1400 and 1750°C for a period of two hours. Sintered densities were subsequently determined via immersion in mercury.

Post-sinter microstructure characterization included evaluation using x-ray diffraction (XRD) and field emission scanning electron microscopy (FE-SEM). The volume fraction of β-Si_3N_4 formed, R_β, relative to residual α-Si_3N_4, was determined from XRD using the following relationship[13]:

$$R_\beta = 1.4434 \times \frac{I_\beta(101)}{I_\beta(101) + I_\alpha(201)} - 0.4434 \left[\frac{I_\beta(101)}{I_\beta(101) + I_\alpha(201)} \right]^2 \times 100$$

The aspect ratio of β-Si_3N_4 grains in individual samples was ascertained through room temperature dissolution of small sections of material in concentrated HF solution, followed by filtering of the resulting suspension, dispersion of the particles onto a polished aluminum stub and examination of the dried product in the FE-SEM.

Sintering aid	Purity (at. %)	Supplier
Y_2O_3	99.99	Metall rare Earth Ltd., Shenzhen, China
MgO	99.9	Inframat Advanced Materials, Farmington, CT
CaO	99.95	Alfa Aesar, Ward Hill, MA

Table I. Summary of the oxide sintering additives used, their purity and supplier.

Sample ID	Composition (wt. %)			
	Si_3N_4	Y_2O_3	MgO	CaO
SN5Y	95	5	-	-
SN5YM	97	2.5	0.5	-
SN2YM	97	2	1	-
SNYM	97	1.5	1.5	-
SN5YM01C	97	2.5	0.5	0.1
SN5YM05C	97	2.5	0.5	0.5

Table II. Compositional designations for the porous Si_3N_4 samples prepared in the present work. Note that samples with CaO additions are prepared with a baseline composition of SN5YM.

RESULTS AND DISCUSSION

The influence of additive composition upon densification behavior is shown in Figure 2 for samples prepared with either 5 wt. % Y_2O_3 additions or varying Y_2O_3:MgO ratios. It is apparent that Y_2O_3 alone does not promote densification, as anticipated from prior studies.[1] When examining the mixed Y_2O_3:MgO additions, it can be seen that increasing MgO content increases the level of densification. The additive levels for these samples are significantly lower than used for conventional pressureless-sintering of Si_3N_4-based ceramics to high density.

The use of pure Y_2O_3 as a sintering aid promotes the transformation of α- to β-Si_3N_4, particularly above 1600°C, with complete transformation observed at 1700°C (Figure 3). It is apparent that the addition of MgO to the mixture significantly lowers the transformation temperature, such that complete transformation is observed for the 5:1 ratio (SN5YM) when sintering at 1600°C and above. Increasing the relative content of MgO results in a slight reduction in the extent of α- to β-Si_3N_4 transformation for a given sintering temperature. Previous studies of the rheological behavior of Si-Y-Mg-O-N glasses demonstrates that increasing the Y:Mg ratio results in an increase in the glass transition and softening temperatures.[14] Such a response may be expected to be reflected in the relative viscosity of such glasses at any given temperature. An increase in viscosity may then be expected to decrease sintered density and increase the α- to β-Si_3N_4 transformation, as noted in the present case, based on the prior arguments of Hampshire and Jack.[15]

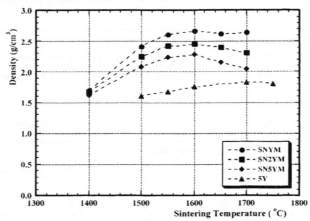

Figure 2. Density as a function of sintering temperature for samples prepared with varying ratios of Y_2O_3:MgO, as well as with single Y_2O_3 additions.

Figure 4 highlights the microstructural changes occurring when sintering the high Y_2O_3:MgO ratio SN5YM composition. The transition from the fine, equiaxed α-Si_3N_4 to the elongated β-Si_3N_4 structure is clearly apparent at 1600°C and above. As noted earlier, these compositions are essentially fully transformed to β-Si_3N_4 at this temperature. The effect of Y_2O_3:MgO ratio upon the sintered microstructure, for a fixed sintering temperature of 1700°C, is shown in Figure 5. Qualitatively, is it is apparent that the β-Si_3N_4 aspect ratio decreases slightly with increasing MgO content in the sintered material. As a consequence of this, the higher Y_2O_3:MgO ratio composition was selected for further study through small additions of CaO which are designed to promote further anisotropic β-Si₃N₄ grain growth, as outlined in Figure 1.

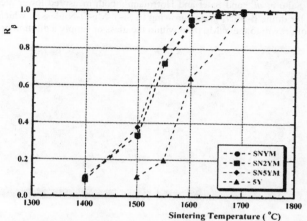

Figure 3. The fraction of β-Si₃N₄ formed during sintering of Y_2O_3/MgO additive compositions between 1400 and 1700°C, and Y_2O_3 additive compositions between 1500 and 1750°C.

The effects of small CaO additions on the sintering behavior of the highest Y_2O_3:MgO ratio composition, when sintering at 1650°C for 2 hours, are summarized in Table III. Even small CaO contents produce a significant increase in densification, although the porosity content after sintering still exceeds 25 vol. %. It may be expected that CaO additions will further lower the glass viscosity at typical sintering temperatures, resulting in increased densification.

Composition	Weight loss (%)	Sintered density (g.cm⁻³)	Nominal retained porosity (vol. %)
SN5YM	2.52	2.15	33.2
SN5YM01C	1.33	2.28	28.9
SN5YM05C	1.37	2.29	28.8

Table III. The effects of CaO additions upon the sintering behavior of the baseline composition SN5YM at 1650°C.

The primary aim of incorporating small CaO additions is to promote the growth of anisotropic β-Si₃N₄ grains during sintering. Measurement of grain dimensions, following their extraction from samples sintered at 1650°C, demonstrates that CaO incorporation does increase the grain aspect ratio (Figure 6). It is apparent from this figure that aspect ratios greater than 25:1 can be achieved with additions of CaO to the 5:1 ratio Y_2O_3:MgO composition. It can also be seen that there is an increase in the maximum grain widths and lengths, which is further reflected in a quantitative manner in Table IV. Table IV demonstrates a mean aspect ratio of ~10.8:1 is obtained for the CaO-free material, which is increased by approximately 12 % when 0.5 wt. % CaO is added. Saito et al[16] and Satet and Hoffmann[17] both measured lower mean grain aspect ratios of 4-6:1 for dilute β-Si₃N₄ grains growing in Si-Y-Mg-O-N glasses, which may represent the sensitivity of growth to the Y:Mg ratio within the glass or simply a dilution or porosity effect.

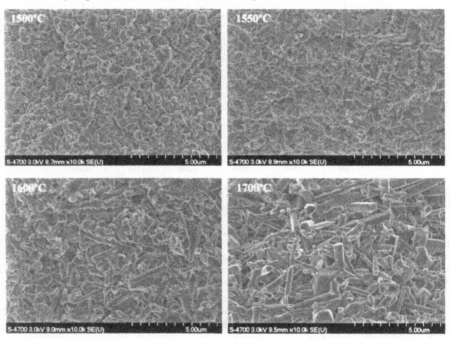

Figure 4. SEM images of microstructural evolution for the composition SN5YM (i.e. Y₂O₃:MgO ratio of 5:1) at sintering temperatures between 1500 and 1700°C. The change from equiaxial α-Si₃N₄ to elongated β-Si₃N₄, with increasing temperature, is readily apparent.

Composition	Length (μm)	Width (μm)	Aspect ratio
SN5YM	3.84	0.355	10.8
SN5YM05	4.17	0.345	12.1

Table IV. The mean β-Si₃N₄ length, width and aspect ratio for β-Si₃N₄ samples prepared with a Y₂O₃:MgO ratio of 5:1, both without and with 0.5 wt. % CaO addition. Mean values are the average of measurements on 100 extracted β-Si₃N₄ grains for each composition

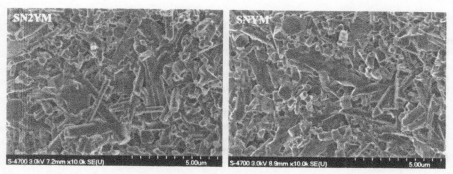

Figure 5. SEM micrographs of the structure of compositions SN2YM (i.e. Y_2O_3:MgO ratio of 2:1) and SNYM (i.e. Y_2O_3:MgO ratio of 1:1) sintered at 1700°C. From the figure, it is qualitatively apparent that increasing the Y_2O_3:MgO ratio increases the grain aspect ratio (c.f. SN5YM sintered at 1700°C in Figure 4).

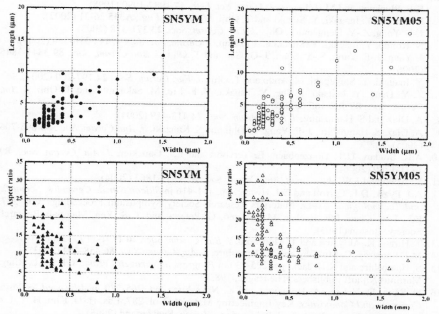

Figure 6. The measured length-width and aspect ratio-width distributions for β-Si₃N₄ grains in compositions SN5YM and SN5YM05C sintered at 1650°C. The measured distributions are plotted on the same scale for direct comparison.

CONCLUSIONS

The current study has demonstrated that porous β-Si_3N_4 based ceramics can be prepared using a low volume fraction of multiple sintering additives (i.e. Y_2O_3, MgO and CaO). This approach has potential benefits over prior methods to prepare porous Si_3N_4 ceramics in that the additives can be selected based on their specific "functionality" during sintering (e.g. densification aid, transformation aid, whisker growth agent, etc.). It has been shown that through this approach, the density (porosity) can be tailored, and that high aspect ratios can be achieved through the addition of oxides known to promote β-Si_3N_4 whisker growth. This work is currently being extended to assess the effects of alternate lanthanide/Group IIIB oxides, as replacements for Y_2O_3, and to evaluate the resulting mechanical behavior of these materials. Combined compositional and microstructural design approaches are also being evaluated, in order to further promote high aspect ratio β-Si_3N_4 whisker growth.

ACKNOWLEDGEMENTS

The authors would like to acknowledge NSERC for provision of funding through the program Inter-American Research in Materials (CIAM).

REFERENCES

1. K.P. Plucknett and M.H. Lewis, *J. Mater. Sci. Lett.*, **17** 1987-1990 (1998).
2. N. Kondo, Y. Inagaki, Y. Suzuki and T. Ohji, *Mater. Sci. Eng.*, **A335** 26-31 (2002).
3. J.-F. Yang, Z.-Y. Deng and T. Ohji, *J. Eur. Ceram. Soc.*, **23** 371-378 (2003).
4. N. Kondo, Y. Inagaki, Y. Suzuki and T. Ohji, *J. Ceram. Soc. Japan*, **112** 316-320 (2004).
5. J. Yang, J.-F. Yang, S.-Y. Shan, J.-Q. Gao and T. Ohji, *J. Am. Ceram. Soc.*, **89** 3843-3845 (2006).
6. Y. Inagaki, Y. Shigegaki, M. Ando and T. Ohji, *J. Eur. Ceram. Soc.*, **24** 197-200 (2004).
7. Z.-Y. Deng, Y. Inagaki, J. She, Y. Tanaka, Y.-F. Liu, M. Sakamoto and T. Ohji, *J. Am. Ceram. Soc.*, **88** 462-465 (2005).
8. A. Diaz and S. Hampshire, *J. Eur. Ceram. Soc.*, **24** 413-419 (2004).
9. A. Diaz, S. Hampshire, J.-F. Yang, T. Ohji and S. Kanzaki, *J. Am. Ceram. Soc.*, **88** 698-706 (2005).
10. A.R. Studart, U.T. Gonzenbach, E. Tervoort and L.J. Gauckler, *J. Am. Ceram. Soc.*, **89** 1771-1789 (2006).
11. K. Araki and J.W. Halloran, *J. Am. Ceram. Soc.*, **87** 1859-1863 (2004).
12. A.J. Pyzik, D.F. Carroll and C.J. Hwang, pp. 411-416 in *Silicon Nitride Ceramics: Scientific and Technological Issues*, Materials Research Society Symposium Proceedings Vol. 287, Eds. I-W. Chen, P.F. Becher, M. Mitomo, G. Petzow and T.-S. Yen, Materials Research Society, Boston (1993).
13. T. Hirata, K. Akiyama and T. Morimoto, *J. Eur. Ceram. Soc.*, **20** 1191-1195 (2000).
14. M.J. Pomeroy, C. Mulcahy and S. Hampshire, *J. Am. Ceram. Soc.*, **86** 458-464 (2003).
15. S. Hampshire and K.H. Jack, pp. 225-230 in *Progress in Nitrogen Ceramics*, Ed. F.L. Riley, Martinus Nijhoff Publishers, The Hague, 1983.
16. N. Saito, D. Nakata, A. Umemoto and K. Nakashima, pp. 265-270 in *Advanced Si-Based Ceramics and Composites*, Key Engineering Materials Vol. 287, Eds. H.-D. Kim, H.-T. Lin and M.J. Hoffmann, Trans Tech Publications, Zurich, Switzerland (2005).
17. R.L. Satet and M.J. Hoffmann, *J. Eur Ceram. Soc.*, **24** 3437-3445 (2004).

ALIGNED PORE CHANNELS IN 8MOL%YTTRIA STABILIZED ZIRCONIA BY FREEZE CASTING

Srinivasa Rao Boddapati and Rajendra K. Bordia
Department of Materials Science and Engineering
University of Washington
Box 352120
Seattle, WA, 98195

ABSTRACT

Freeze casting is a simple and cost effective way of making porous ceramics with aligned pore channels. The structure and volume of the pores in the freeze cast ceramic are mainly governed by the temperature gradient driving the solidification, volume fraction and size of the particles in the slurry, viscosity of the slurry and sintering temperature. Microstructural evolution of freeze cast 20vol% yttria stabilized zirconia (YSZ) as a function of sintering temperature has been investigated. The change in pore orientation from purely axial at the base of the pellet to purely radial at the top portion of the pellet (25 mm in height) has been attributed to the change in the direction of temperature gradient. The inter-particle porosity in the pore walls was completely eliminated when the pellet was sintered at 1500°C unlike 1200°C and 1300°C samples.

INTRODUCTION

Freeze casting is a simple and cost effective way of making porous ceramics. Freeze casting involves freezing of a ceramic slurry followed by sublimation of the solidified liquid, for example, ice in the case of water based slurry. Freeze casting is suitable for producing aligned and/or interconnected pore channels. The size and shape of the pore channels is a function of temperature gradient applied during freezing, volume fraction of the particles, particle size and shape, and sintering temperature. Porous ceramics have been processed by freeze casting using both aqueous and non-aqueous slurries. Fukasawa[1-2] and co-workers have investigated the formation of complex pore structures in alumina and aligned pore channels in Si_3N_4[3]. Sofie and Dogan[4] have investigated the freeze casting of aqueous alumia slurries with glycerol. They found out that the presence of glycerol reduced the defects formed during freeze casting, and reduced the viscosity of the slurry thereby making it possible to increase the particle loading in the slurry. Araki and Halloran[5-7] have developed a new freeze casting technique based on sublimable vehicles such as camphene, and naphthalene-camphor eutectic system, which make it possible to freeze cast ceramic slurries near room temperature. Koch et al.[8] carried out freeze casting with alumina slurry containing silica sol which transforms to a gel during freezing. The gel formed during freeze casting retains the shape of the cast body and the solvent is either sublimated or evaporated. They also found out that lower cooling rates, observed at distances far away from the cooling surface, resulted in bigger pores. Freeze casting has also been used to build composites with complex microstructures which mimic the microstructure of natural materials such as nacre[9]. Freeze casting of silica sols on Si substrates was carried out by Kisa et al.[10]. Koh et al.[11-12] investigated the effect of polystyrene addition on the freeze casting of ceramic/camphene slurries and freeze casting of dilute ceramic slurries to form interconnected pore networks. Shanti et al.[13] studied the particle redistribution during dendritic solidification by conducting freeze casting experiments with Alumina-camphene system.

In the present work, freeze casting is used to form aligned pore channels in 8mol% yttria stabilized zirconia, and to investigate the effect of sintering temperature on the microstructural evolution of the freeze cast samples.

EXPERIMENTAL PROCEDURE

Porous YSZ samples have been prepared by freeze casting aqueous YSZ slurry. 8mol%YSZ particles (Tosoh), a dispersant (Duramax[TM] D-3005; 5wt% of the solids) and a binder (Duramax[TM] B-1020; 5wt% of the solids) and de-ionized water were used to prepare the slurry. The dispersant was dissolved in water and the YSZ powder was added to the solution while stirring. The slurry was ultrasonicated for 3 min. Later binder was added and ball milled for 24h. The ratio of the slurry weight to the milling media (MgO stabilized zirconia) weight was kept at 1:1. The experimental set-up used to freeze cast the samples is shown in Fig.1. To produce aligned pore channels, directional solidification was induced by attaching styrene cylindrical molds to aluminum metal block. The block was immersed in a Pyrex dish containing liquid nitrogen. Ball milled slurry was poured into the styrene molds and allowed to solidify. The freeze cast samples were dried in a freeze drier for 48 hours. The diameter and height of the pellets are 11.2 mm and 30 mm, respectively. After freeze drying the samples were sintered in air at 1200°C, 1300°C and 1500°C for 2h. The microstructure of both the green stage and sintered samples, both transverse and longitudinal sections, were examined using a field emission scanning electron microscope (JEOL JSM 7000).

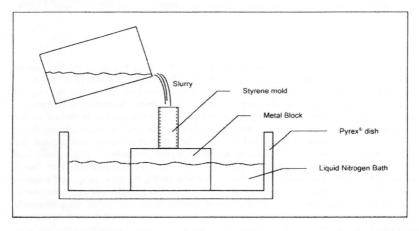

Fig.1: Experimental set-up used for freeze-casting of YSZ

RESULTS AND DISCUSSION

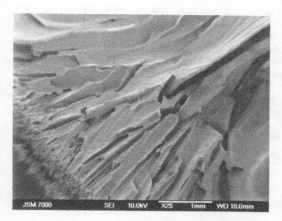

Fig.2: Top portion of the 20vol% YSZ green stage pellet. The channels are all oriented in radial directions.

Fig.3: Near close packing of 8mol%YSZ particles achieved by rejection of the particles by 20vol% YSZ aqueous slurry

The green stage pellets did not exhibit enough handling strength. As a result most of the samples broke during removing the samples from the styrene molds. Although the binder content was around 5 wt% of the solids weight, it was not sufficient to provide enough strength to the green pellets. However, addition of higher amounts of binder increased the viscosity of the slurry and the slurry had to be rolled without milling media for about 18h to remove the entrapped air. Macroscopic observations of the broken green pellets revealed that the pore channels are predominantly oriented in axial direction up to about 20 mm length and in the rest 10 mm the pores are predominantly oriented in radial direction. This change in the pore orientation appears to be caused by the change in the temperature gradient along the height. Although the temperature at different locations was not measured, it is highly likely that radial direction is preferred at top portion of the pellet and axial direction is preferred at the bottom portion of the pellet due to proximity of the bottom portion of the pellet to the liquid nitrogen bath. The microstructures of the 20vol% green pellets are shown in Figures 2 and 3. Fig.2 shows large radial channels (100-200 μm) at the top portion of the pellet. Near closing of particles achieved in the wall of a pore channel can be seen in Fig.3. The transition in the orientation of the pore channels in 1200°C sample are shown in Figures 4 and 5.

The arrow points towards the top portion of the pellet

Fig.4: Narrowing down of axial pore channels and increase in the radial pore channels at the top portion (sintered at 1200°C)

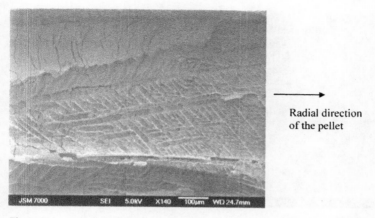

Radial direction
of the pellet

Fig.5: Intersection of both radial and axial pore channels near the top portion of the pellet (sintered at 1200°C)

Fig.6: The walls of the pore channels in a sample sintered at 1200°C. There is still some inter-particle porosity left in the sample.

Axial direction
of the pellet

Fig.7: Inner surface region of the pore channels. The pore channels are around 20-30µm (sintered at 1300°C)

Fig.6 shows the sintering of YSZ particles at 1200°C. There is still some inter-particle porosity left in the sample after 2h of holding. The width of the axial pore channels is about 10 µm in the sample sintered at 1300°C. Rough surface seen on the channels was created by the particles pushed in to the interdendritic arms of the ice crystals during freeze casting. Axial pore channels and dense microstructure of the pore walls in the sample sintered at 1500°C can be seen in Figures 8 and 9. The inter-particle porosity was completely eliminated due to higher sintering temperature. The cracks that were formed in the green stage can still be seen in the sintered samples (see Fig.8).

Fig.8: Axial pore channels in the pellet sintered at 1500°C

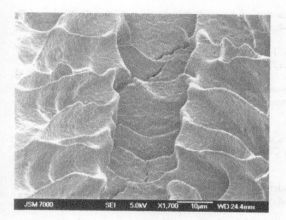

Fig.9: The inter-particle porosity is completely eliminated in the pore walls (sintered at 1500°C)

SUMMARY AND CONCLUSIONS

Porous YSZ green pellets have been processed by freeze casting aqueous 20vol% YSZ slurry. Binder content equivalent to 5wt% of the solids weight was not sufficient to provide sufficient handling strength. The orientation of the pore channels changed from axial at the bottom to radial at the top due to change in the direction of temperature gradient. Higher sintering temperatures led to complete removal of inter-particle porosity in the pore walls.

ACKNOWLEDGEMENTS

Mr. Aaron Feaver's help in conducting the freeze drying experiments and Prof. Guozhong Cao's permission to use the freeze dryer are gratefully acknowledged.

REFERENCES

[1]T. Fukasawa, Z.-Y. Deng, M. Ando, Pore structure of porous ceramics syntehsized from water-based slurry by freeze-dry process, J. Mater. Sci., 36 (2001), 2523-2527.

[2]T. Fukasawa, M. Ando, T. Ohji and S. Kanzaki, Synthesis of porous ceramics with complex pore structure by freeze-dry processing, J. Am. Ceram. Soc., 84 [1], (2001), 230-232.

[3]T. Fukasawa, Z.-Y. Deng, M. Ando, T. Ohji and S. Kanzaki, Synthesis of porous silicon nitride with unidirectionally aligned channels using freeze-drying process, J. Am. Ceram. Soc., 85 [9], (2002), 2151-2155.

[4]S. W. Sofie and F. Dogan, Freeze casting of aqueous alumina slurries with glycerol, J. Am. Ceram. Soc., 84 [7], (2001), 1459-1464.

[5]K. Araki and J. W. Halloran, New Freeze-casting technique for ceramics with sublimable vehicles, J. Am. Ceram. Soc., 87 [10], (2004), 1859-1863.

[6]K. Araki and J. W. Halloran, Room-temperature freeze casting for ceramics with nonaqueous Sublimable vehicles in the naphthalene-camphor eutectic system, J. Am. Ceram. Soc., 87 [11], (2004), 2014-2019.

[7]K. Araki and J. W. Halloran, Porous ceramic bodies with interconnected pore channels by a novel freeze casting technique, J. Am. Ceram. Soc., 88[5], (2005), 1108-1114.

[8]D. Koch, L. Andresen, T. Schmedders and G. Grathwohl, Evolution of porosity by freeze casting and sintering of sol-gel derived ceramics, J. of sol-gel Sci and Tech., 26 (2003), 149-152.

[9]S. Deville, E. Saiz, R. K. Nalla and A. P. Tomsia, Freezing as a path to build complex composites, Science, 311, 27 January 2006, 515-518.

[10]P. Kisa, P. Fisher, A. Olszewski, I. Nettleship, N. G. Eror, Mat. Res. Soc. Symp. Proc., 788 (2004), L8.7.1-L8.7.5

[11]Y.-H. Koh, E.-J. Lee, B.-H. Yoon, J.-H. Song, H.-E. Kim and H-W. Kim, "effect of polystyrene addition on freeze casting of ceramic/camphene slurry for ultra-high porosity ceramics with aligned pore channels," J. Am. Ceram. Soc., 89[12] (2006), 3646-3653.

[12] Y.-H. Koh, J.-H. Song, , E.-J. Lee and H.-E. Kim, "Freezing dilute ceramic/camphene slurry for ultra-high porosity ceramics with completely interconnected pore networks," J. Am. Cera. Soc., 89[10] (2006) 3089-3093.

[13]N. O. Shanti, K. Araki and J. W. Halloran, "Particle redistribution during solidification of particle suspensions," J. Am. Ceram. Soc., 89[9] (2006), 2444-2447.

PREPARATION OF A PORE SELF-FORMING MACRO-/MESOPOROUS GEHLENITE CERAMIC BY THE ORGANIC STERIC ENTRAPMENT (PVA) TECHNIQUE

Dechang Jia*, DongKyu Kim and Waltraud M. Kriven
Department of Materials Science and Engineering, University of Illinois at Urbana-Champaign, Urbana, IL 61801, USA

ABSTRACT

The organic steric entrapment (PVA) method has been successfully used to develop a pore self-forming, macro-/mesoporous gehlenite ($2CaO \cdot Al_2O_3 \cdot SiO_2$ or "C_2AS") ceramic. The C_2AS ceramic manifests a very good pore self-forming ability, no vesicants and additives need to be incorporated, and no hydrothermal treatment is needed, as is generally the case in other kinds of porous ceramics or glasses. The pore self-forming behavior is accompanied by a significant volume expansion (~134%) of the samples. The pores can be either open or closed, and the total porosity attains ~80%. The porous ceramic "sintered" at 1400 °C for 1 h has a flexure strength value of 11.2 ± 3.4 MPa, after being further thermally etched at 1200 °C for 1h, its bending strength value having increased to 15.4 ± 4.2 MPa. The mechanism for the pore self-forming behavior is discussed.

INTRODUCTION

Porous ceramics and glasses have received much attention recently because of their unique properties and have found many applications in a variety of areas. Many processing routes have been developed for the preparation of macroporous ceramics, and they can be classified into three types, viz., replica, sacrificial template and direct foaming methods, according to Studart et al.[1] For specific methods, partial densification of green compacts, bubble generation into a slurry or at a green state during a specific thermal treatment, stacking of pre-sintered granules or fibers, aero gel and sol-gel methods, pyrolysis of various organic additives, the polymeric sponge method, etc, have been reported.[2] For most of the methods above, however, vesicants and/or additives need to be incorporated, which in turn make the process complex, less controllable and more expensive, etc. Until now, many kinds of engineering ceramics or ceramic composites, such as SiO_2,[3-5] Al_2O_3,[6-10] PSZ, cordierite,[2] mullite,[11-13] ZTA, SiC-Al_2O_3, mullite-zirconia,[1,2] SiC_w/SiO_2,[14] as well as SiO_2-ZrO_2, mullite and zirconia matrix composites with Al_2O_3 platelets[15] have been fabricated into porous ceramics. Porous gehlenite C_2AS ceramics, however, have not been reported as yet. In this paper, therefore, the organic steric entrapment (PVA) method has been used to develop a pore self-forming macro-/mesoporous, gehlenite ceramic.

---

* On sabbatical leave from Institute for Advanced Ceramics, Harbin Institute of Technology, 150001, Harbin, China[1]

EXPERIMENTAL PROCEDURES

The organic steric entrapment (PVA) method, invented in our laboratory, has been used for the preparation of many kinds of oxide powders[15-20]. In this paper, it was also used to synthesize the starting ceramic powders for the preparation of the C_2AS porous ceramic. Reagent grade, calcium nitrate tetrahydrate ($Ca(NO_3)_2 \cdot 4H_2O$) and aluminum nitrate nonahydrate ($Al(NO_3)_3 \cdot 9H_2O$) were used as starting materials to obtain calcium and aluminum oxides. A Ludox SK colloidal SiO_2 (25 wt% SiO_2 sol) was used to supply the SiO_2. These cation sources were dissolved in stoichiometric ratios for gehlenite, or "C_2AS" in cements terminology, in deionized water. To improve the solubility of the Ludox SK, the pH value of the solution was adjusted by addition a few drops of nitric acid. Once the cation sources were completely dissolved, the 5 wt% PVA solution (Polyvinyl alcohol, 205S) was added. Water was evaporated by continuous magnetic stirring and later, hand stirring with glass rods during heating on a hot plate (~150 °C), until the decomposition of the organic/inorganic precursors had fully taken place. The products of the decomposed precursors were ground and calcined at a heating rate of 5 °C/min in a common box furnace at 800 °C for 1 h, and pure white powders were obtained.

After synthesis, the powder was subjected to wet attrition milling at 50 rpm for 1 h in a jar, 85 mm in diameter by 105 mm in height. For one batch of powders (30 g), 700 g of ZrO_2 balls 5 mm in diameter were used as milling media, and 250 mL of isopropyl alcohol was the solvent. The solvent was evaporated at 125 °C in an oven, and the powders obtained were sieved through a 30~40 mesh sieve. Then, the powders were uniaxially pressed at 10 MPa, followed by cold isostatic pressing at 400 MPa. The flat-shaped, green compacts or billets obtained were directly sintered in air atmosphere at 1400 °C, using a heating rate of 5 °C/min, holding for 1 h and cooling down at 5 °C/min.

The particle size distribution and specific surface area of the as-milled powders were examined in a particle size analyzer. Phases of the starting powder and sintered bulk samples were determined by an X-ray diffractometer using $CuK_{\alpha 1}$ radiation (40 kV, 20 mA) at a scanning speed of 0.4 °/min.

The morphology of the as-milled powder particles, and pore microstructure of the calcined samples were examined by scanning electron microscopy (SEM). Open pore size distributions, average diameters of open pores and the apparent porosity of the porous ceramics were measured by mercury infiltration porosimetry (MIP) system. The density values were also calculated by the Archimedes weight/volume method, and the total porosity of the samples was calculated using 2.98 g/cm^3 as the theoretical density for gehlenite (C_2AS).

The bend strength of the porous C_2AS gehlenite was tested by the three-point-bending method in an Universal Instron Testing machine (Instron, 4502, Japan) with a crosshead lowering speed of 0.5 mm/min, using rectangular bars (4 mm × 6 mm × 40 mm in dimension) as samples.

RESULTS

(1) Particle size distribution and morphology of the starting powders

The particle size distribution and particle morphology of the as-attrition milled C_2AS powders are shown in Figs. 1 and 2, respectively. As indicated, the particle sizes showed a single peak distribution. Particles in larger sizes can be seen to have a flake shape resulted from the attrition milling of the calcined PVA foams. According to the particle size analysis, the median and average diameters were 5.5 and 5.7 μm, respectively, and the specific area attained 1.2 m²/cm³.

(2) Pore self-forming behavior and accompanying volume expansion

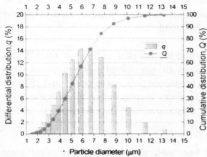

Fig. 1. Particle size distribution of the as-attrition milled powder synthesized by the PVA technique according to the stoichiometric ratio for gehlenite (C₂AS)

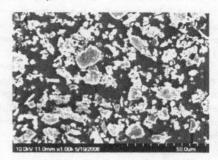

Fig. 2. SEM images of the as-attrition milled powder particles synthesized by PVA technique according to the stoichiometric ratio for C₂AS

Fig. 3. Optical micrographs of the gehlenite C₂AS ceramic bend bars before (a) and after calcination showing a significant increase in dimensions

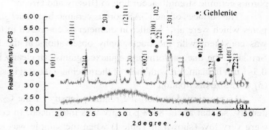

Fig. 4. XRD patterns of as-attrition milled powder designed according to the stoichiometric ratio of C₂AS and synthesized by PVA technique (a), and bulk samples sintered at 1400 °C for 1h (b)

Initially we had planned to prepare densified gehlenite ceramic samples. However, no shrinkage behavior was observed when the cold isostatically pressed (CIP) green beams were calcined at 1400°C for 1 h in air atmosphere, even though the temperature was very close to 1623°C, the melting point of pure C₂AS. Instead, significant expansion of the samples was unexpectedly observed (seen in Fig. 3). The maximum linear dimensions increased by as much as 47%, and the total volume expansion rate attained 134%. The original shapes of the green bend bars were still retained except that there was a slight swelling in the middle part of the bars. The sample surfaces were smooth and free from any macro-cracks (also shown in Fig. 3). It was found that the relative density was only about 20%, which meant that nearly 80% porosity was introduced. The interesting fact is that this was achieved without addition of any kinds of vesicants or special treatments such as hydrothermal treatment. Thus, a novel kind of porous ceramic or a new route for preparing porous ceramics was found.

(3) Phase composition

XRD patterns of the as-attrition milled powders and the samples calcined at 1400 °C are presented in Fig. 4. As seen, the as-attrition milled powders still maintained their amorphous state. For the sample "sintered" at 1400 °C, the original wide, broadened peak between 25 ° and 35 ° in 2 theta disappeared, and instead, all the main diffraction peaks corresponding to gehlenite were observed. They became sharp, indicating that complete crystallization of the amorphous C_2AS had taken place.

(4) SEM microstructure of the porous ceramic

SEM morphology of the polished, fracture and free "sintered" surfaces of the samples are provided in Fig. 5. As indicated in Fig. 5 (a), the sample had a good uniformity in pore size. Most of the pores were smaller than 10 μm in diameter. From the fractographs taken near the surface of the samples (shown in Fig. 5 (b)), the pore network structure could clearly be seen. Additionally, a relatively dense surface layer of about 10-20 μm in thickness was also observed. The pore walls could be as thin as the width of just one single grain to several grains. Additionally, no microscopic or macroscopic flaws were observed in the pore walls or struts of the samples, which is beneficial for porous ceramic strength according to Brezny and Green.[21]

The surfaces of the C_2AS porous ceramic bend bars shown in Fig. 5 (c) had a lower amount of pores which were also smaller in diameter as compared to the interior structure of the samples. This was consistent with the fractography observations at the free surface layer shown in Fig. 5 (b). Similarly, the fine, uniform and equiaxed grains can clearly be seen in Fig. 5 (d). They were between 0.5~1 μm in diameter, with most of them being ~ 0.6 μm in size.

(5) Pore size distribution

As mentioned earlier, the relative density of the C_2AS porous was only about 20% of full density, so that about 80% porosity was introduced. In comparison, the apparent porosity values were very low (shown in Table I) when the sample was tested by mercury intrusion porosimetry (MIP) system. Moreover, it should be noted from the data provided in Table I that the apparent (or skeletal) density values of the two samples (0.80 and 0.75 g/cm³, respectively) were still much lower than the theoretical density of gehlenite, 2.98 g/cm³. Therefore, it can be concluded that most of the pores in the C_2AS porous gehlenite were of the closed type or simply disconnected from the outside due to the self sealing effect during, which was consistent with the SEM observations.

When the C_2AS porous ceramic samples were placed into water, all of them floated on the surface of the water for a period of time. Small air bubbles appeared only on certain parts of the sample surfaces. Eventually, most of the samples sank to the bottom of the beaker. However, some samples could even float on the water without sinking, as shown in Fig. 6.

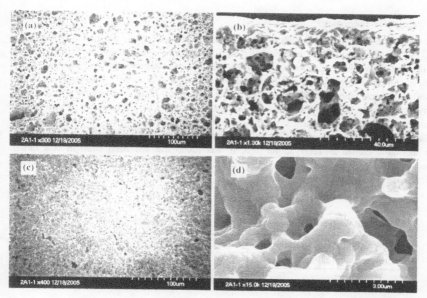

Fig. 5. SEM micrographs of (a) polished surface, (b) fractograph near sample surface, (c) sample surface morphology, and (d) grain morphology on sample surface, respectively, showing the pore size, network structure, the relatively denser surface layer and grain size of the C_2AS porous ceramic samples sintered at 1400 °C for 1 h.

Table I. Mercury intrusion porosimetry (MIP) analysis of C_2AS ceramic sintered at 1400 °C

Properties	1#	2#
Total pore area (m²/g)	0.24	-*
Median pore diameter (Volume) (μm)	9.8	-
Median pore diameter (Area) (μm)	0.04	-
Average pore diameter(4V/A) (μm)	1.01	-
Bulk density (g/cm³)	0.76	0.75
Apparent (skeletal) density (g/cm³)	0.80	0.75
Apparent porosity (%)	4.65	0

* The tested value, 0, is omitted due to its clearly being incorrect.

Fig. 6. A piece of the C_2AS porous gehlenite sample floating on water

(6)   Mechanical Properties

Three-point-bend strength testing results of the C_2AS porous ceramic are given in Table II. As shown, the strength values were unexpectedly high for such a high porosity, particularly for the gehlenite ceramic whose bulk mechanical properties for dense samples is low. The good bend strength of the porous gehlenite could be attributed to the fine and uniform pore structure, robust pore walls and struts free of any flaws, as well as the fine grains. If the compressive strength value were hypothesized to be double the flexural strength value, they would basically lie in the zone between the theoretical compressive strength lines of the open-cell and closed-cell porous ceramics, according to the model proposed by Gibson and Ashby [22]. Additionally, when the specimens were further thermally etched at 1200 °C for 1 h, it was found that the bend strength was further unexpectedly increased by nearly 38%, which could be principally attributed to the self healing behavior, leading to a much denser surface layer.

Table II. Three-point-bending strength of the C_2AS porous ceramics

Sintering and treatment condition	1400°C/1h	1400°C/1h+1200 °C/1h
Bending strength (MPa)	11.2 ± 3.4	15.4 ± 4.2

(7)   TG/DSC analysis of the starting powder

Thermogravimetric (TG) and differential scanning calorimetric (DSC) curves of the as-attrition milled C_2AS starting powders were shown in Fig. 7. From the TG curve in this figure it is clearly seen that a powder mass loss of ~1.66% took place between 890 °C and 960 °C. In the corresponding DSC curve, an exothermic reaction peak appeared but having a hysteresis of 30 °C, i.e, from 920 to 990 °C. Considering the amorphous state of the starting powder calcined at 800 °C, the exothermic peak of the DSC curve could be attributed to the crystallization of the amorphous C_2AS powders, which may perhaps be triggered by the loss of the atmospheric water of hydration.

DISCUSSION

The expansion behavior during calcination was reported recently in glasses.[23] It was concluded that water incorporated into the glass structure by the hydrothermal reaction acted as a vesicant in the latter calcinations, resulting in the formation of macropores. In C-S-H [23,24] and CaO-SiO_2-Al_2O_3-H_2O [25] systems, hydroxides, such as hydrogarnet ($Ca_3Al_2(SiO_4)\cdot(OH)_8$), tobermorite ($Ca_5(Si_6O_{18}H_2)\cdot(4H_2O)$) were reported and strength development was found after hydrothermal treatment.

In this work, we deduced that the expansion of the gehlenite green compacts during calcination might be related to atmospheric hydration with water because gehlenite is in the CaO-Al_2O_3-SiO_2 ternary system. This ternary phase diagram contains cementitious compounds some of which are known to be hydrophilic. Moreover, it was also found that some strength development took place for the CIPped C_2AS samples placed in air. Hence, it can be deduced from the above that hydration which took place at least to some degree can somewhat account for the pore self-formation of the C_2AS gehlenite. This hypothesis was supported by the TG/DSC analysis result shown in Fig.7. Thus, it was concluded that the automatic hydration or water adsorption of the C_2AS powders may be responsible for the mechanisms of the pore self-formation of C_2AS porous ceramics. A more detailed study of self-forming porosity in gehlenite (C_2AS) ceramics is the subject of two upcoming papers.[26,27]

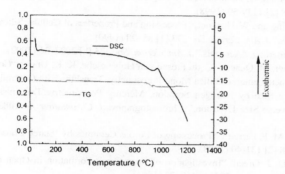

Fig. 7. TG and DSC curves with increasing temperature of the as-attrition milled C_2AS powder

CONCLUSIONS

The organic steric entrapment (PVA) method has been used to develop a pore self-forming macro-/mesoporous gehlenite ceramic having a total porosity of ~80%, and a significant volume expansion as large as 134%. No vesicants and additives were required and no hydrothermal treatment was needed.

The automatic hydration or water adsorption of the C_2AS powders were hypothesized to be responsible for the mechanisms of the pore self-forming behavior, and it might be a universal characteristic for all hydrophilic ceramic powders prepared by the organic steric entrapment (PVA) technique.

The C_2AS porous ceramics manifested relatively high flexural strength, which could mainly be attributed to the robust pore walls and struts free of any flaws, the fine and uniform pores in the sintered samples, as well as to a relatively denser surface layer formed on thermally treated samples.

ACKNOWLEDGEMENTS

Part of this work was carried out in the Center for Microanalysis of Materials, in the Frederick Seitz Materials Research Laboratory, which is partially supported by the U.S. Department of Energy under grant DEFG02-91-ER45439. The authors thank Xiaoming Duan and Boyang Liu at the Institute for Advanced Ceramics, Harbin Institute of Technology, for particle size measurements as well as TG and DSC analyses. Dechang Jia also gratefully acknowledges the

China Scholarship Council (CSC) for awarding a scholarship for pursuing research in the United States as a visiting scholar.

REFERENCES

1. A. R. Studart, U. T. Gonzenbach, E. Tervoort and L. J. Gauckler, "Processing Routes to Macroporous Ceramics: A Review," *J. Am. Ceram. Soc.*, **89** [6] 1771-89 (2006).
2. L. Montanaro, Y. Jorand, G. Fantozzi and A. Negro, "Ceramic Foams by Powder Processing," *J. Eur. Ceram. Soc.*, **18** [2] 1339-50 (1998).
3. T. Fujiu, G. L. Messing and W. Huebner, "Processing and Properties of Cellular Silica Synthesized by Foaming Sol Gels," *J. Am. Ceram. Soc.*, **73** [1] 85–90 (1990).
4. J. Fricke and A. Emmerling, "Aerogels," *J. Am. Ceram. Soc.*, **75** [8] 2027–36 (1992).
5. D. Lubda, W. Lindner, M. Quaglia, C. du Fresne von Hohenesche, K. K. Unger, "Comprehensive Pore Structure Characterization of Silica Mono Porosimetry, Transliths with Controlled Mesopore Size and Macropore Size by Nitrogen Sorption, Mercury Porosimetry, Transmission Electron Microscopy and Inverse Size Excusion," Chromatography, *J. Chromatogr A.*, **1083** [1-2] 14-22 (2005).
6. O. Lyckfeldt and J. M. F. Ferreira, "Processing of Porous Ceramics by 'Starch Consolidation'," *J. Eur. Ceram. Soc.*, **18** [2] 131-40 (1998).
7. D. D. Brown and D. J. Green, "Investigation of Strut Crack Formation in Open Cell Alumina Ceramics," *J. Am. Ceram. Soc.*, **77** [6] 1467–72 (1994).
8. T. Fukasawa, M. Ando, T. Ohji, and S. Kanzaki, "Synthesis of Porous Ceramics with Complex Pore Structure by Freeze–Dry Processing," *J. Am. Ceram. Soc.*, **84** [1] 230–2 (2001).
9. K. Araki and J. W. Halloran, "Porous Ceramic Bodies with Interconnected Pore Channels by a Novel Freeze Casting Technique," *J. Am. Ceram. Soc.*, **88** [5] 1108–14 (2005).
10. S. Krenar, L. M. Matthew, Y. Di, and V. Henk, "Preparation and Properties of Porous α-Al_2O_3 Membrane Supports," *J. Am. Ceram. Soc.*, **89** [6] 1790–4 (2006).
11. J. M. Tulliani, L. Montanaro, T. J. Bell, and M. V. Swain, "Semiclosed-Cell Mullite foams: Preparation and Macro- and Micromechanical Characterization," *J. Am. Ceram. Soc.*, **82** [4] 961–8 (1999).
12. E. Roncari, C. Galassi and C. Bassarello, "Mullite Suspensions for Reticulate Ceramic Preparation," *J. Am. Ceram. Soc.*, **83** [12] 2993–8 (2000).
13. R. Barea, M. I. Osendi, P. Miranzo and J. M. F. Ferreira, "Fabrication of Highly Porous Mullite Materials," *J. Am. Ceram. Soc.*, **88** [3] 777–9 (2005).
14. K. Yanagisawa, N. Z. Bao, L. M. Shen, A. Oada, K. Kajiyoshi, Z. Matamoras-Veloza, J. C. and Rendón-Angeles, "Development of A Technique to Prepare Porous Materials from Glasses," *J. Eur. Ceram. Soc.*, **26** [4-5], 761-5 (2006).
15. M. A. Gülgün and W. M. Kriven, "A Simple Solution-Polymerization Route for Oxide Powder Synthesis," *Ceramic Transactions*, **62**, 57-66 (1995).
16. S. J. Lee and W. M. Kriven, "Crystallization and Densification of Nano-size, Amorphous Cordierite Powder Prepared by a Solution-Polymerization Route," *J. Am. Ceram. Soc.*, **81** [10] 2605-12 (1998).

17. M. A. Gülgün, M. H. Nguyen and W. M. Kriven, "Polymerized Organic-Inorganic Synthesis of Mixed Oxides," *J. Am. Ceram. Soc.*, **82** [3] 556-60 (1999).
18. M. H. Nguyen, S. J. Lee and W. M. Kriven, "Synthesis of Oxide Powders via a Polymeric Steric Entrapment Precursor Route," *J. Mater. Res.*, **14** [8] 3417-26 (1999).
19. M. A. Gülgün, W. M. Kriven and M. H. Nguyen, "*Processes for Preparing Mixed-Oxide Powders,*" US patent number 6,482,387 issued Nov 19th 2002
20. S. J. Lee, Ch. H. Lee and W. M. Kriven, "Synthesis of Low-firing Anorthite Powder by the Steric Entrapment Route," *Ceram. Eng. and Sci. Proc.*, **23** [3] 33-40 (2002).
21. R. Brezny and D. J. Green, "Fracture Behavior of Open-Cell Ceramics," *J. Am. Ceram. Soc.*, **72** [7] 1145–52 (1989)
22. G. L. J. Gibson and M. F. Ashby, *Cellular Solids: Structure and Properties*, p.510, 2nd edition, Cambridge University Press, Cambridge, 1997
23. G. L. Kalousek, Crystal Chemistry of Hydrous Calcium Silicates: I, Substitution of Aluminum in the Lattice of Tobermorite, *J. Am. Ceram. Soc.*, **40** [3] 74–80 (1957).
24. H. Maenami, O. Watanabe, H. Ishida and T. Mitsuda, "Hydrothermal Solidification of Kaolinite–Quartz–Lime Mixtures," *J. Am. Ceram. Soc.*, **83** [7] 1739–44 (2000).
25. R. Sarkar and S. K. Das, Hydrothermal Treatment of $CaO-SiO_2-Al_2O_3-H_2O$ System, *Am. Ceram. Soc. Bull.*, 9201-10, (2006).
26. "Sintering Behavior of Gehlenite. Part I. Self-forming, Macro-/Mesoporous Gehlenite – Pore Forming Mechanism, Microstructure, Mechanical and Physical Properties," D. Jia, D.-K. Kim and W. M. Kriven, *J Amer. Ceram. Soc.*, in press (2007).
27. "Sintering Behavior of Gehlenite. Part II. Microstructure and Mechanical Properties" D. Jia and W. M. Kriven, *J Amer. Ceram. Soc.*, in press (2007).

NONLINEAR STRESS-STRAIN BEHAVIOR EVALUATION OF POROUS CERAMICS WITH DISTRIBUTED-MICRO-CRACK MODEL

Hidenari Baba
Ishikawajima-Harima Heavy Industries, Co., Ltd.
1, Shin-Nakahara-Cho, Isogo-Ku
Yokohama, Kanagawa, 235-8501, JAPAN

Akihiko Suzuki
Graduate School of Science and Engineering, Saitama University
255, Shimo-Okubo, Sakura-Ku
Saitama, Saitama, 338-8579, JAPAN

ABSTRACT

To evaluate the extent of damage tolerance of porous ceramics, we have proposed Distributed-Micro-Crack (DMC) model and applied it to the nonlinear stress-strain behavior analysis of porous ceramics. In our previous studies with DMC model, a uniform initial crack size distribution was used. Though a uniform initial crack size distribution can predict the initial stage of deformation (i.e. the quasi-linear stress-strain region) of porous ceramics, it cannot evaluate the final stage of deformation near fracture. It can be considered that this discrepancy is due to the localized damage mechanism that cannot be estimated by DMC model with a uniform initial crack size distribution. Therefore, a nonuniform initial crack size distribution is introduced into the model in present studies. The modified DMC model with nonuniform initial crack size distribution can simulate localized damage. Not only initial stage of stress-strain relationship but also near fracture stage of it obtained by the model shows good agreement with experimental data.

1. INTRODUCTION

Porous ceramics have been applied to dust collecting equipment, gas separator, and adsorbent, etc. To maintain structural integrity during operation, strength assessment of these components is required. Since porous ceramics are basically categorized as brittle materials, strength assessment has been performed based on linear elastic fracture mechanics. However, such ceramics show a little nonlinear stress-strain behavior and have damage tolerance. Therefore, it is considered that strength assessment utilizing these properties can predict porous ceramic deformation and fracture behavior more precisely. To establish the evaluation method of the extent of damage tolerance of porous ceramics, DMC model [1] has been proposed and has been applied to prediction of nonlinear stress-strain behavior. In this model, pores in the material are modeled as penny shape cracks with arbitrary orientation. The crack propagation behavior in the model depends on R-curve property [2] obtained from notched specimen bending test of the same porous ceramics. In previous studies, the distribution of initial pore size was not considered. A uniform crack length was assumed as initial pore size. Although such a model can predict the initial stage of deformation (i.e. quasi-linear stress-strain region) of porous ceramics, it cannot evaluate the final stage of deformation near fracture. It can be considered that this discrepancy is due to the localized damage mechanism that cannot be estimated by the model with a uniform initial crack size distribution. Therefore, a nonuniform initial crack size distribution is introduced into the model.

In the present study, nonlinear stress-strain behavior study by DMC model with initial crack size distribution is presented. Weibull distribution is employed as the initial crack size distribution profile. The strength of porous ceramics is discussed in relation to nonlinear behavior and loading type.

2. DISTRIBUTED-MICRO-CRACK MODEL

In this study, porous ceramics were assumed to be brittle materials which have numerous microcracks. The microcrack propagation behavior was controlled by R-curve property of this material. This leads to macroscopic nonlinear stress-strain deformation under loading. To simplify the numerical analysis, increasing the number of cracks and crack joining have not been taken into consideration.

Schematic illustration of the model that has a penny shape crack in the material is shown in Fig.1(a).

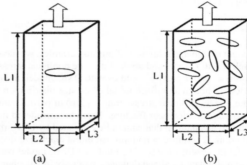

(a)   (b)

Figure 1   Nonlinear stress-strain behavior model under uniaxial tension.
(a) Component with a single microcrack
(b) Component with a large number of microcracks

The loading point displacement and strain of this model can be obtained as

$$u = u_0 + \int_0^A \frac{\partial G}{\partial P} dA \qquad (1)$$

$$\varepsilon = \frac{u}{L_1} = \frac{\sigma}{E} + \frac{4\pi^2 F^2}{3E'} \frac{\sigma a^3}{L_1 L_2 L_3} \qquad (2)$$

where u :loading point displacement, G :strain release energy, P :load, A :area of crack,

ε :average strain, σ :stress, F :shape coefficient, E :Young's modulus, E': $E/(1-v_{\cdot}^{2})$, v : Poisson's ratio, a : crack radius (mm),

In practice, numerous microcracks exist in porous materials. If the microcracks are distributed as shown in Fig.1(b) and the number of cracks equal to N, stress-strain relationship can be expressed as equation (3).

$$\varepsilon = \frac{\sigma}{E} + \frac{16}{3}\frac{N}{L_1 L_2 L_3}\frac{\sigma}{E'}a^3 \qquad (3)$$

If microcracks are distributed uniformly in the material, we can define crack density as given in equation (4) and equation (3) becomes equation (5)

$$q = \frac{N}{L_1 L_2 L_3} \qquad (4)$$

$$\varepsilon = \frac{\sigma}{E} + \frac{16}{3}q\frac{\sigma}{E'}a^3 \qquad (5)$$

where q :crack density (1/m3).

In equation (5), second term of right side is corresponds to an additional strain component due to existence of microcracks in the model. If the crack length increases in accordance with the R-curve property, the relative amount of second term in equation (5) will be higher. This leads to post peak behavior and the nonlinear stress-strain relation can be simulated. The R-curve property used in this study can be written as equation (6),

$$K_R = B(\Delta a)^n \qquad (6)$$

where K_R :crack growth resistance, Δa : increment of crack propagation, B, n :material constants. When the constitutive equations (5) and (6) are substituted into a FEM program, numerical simulation of stress-strain relationship with DMC model can be performed.

The model stated above can be applied to the uniaxial stress condition. However, in almost all practical structures or machines, stress concentrations exist. In general, these portions become multiaxial stress fields. And also, each microcrack that existed in the material has an arbitrary orientation. The effects of the multiaxial stress field and crack orientation on stress-strain relationship must be considered. To describe these effects, DMC model was modified to apply to the multiaxial stress field and to consider crack orientation. [1]

3. EXPERIMENT PROCEDURE

3.1 Material and Specimen

To measure the nonlinear stress-strain relationship of porous ceramics, tensile tests, three-point bending tests, and four point bending tests were conducted. The material was porous silicon carbide as used for filtration of food, gas analyzing devices, filtration of hot gases, and as a coalescer for oil aerosols in hot gases, as fabricated by Pall Corporation (SL05). The specimen was cut from a tube having an outer diameter is 70 mm and an inner diameter is 40 mm. The porosity was 35 % and the bulk density was 1.9 g/cm^3. Tensile test specimen is shown in Fig.2(a). The thickness of the tensile specimen is 3 mm. Prismatic bar type specimen was used for three point and four point bending test and its cross section size is 8 mm (height) x 6 mm (width) and length is 70 mm.

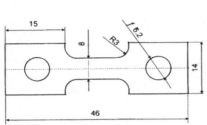

Figure 2(a) Tensile specimen geometry

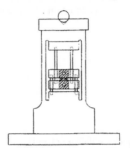

Figure 2(b) Crack Stabilizer

3.2 Tensile and Bending Test

The tensile test was carried out with a screw type material test machine and the load cell capacity of the machine is 10 kN. The cross head speed for tensile test was 0.05mm/min. Strain was measured by uniaxial strain gauges attached in the center of both sides of the specimen. To achieve stable crack propagation, crack stabilizer [3] was used. Figure 2(b) shows schematically the whole view of the stabilizer and specimen installed in the stabilizer through the pins (specimen is shown in the hatch). When the compression load is applied at the top of the stabilizer by testing machine, the tensile loading is applied to the specimen. Since the rigidity change of the columns of the stabilizer is much greater than that of the specimen, the influence of the rigidity change caused by the fracture or damage of the specimen is small on the rigidity of the testing system. As a result, an almost complete displacement control was achieved and the stable fracture test was conducted.

Three and four point bending tests were carried out with a hydraulic servo material test machine and the load cell capacity of the machine is 5 kN. The cross head speed for bending tests were 0.05 mm/min. The outer span of both bending tests are 60 mm, inner span of four point bending test was 30 mm. Strain of bending tests were also measured by uniaxial strain gauges attached to the tension side bottom surface of the specimen.

4. DMC MODEL MATERIAL PARAMETER

Prior to porous ceramics nonlinear stress-strain behavior analysis with DMC model, some material constants summarized in Table 1 had to be determined.

Table 1   Material constants of DMC model

Young's modulus, Poisson's ratio	E =260GPa, v =0.17	
Crack density	q =6 x 10^{15}	
R-curve parameter	B =0.4, n =0.109	
	Uniform	Nonuniform
Initial crack length distribution	$2a$ =10 μ m	$a_0 = a_0^* \cdot \dfrac{1}{\left\{ \ln\left(\dfrac{1}{1-P} \right) \right\}^{\frac{2}{m}}}$

Because the average Young's modulus of monolithic SiC is approximately 400 GPa and the porosity of the porous ceramics is 35% as mentioned before, the Young's modulus of the base material without the crack portion in DMC model was set to 260 GPa (=400*(1.0-0.35)). Regarding Poisson's ratio of base material, 0.17 was used. (Refer to ceramics manufacturer material data sheet [4]) The crack density q was equal to 6x10^{15}. The gradient of the stress-strain curve's linear portion as calculated from DMC model with the crack density was consistent with actual porous ceramics (SL05) Young's modulus.

R-curve parameters were obtained by four point bending test with a notched specimen. The notched specimen was made from same lot for both tensile and bending test specimens. The compliance of this material is defined as equation (7)

$$\lambda = \varepsilon / P \qquad (7)$$

where λ : compliance, ε : strain, P : load.
The relation between compliance and crack propagation length can be determined by FEM analysis [5]. With this manner, R-curve parameter B =2.34 and n =0.109 were obtained. However, this R-curve parameter could not simulate nonlinear behavior of porous ceramics in preliminary studies of DMC analysis. Therefore, R-curve parameter B was changed to 0.4.

Though initial crack length was assumed as uniform and set to 10 μ m that corresponds to an average grain size ($2a_0$ =10 μ m) in previous studies [1],[6], nonuniform distribution of initial crack length also has taken into consideration in this study. However, since actual initial crack length distribution is very difficult to measure, we assumed pseudo initial crack length profile function as

$$P = 1 - \exp\left\{ -\left(\frac{a_0^*}{a_0} \right)^{\frac{m}{2}} \right\} \qquad (8)$$

where P : cumulative existing probability of crack and its length is larger than a_0, a_0^*, m are material constants. This function can be applied to monolithic ceramics crack length profile and it leads to the strength of a two-parameter Weibull distribution with Weibull parameter m. Equation (8) becomes

$$a_0 = a_0^\cdot \frac{1}{\left\{\ln\left(\frac{1}{1-P}\right)\right\}^{\frac{2}{m}}} \qquad (9)$$

and

$$\frac{1}{\bar{a}_0} = \frac{1}{a_0^\cdot}\left(\frac{2}{m}!\right) \qquad (10)$$

where $\bar{a}_0$: average of a_0. Weibull modulus m in equation (10) was set to 15. This value was based on results of round robin tests with same material. With a random number P ($0<P<1.0$) substituted into Equation (9), a nonuniform initial crack distribution can be generated.

5. EXPERIMENTAL RESULTS AND DISCUSSION
5.1 Tensile Test

Figure 3 shows the stress-strain curve for tensile test. The solid line is the experimental result and the dotted line is DMC simulation. A uniform initial crack length assumption was used for DMC simulation in Fig.3. Regarding the maximum stress, there is large discrepancy between experiment and DMC simulation in Fig.3.

Figure 4 shows the stress-strain curve for the tensile test and for DMC analysis with a nonuniform initial crack length distribution as explained in section 3. Four different nonuniform initial crack distributions were created by use of different random number sets. Hence, four DMC analysis results are shown in Fig.4. A comparison of DMC simulations with nonuniform initial crack length between Fig.3 and Fig.4 shows an earlier nonlinear behavior and decreased maximum stress. The maximum stress occurred around of 7.5×10^{-3}, then decreasing stress phenomena so-called "strain softening" can be simulated. In DMC model with nonuniform initial crack length distribution, some larger cracks relative to uniform initial crack length were stochastically exist. Therefore, it can be considered that these cracks initiate localized damage phenomena and lead to earlier nonlinear behavior and decreased maximum stress.

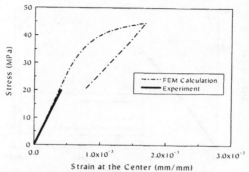

Figure 3   Stress-strain curve for flat plate under tension calculated by DMC model with a
uniform initial crack size distribution

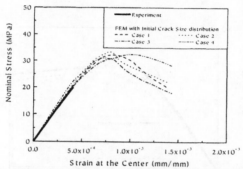

Figure 4   Stress-strain curve for a flat plate under tension calculated by DMC model with a
nonuniform initial crack size distribution (nominal stress vs strain at the center)

5.2 Three and Four Point bending Test

Figure 5 and 6 show the three point bending test results. DMC analysis with a uniform initial
crack length is shown in Fig.5, and the nonuniform result is shown in Fig.6. A comparison
between Fig.5 and Fig.6 shows that the nonuniform initial crack length distribution decreases
maximum bending strength. Also, maximum bending strength of the nonuniform initial crack
length distribution is in good agreement with the experiment.

Regarding the four point bending test and DMC simulation, these results are shown in Fig.7
and 8. As same as above, maximum strength of DMC simulation with nonuniform initial crack
length shows good agreement with experiment.

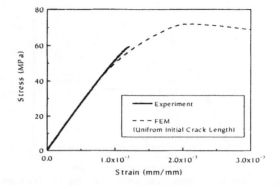

Figure 5 Stress-strain curve for 3-point bending of a prismatic bar calculated by DMC model with a uniform initial crack size distribution

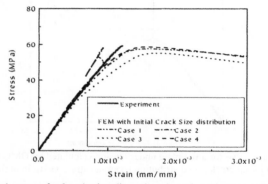

Figure 6 Stress-strain curve for 3-point bending of a prismatic bar calculated by DMC model with a nonuniform initial crack size distribution

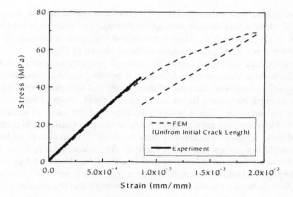

Figure 7   Stress-strain curve for 4-point bending of a prismatic bar calculated by DMC model
with a uniform initial crack size distribution

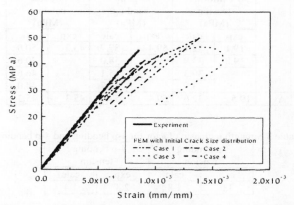

Figure 8   Stress-strain curve for 4-point bending of a prismatic bar calculated by DMC model
with a nonuniform initial crack size distribution

6. DISCUSSION

Though porous ceramics are categorized as brittle materials, a slight nonlinearity of stress-
strain relationship is observed prior to fracture. This nonlinearity is due to growth and/or
initiation of microcracks. This means that porous ceramics have a damage tolerance property
based on microcrack propagation prior to the whole destruction of the system. DMC analysis of

porous ceramics to evaluate quantitative damage tolerance revealed that the fracture of porous ceramics is unstable destruction associated with the damage localization caused by the softening behavior of stress-strain relationship. Also, it was found that stress relaxation by the nonlinear stress-strain behavior was caused prior to destruction as shown in Fig.8.

To evaluate the effects of such behavior on the strength property of porous ceramics, the maximum stress for tensile, three point and four point bending test and simulation results summarized in Table2. Moreover, the ratio of three point and four point bending strength to tensile defined as strength increasing factor is shown in Table3. Both of the calculated ratios are smaller than experimental ones. This is because the experimental value of tensile is smaller than the calculated one. However, the ratio of three point bending to tension is higher than the ratio of four point bending to tension. The ratio of three point bending to four point bending from experiment and calculation are also shown in Table 4. As shown in Table 4, the strength increasing factor between three point bending and four point bending is independent of the method and the three point bending strength is higher than four point one. This means that stress-strain nonlinear behavior causes an increase of strength and the amount is dependent on loading type.

Table 2 Maximum nominal stress of SiC porous ceramics (SL#5) for tension, 3-p bending and 4-p bending

	Uniform tension (MPa)		3-point bending (MPa)		4-point bending (MPa)	
	exp	calc	exp	calc	exp	calc
	19.1	32.1	59.4	57.7	45.3	50.0
	19.8	33.9		58.0		40.6
		30.3		55.1		46.6
		33.9		58.7		44.0
Av.	19.5	32.6	59.4	57.4	45.3	45.3

Table 3 Strength increasing factor for 3-p bending and 4-p bending

3-p bending / tension		4-p bending / tension	
exp	calc	exp	calc
3.0	1.8	2.3	1.4

Table 4 Strength ratio for 3-p bending to 4-p bending

3-p bending / 4-p bending	
exp	calc
1.3	1.3

7. CONCLUSION

- Distributed-Micro-Crack (DMC) model with a nonuniform initial crack size distribution has been developed. DMC model can predict fairly well maximum bending strength of porous ceramics.

- DMC model analysis revealed that unstable fracture behavior of porous ceramics is caused by damage localization due to strain softening.

- Porous ceramics show a nonlinear stress-strain relationship and the extent of the nonlinearity is dependent on not only on material property such as initial crack length but also on loading type.

8. REFERENCES

1 A.Suzuki, F.Takemasa and H.Baba, "Damage tolerance assessment of synergy ceramics", Proceedings of International Conference on Advanced Technology in Experimental Mechanics 2003(ATEM' 03), CD-ROM, Manuscript No. OS08W0038 (6pages) (2003), The Japan Society of Mechanical Engineers.

2 For example, A.G.Evans and K.T.Faber, "Crack-Growth Resistance of Microcracking Brittle Materials", J.Am.Ceram.Soc., **67**, No.2, 255-260(1984)

3 A.Suzuki and H.Baba, "Assessment of damage tolerance and reliability for ceramics", Mat. Sci. Res. Int., **9**, No.2, 160-166 (2003).

4 For example, "Characteristics of KYOCERA technical ceramics", http://www.kyocera.co.jp/prdct/fc/product, KYOCERA Corporation, (2004).

5 K.Tanaka, Y.Akiniwa, H.Kimachi, Y.Kita, "R-curve behavior in fracture of notched porous ceramics", Engng. Fract. Mech., **70**, 1101-1113 (2003)

6 A.Suzuki, H.Baba and F.Takemasa, "Assessment of damage tolerance for porous ceramics", Ceramics Engineering & Science Proceedings (Proceedings of 27th international Cocoa beach conference on advance ceramics and composites), **24**, Issue 3, 165-170(2003)

THERMAL SHOCK BEHAVIOR OF NITE-POROUS SiC CERAMICS

YI-HYUN PARK
Graduate School of Energy Science, Kyoto University
Uji, 611-0011, Japan

JOON-SOO PARK, TATSUYA HINOKI and AKIRA KOHYAMA
Institute of Advanced Energy Science, Kyoto University
Uji, 611-0011, Japan

ABSTRACT

Recently, there has been an increasing interest in the applications of porous SiC ceramics as functional materials of sustainable and advanced energy systems because of their low thermal-expansion coefficient and good thermal-shock resistance as well as excellent physical and chemical stability at elevated temperature. In most of their applications, porous SiC ceramics are subjected to large temperature difference, which may cause transient thermal stress. Therefore, from the view point of safety and stability, it is necessary to investigate the thermal shock behavior of porous SiC ceramics. The purpose of the present work is to investigate the mechanical and thermal shock behaviors of NITE-Porous SiC ceramics. Porous SiC ceramics have been manufactured based on the Nano Infiltration Transient Eutectic process (NITE process). The flexural strength and critical temperature difference of NITE-Porous SiC ceramics was measured by the three-point bending test and the water-quenching test. The NITE-porous SiC ceramics exhibited a substantially high strength in comparison with other conventional porous SiC ceramics, due to its robust microstructure consisted of spherical pores. The tendency that critical temperature difference increased with increasing porosity and decreasing specimen thickness was confirmed.

INSTRODUCTION

Recently, there has been an increasing interest in the applications of porous ceramics as hot-gas or molten-metal filters, catalyst supports, battery electrodes, heat insulators, ion exchangers, gas sensors, and water cleaners. In particular, porous SiC ceramics are considered as functional materials of sustainable and advanced energy systems, such as perforated containment wall or flow channel inserts for blanket module of fusion reactor, and inner/outer tube of a coated particle type fuel compartment for horizontal flow cooling concept with directly cooling system on Gas-Cooled Fast Reactor, because of their low thermal-expansion coefficient, low thermal conductivity and good thermal-shock resistance as well as excellent physical and chemical stability at elevated temperature[1,2]. Flow channel inserts made of a SiC fiber reinforced SiC matrix composite were first proposed by Tillack and Malang[3] as a means for electrical and thermal insulation between the flowing liquid metal and the load-carrying channel walls to reduce the magnetohydrodynamic pressure drop in the dual-coolant lead lithium blanket channels of a fusion power reactor. The main attraction of the flow channel inserts is that SiC fiber reinforced SiC matrix composite has relatively low thermal conductivity, allowing for the reduction of heat losses from the breeder and therefore high bulk temperatures at the blanket exit, making the overall thermal efficiency of the blanket higher[4]. However, manufacturing method for a SiC fiber reinforced SiC matrix composites is very complicated and time consuming work. On the other hand, manufacturing of porous SiC is simpler compared with those of SiC/SiC composites. In addition to, porous SiC ceramics has relatively low electrical and thermal conductivity. For these reasons, porous SiC ceramics is one of candidate materials as the flow channel insert in the dual-coolant lead lithium blanket module.

A number of manufacturing approaches have been applied to fabricate porous SiC including polymer pyrolysis, oxidation bonding, and reaction bonding[5-8]. However, their processes are complicated and conventional porous SiC shows insufficient physico-chemical stability under high temperature environment. Therefore, from the view point of safety and stability, it is necessary to develop an uncomplicated manufacturing method and to investigate mechanical and thermal properties of porous SiC ceramics. In this study, porous SiC ceramics have been manufactured based on the Nano Infiltration Transient Eutectic process (NITE process), which is developed as a processing technique for high performance a SiC fiber reinforced SiC matrix composite[9]. Moreover, the effects of porosity on the mechanical properties and thermal shock resistance are investigated.

EXPERIMENTAL PROCEDURE
Materials

SiC nano powder (β-SiC, purity >99.0%, mean particle size 50 nm, HeFei, China) and carbon powder (mean particle size 80 nm, Cancarb, Canada) were used as the starting materials. For sintering additives, we used Al_2O_3 (purity >99.99%, mean particle size 0.3 μm, Kojundo Chemical Lab. Co., Ltd., Japan) and Y_2O_3 (purity >99.99%, mean particle size 0.4 μm, Kojundo Chemical Lab. Co., Ltd., Japan). Powders were wet-milled in isopropanol for 12h using Planetary Ball Mill (Fritsch, Germany). The slurry was dried and then passed through a sieve with 100 μm screen. Mixed powders were hot-pressed under a pressure of 20 MPa at 1900°C for 1h with a heating rate of 10°C/min. The sintered SiC including carbon particles was subsequently annealed in the air at 700 °C for decarburization process. Carbon particles can be removed by reaction with oxygen according to the following equation:

$$C + O_2 \rightarrow CO_2 \quad (1)$$
$$2C + O_2 \rightarrow CO \quad (2)$$

Microstructure and mechanical properties

Bulk density of NITE-Porous SiC ceramics was measured by the Archimedes method, using distilled water. Porosity was calculated from relative density and theoretical density, which were obtained by the rule of mixture. The pore shape and size distribution were examined by optical and scanning electron microscopy (FE-SEM, Model JSM-6700F, Jeol, Japan). Flexural strength of NITE-porous SiC ceramics was investigated using three-point bend testing. Rectangular bar 25.0 mm × 4.0 mm × 2.5 mm were prepared by grinding and cutting for porosity measurement and bending test. The tensile faces of the bars were subsequently polished down to 1 μm diamond polish. The tensile edges were beveled to decrease the effect of edge cracks. All tests were conducted on a conventional screw-driven loading frame (Model 5581, Instron, U.K.), with a crosshead speed of 0.5 mm/min, using a three-point bending jig of 18 mm support span, at room temperature. Elastic modulus of the NITE-Porous SiC ceramics was measured using tensile testing which was also conducted on the Instron machine.

Thermal shock test

Thermal shock resistance of the NITE-Porous SiC ceramics was estimated by the water quenching method. The specimens were heated at a rate of 10°C/min to a preset temperature and held at this temperature for 20 min. And then, the heated specimens were then dropped into a water bath, which was maintained at 20°C. After drying at 100°C for 1h, the residual strengths of the quenched specimens were determined in a three-point bend testing. All tests were conducted on a conventional screw-driven loading frame (Model 5581, Instron, U.K.), with a crosshead speed of 0.5 mm/min, using a three-point bending jig with 18 mm support span, at room temperature.

RESULTS AND DISCUSSION
Microstructure and pore-size distribution

Fig. 1 shows the microstructures and pore-size distributions of the NITE-Porous SiC ceramics. Microstructures are cross-section surface of the specimens. For all specimens, we were

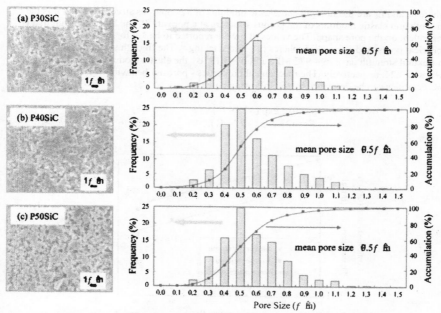

Fig. 1 Microstructures of NITE-Porous SiC ceramics and pore-size distribution

not able to observe residual carbon particles after decarburization process. It is means that most of the pores in the NITE-Porous SiC ceramics are formations of the open pore. Furthermore, as shown in Fig. 1, it was confirmed that mean pore size of the NITE-Porous SiC ceramics is 0.5 μm. The mean pore size of the NITE-Porous SiC ceramics was not changed by the porosity. Accordingly, pore-size of the NITE-Porous SiC ceramics depends on the size of carbon particles instead of quantity of the carbon particles.

Mechanical properties

Flexural strength and elastic modulus of the NITE-Porous SiC ceramics are shown in Fig. 2. Flexural strength of the NITE-porous SiC ceramics, which has porosity of 30 %, 40 %, and 50 % was about 410 MPa, 305 MPa, and 170 MPa, respectively. Moreover, elastic modulus of the NITE-Porous SiC ceramics, which has porosity of 30 %, 40 %, and 50 % was about 190 GPa, 120 GPa, and 75 GPa, respectively. Generally, the mechanical properties of porous ceramics depend strongly on porosity. The porosity dependency of mechanical properties can be approximated by an exponential equation as follows,

$$\sigma = \sigma_0 \exp(-bP) \qquad (3)$$
$$E = E_0 \exp(-bP) \qquad (4)$$

where σ_0 and E_0 are the strength and elastic modulus of a nonporous structure, σ and E is the strength and elastic modulus of the porous structure at a porosity P, and b is a constant that is dependent on the pore shape. The values of b were reported to be 1.4 for cylindrical pores, 3 for spherical pores, and 5 for solid spheres in cubic stacking[10]. The fit of this equation to the results in flexural strength gave $\sigma_0 = 972$ MPa and b = 3.19. By the elastic modulus, E_0 and b was 423 GPa and 3.24, respectively. The mechanical properties - porosity behavior of the NITE-porous SiC ceramics can be described by the following equation,

$$\sigma = 972 \exp(-3.19P) \qquad (5)$$
$$E = 423 \exp(-3.24P) \qquad (6)$$

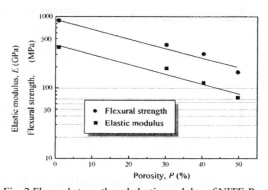

Fig. 2 Flexural strength and elastic modulus of NITE-Porous SiC

As discussed above, the main shape of the pores in the NITE-Porous SiC ceramics can be viewed as a spherical pore. Consequently, the NITE-porous SiC ceramics exhibited a substantially higher strength in comparison with other conventional porous SiC ceramics, due to its robust microstructure consisting of spherical pores.

Thermal shock resistance

Fig. 3 shows the relative residual strengths of the quenched NITE-Porous SiC ceramics with results for dense NITE-SiC ceramics included for comparison. All of the specimens show a sharp drop in retained strength at critical temperature difference, which is the value of temperature difference at the 50% of residual strength. The critical temperature difference of the NITE-porous SiC ceramics which had porosity of 30%, 40%, and 50% was estimated to be about 510°C, 550°C, and 630°C, respectively. In other words, the critical temperature difference of the NITE-Porous SiC ceramics increased with increasing porosity.

The effect of specimen thickness on thermal shock resistance of the NITE-Porous SiC ceramics is shown in Fig. 4. The critical temperature difference of the NITE-porous SiC ceramics, which has thickness of 1.19 mm, 1.90 mm, and 2.90 mm was about 610 °C, 550 °C, and 450 °C, respectively. To reflect the importance of specimen size and heat-transfer rate in determining the critical temperature difference (ΔT_c) under a water-quenching thermal shock condition, an empirical equation has been suggested by Becher et al.[11].

$$\Delta T_c = \frac{\sigma_f (1-v)}{\alpha E} \left(1 + \frac{Bk}{ah} \right) \quad (7)$$

Here, B is a shape factor dependent on specimen geometry, k the thermal conductivity, a the characteristic heat-transfer length, and h the surface heat-transfer coefficient. From Eq. (7), the critical temperature difference value is inversely proportional to the heat-transfer length.

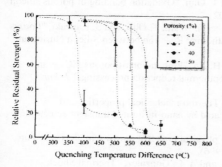

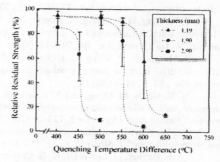

Fig. 3 Relative retained strength versus thermal shock ΔT for quench test. (specimen thickness : 1.90mm)

Fig. 4 Relative residual strength as a function of quenching temperature difference for an NITE-Porous SiC ceramics with a porosity of 40 %

Increasing of specimen thickness should increase the heat-transfer length, hence decreasing the critical temperature difference. Such a tendency accords with experimental results.

CONCLUSION

In this work, the mechanical properties and thermal shock behaviors of the high performance NITE-Porous SiC ceramics were investigated by three-point bending and water-quenching test. The NITE-porous SiC ceramics exhibited a substantially high strength in comparison with other conventional porous SiC ceramics, due to its robust microstructure consisted of spherical pores. From the results of thermal shock test, it appeared that increasing the porosity and decreasing the specimen thickness increased the critical temperature difference of the NITE-Porous SiC ceramics. The application of NITE-Porous SiC ceramics in the functional materials of sustainable and advanced energy systems seems to hold considerable promise.

REFERENCES

[1] A. Kohyama, M. Seki, K. Abe, T. Muroga, H. Matsui, S. Jitsukawa, and S. Matsuda, "Interactions between fusion materials R&D and other technologies," *J. Nucl. Mater.*, **283-287**, 20-7 (2000).

[2] P. Fenici, A. J. Frias Rebelo, R. H. Jones, A. Kohyama and L. L. Snead, "Current status of SiC/SiC composites R&D," *J. Nucl. Mater.*, 258-263, 215-25 (1998).

[3] M. S. Tillack and S. Malang, "High Performance PbLi Blanket," Proceedings of the 17th IEEE/NPSS Symposium Fusion Engineering, **2**, p. 1000 (1997).

[4] S. Smolentsev, N. B. Morley, and M. Abdou, "Magnetohydrodynamic and thermal issues of the SiC$_f$/SiC flow channel insert," *Fusion Sci. Tech.*, **50**, 107-19 (2006).

[5] S. Zhu, S. Ding, H. Xi, and R. Wang, "Low-temperature fabrication of porous SiC ceramics by preceramic polymer reaction bonding," *Mater. Lett.*, **59**, 595-7 (2005).

[6] J. H. She, Z. Y. Deng, J. Daniel-Doni, and T. Ohji, "Oxidation bonding of porous silicon carbide ceramics," *J. Mater. Sci.*, **37**, 3615-22 (2002).

[7] J. F. Yang, G. J. Zhang, N. Kondo, J. H. She, Z. H. Jin, T. Ohji, and S. Kanzaki, "Porous 2H-Silicon Carbide Ceramics Fabricated by Carbothermal Reaction between Silicon Nitride and Carbon," *J. Am. Ceram. Soc.*, **86**, 910-4 (2003).

[8] J. M. Qian, J. P. Wang, G. J. Qiao, and Z. H. Jin, "Preparation of porous SiC ceramic with a woodlike microstructure by sol-gel and carbothermal reduction processing," *J. Eur. Ceram. Soc.*, **24**, 3251-9 (2004).

[9] Y. Katoh, S. M. Dong, and A. Kohyama, "Thermo-mechanical properties and microstructure of silicon carbide composites fabricated by nano-infiltrated transient eutectoid process," *Fusion Eng. Des.*, **61-62**, 723-31 (2002).

[10] R. W. Rice, "Evaluation and extension of physical property-porosity models based on minimum solid area," *J. Mater. Sci.*, **31**, 102-18 (1996).

[11] P. F. Becher, D. Lewis III, K. R. Carman, and A. C. Gonzalez, "Thermal Shock Resistance of Ceramics: Size and Geometry Effects in Quench Tests," *Am. Ceram. Soc. Bull.*, **59** [5], 542-45 (1980).

OPTIMIZATION OF THE GEOMETRY OF POROUS SiC CERAMIC FILTERING MODULES USING NUMERICAL METHODS

S. Alexopoulos, G. Breitbach, B. Hoffschmidt, FH Aachen, Jülich, Germany; P. Stobbe, Stobbe Tech Ceramics A/S, Holte, Denmark

ABSTRACT

An optimization of the geometrical shape of porous ceramic filtering modules of SiC is presented. These modules consist of channel crossed carriers where the channels are covered by thin ceramic coatings. The analysis method is based on finite element modeling concerning the fluid flow in porous media for different geometrical configurations of the filtering module.

For the characterization of a module different physical and geometrical quantities were chosen. A theoretical model using these variables has been set up and was then implemented in a suitable numerical simulation environment. The geometry optimizations of the modules were carried out aiming in a reduction of the used membrane area and in a higher permeate flow. Different geometrical parameters like wall thickness and channel width of the carrier have been varied. An optimum has been estimated with respect to wall thickness and channel width. The geometry form of a 0.144 m diameter membrane module has been considered. The pressure build up especially in the inner parts of the module depends on the permeability and the size of the carrier. Therefore the inner parts in large porous carriers do not contribute significantly to the filtered permeate flow. In this context measures to improve the efficiency of the modules are demonstrated: an appropriate removal of inefficient regions.

INTRODUCTION

High mechanical strength and corrosion resistance is one of the key issues in utilization of membranes in most of the applications like food & beverage, potable water, medical & pharmaceutical industry and industrial. For such applications a ceramic membrane module is suitable. SiC as a carrier material has turned out to be an excellent choice because of its corrosion resistance, mechanical strength, size potential and possibility to membrane coatings.

A typical ceramic membrane element has multiple parallel passageways that run from one face to the opposite end face. During processing, the feed stream to be treated is introduced under pressure at one end of the element, flows through the passageways over the membrane, and is withdrawn at the downstream end of the element as "retentate". Material which passes through the membrane ("permeate") flows into the cell walls of the monolith. The combined permeate from all the passageways flows toward the periphery of the monolith support, and is removed through an integral, "skin" at the exterior of the monolith. Ceramic membranes are formed on the internal passageway wall surfaces of the ceramic honeycomb monoliths by slip casting porous coatings of ceramic particles[1].

Competitive cost is the other key for implementation of ceramic membranes. It should be obvious that a way for reducing prices of ceramic membranes is by the increase of the membrane module filtration area significantly. This is possibly to achieve by using huge SiC diameters like 0.144 or 0.180 m. For such huge diameters new problems occur. They are related to the choice of the right physical dimensions and of the geometry form. The geometry optimizations of the module are carried out aiming in a reduction of the used membrane area and in a higher permeate flow. Under these conditions different geometrical quantities are parameterized with a chosen numerical software tool. Dolecek and Cakl simulated the permeate flow in porous support of a hexagonal 19-channel ceramic membrane[2]. The configuration of a 0.144 m diameter porous ceramic membrane was chosen for the following study.

THEORETICAL BACKGROUND

In a porous material the volumetric flow rate q [m^3/(m^2/s)] is given by Darcy's law[3]

$$q = -K \cdot grad \; p \tag{1}$$

p is the pressure field inside the porous material, K depends on the porous structure parameters and the viscosity of the fluid. It is usually given in the following form:

$$K = \frac{B}{\eta} \tag{2}$$

B depends on the pore size and of the porosity. There exist several models to determine B by calculations[4].

Assuming incompressible fluids and stationary condition the equation for the calculation of the pressure field is the classical Laplace equation

$$div\left(-\frac{B}{\eta} \cdot grad \; p \right) = 0 \tag{3}$$

or

$$\Delta p = 0 \tag{4}$$

It should be mentioned that the structure of the governing equation is the same as the equation for conduction of heat in solids. Therefore corresponding computer programs can be used for numerical analysis.

The considered structure has complex boundaries (see Figure 1).

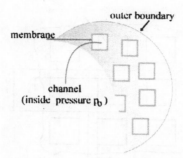

Figure 1. Cross section of a filter module; inside the material are membrane covered channels with fluid pressure p_0

Inside the material there are membrane covered channels. The boundary condition here is given analog to the "radiation" boundary condition in heat transfer[5]:

$$\frac{B}{\eta} \cdot \frac{\partial p}{\partial n} = \frac{K_M}{l_M} \cdot (p_0 - p) \tag{5}$$

The membrane is not integrated into the porous material. It is considered as a boundary flow resistance. It corresponds to the boundary heat transfer coefficient. The condition at the outer boundary is given by the constant outside pressure.

QUANTITIES FOR THE CHARACTERISATION AND OPTIMIZATION OF A CERAMIC MODULE

In general different quantities determine the fluid flow through porous modules and describe the filtration potential of these ceramic membranes. Both the geometrical and membrane quantities are considered here with special emphasis to the geometrical properties of the carrier.

A ceramic module is characterized by the following parameters:
- the radius R, the length L
- the inner channel width a, the basic geometrical cell dimension b equal to the channel width a plus the wall thickness w (see Figure 2)
- relevant material properties of the SiC-carrier: Porosity ε_p, pore diameter d_p
- Properties and structure of the membrane
  - Number and thickness of layers l_M
  - Porosity ε_M and pore diameter of the layers d_M

Based on the parameters the following quantities must be calculated (or measured)
- Membrane area
- Permeability of carrier and membrane
  - Permeability of the SiC grid structure
  - Permeability of the membrane

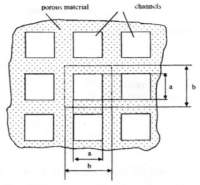

Figure 2. Elementary cell in porous material, $w = b - a$

The properties for an ideal filtering module are now discussed.
The module has to be optimized according to three quantities. These are the module effectivity, the membrane area and the flow of the module.

The module effectivity E is defined as the ratio of the permeate flow Q_{per} to the maximal possible flow Q_{max} by negligible resistance of the porous carrier.

$$E = \frac{Q_{per}}{Q_{max}} \qquad (6)$$

The permeate flow Q_{per} depends on the pressure, the geometrical and the membrane properties of the module:

$$Q_{per} = Q_{per}(a, b, d_M, d_p, \varepsilon_M, \varepsilon_p, l_M, R, p_0) \qquad (7)$$

A maximal possible flow is calculated by:

$$Q_{max} = \pi \cdot R^2 \cdot \frac{4a}{b^2} \cdot \frac{K_M}{l_M} \cdot p_0 \cdot L \qquad (8)$$

The factors K_M and l_M are characteristic for the membrane and describe the hydraulic conductivity and the thickness of the membrane layer. In this paper the values for these membrane properties are assumed constant.
The module effectivity is a function of the following parameters:

$$E = E(a, b, d_M, d_p, \varepsilon_M, \varepsilon_p, l_M, R) \qquad (9)$$

Another characteristic quantity is the membrane area per volume. The goal is to minimize the cost for the production of the module by minimizing the membrane area used.

The third quantity is the permeate flow. A module has a good performance when it can filtrate as much as possible. In an optimization process all these three quantities must be considered.

CHOSEN NUMERICAL SOFTWARETOOL

The optimization of the geometry of the module has been carried out numerically. A numerical model for the fluid flow through a membrane in a porous medium is implemented with finite element methods using the software environment of COMSOL[6]. COMSOL has the capability to generate complicated module geometries, is user friendly and has an integrated numerical equation solver.

During the development of the numerical model the fluid flow through the porous carrier is described with the same transport equation. For each channel the infiltration of the fluid through each membrane is implemented as a boundary condition. The simulation gives reasonable and descriptive results. They provide macroscopic information about the flow rate through the whole module as well as a detailed two-dimensional pressure distribution in the porous material.

OPTIMISATION OF THE CHANNEL GEOMETRY

For computation a porous SiC module with a diameter of 0.144 m is used (see Figure 3). It is assumed that the channels inside are covered by a 50 nm membrane with an effective conductivity K_M of $1.1e^{-13}$ m^4/N/s.

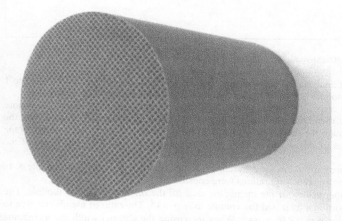

Figure 3. Geometrical shape of a 0.144 m diameter SiC porous module

Figure 4 shows a typical geometry model for analysis. For symmetry reasons a quarter of the module geometry is considered. The mesh of that geometry consists of 6,718 triangular elements.

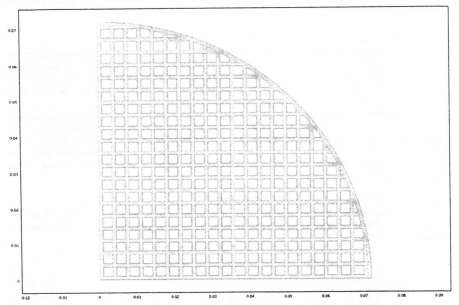

Figure 4. Geometrical shape with mesh of the considered 0.144 m module

In the following analysis the pressure drop from channel to environment is set equal to 1 bar. The length of the module is set to 1 m and the porosity ε_p of the carrier is equal to 0.5.

The membrane consists of three different layers of which the outer layer has a pore diameter of 50 nm. The membrane thickness and porosity of each layer is set constant. Only the fluid flow of water through the module is considered without respect to the effects of membrane fouling. The main focus of this optimization was on the consideration of the resistance of the carrier.

Calculations were carried out for different values of channel width and wall thickness. In total nine different geometrical configurations were considered.

Table I shows the results for the membrane area for the 0.144 m diameter module. The membrane area is proportional to a and the volume to $(a + w)^2$. The ratio of membrane area to volume is then proportional to $a / (a + w)^2$. When increasing the channel width the membrane area per volume of the module decreases. On the other hand, decreasing the wall thickness the membrane area per volume decreases as well. The module with the lowest membrane area is the module with 2.5 mm channel width and 1 mm wall thickness.

Table I. Calculated membrane area by variation of channel width and wall thickness

Membrane area in m²		channel width		
		1.5 mm	2.0 mm	2.5 mm
wall thickness	0.6 mm	22.2	19.3	17.0
	0.8 mm	18.5	16.6	15.0
	1.0 mm	15.6	14.5	13.3

Table II shows the results for the module flow for the 0.144 m diameter module. When decreasing the channel width the module flow increases. The module with the highest permeate output is the module with 0.6 mm wall thickness and 1.5 mm channel width.

Table II. Calculated module flow by variation of channel width and wall thickness

Module flow in l/h		channel width		
		1.5 mm	2.0 mm	2.5 mm
wall thickness	0.6 mm	3,019	2,579	2,214
	0.8 mm	2,974	2,537	2,294
	1.0 mm	2,858	2,562	2,256

There is an interesting observation considering Tables I and II. Increasing the wall thickness at constant channel width, results in a remarkable reduction of membrane area. But in comparison there is only a small change in the module flow. This can be explained as follows. When reducing the membrane area a decreasing flow is expected. But an increasing wall thickness at the same time achieves a smaller flow resistance inside the porous carrier.

Table III shows clearly that the module flux increases strongly as a function of wall thickness.

Table III. Calculated module flux by variation of channel width and wall thickness

Module flux in l/h/m²		channel width		
		1.5 mm	2.0 mm	2.5 mm
wall thickness	0.6 mm	136.0	133.6	130.2
	0.8 mm	160.8	152.8	152.9
	1.0 mm	183.2	176.7	169.6

Table IV shows the results for the effectivity of the module. With increasing wall thickness the effectivity of the module increases. The optimal modules are found in the bottom row belonging to 1 mm wall thickness.

Table IV. Calculated effectivity by variation of channel width and wall thickness

Effectivity in %		channel width		
		1.5 mm	2.0 mm	2.5 mm
wall thickness	0.6 mm	22.3	21.9	21.3
	0.8 mm	26.3	24.9	25.1
	1.0 mm	29.9	28.9	27.7

CONSIDERATIONS TO THE OPTIMIZATION OF THE MODULE GEOMETRY

For the set of the here chosen parameters, a remarkable pressure build up is found in the central region of the module (see Figure 5). The pressure difference which drives the permeate flow through the membrane is small in that region. For this reason the inner parts have a very small contribution to the total permeate flow.

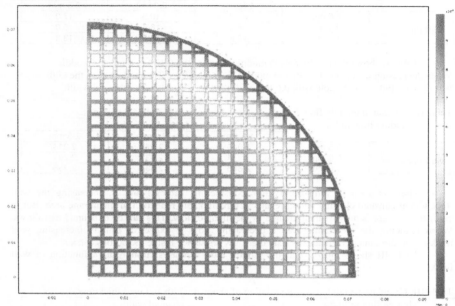

Figure 5. Pressure distribution in the 50 nm membrane filter module with the diameter of 0.144 m, a wall thickness of 1.0 mm and a channel width of 2.5 mm

An improvement of the filtering performance can be reached by creating an additional inside surface from which the filtrate can be discharged. Then the filtering module gets the form of a hollow cylinder with the internal radius R_i and the external radius R_a. As described below the flow through the module is investigated by varying the internal radius for a fixed outer radius.

For the 0.144 m module numerically calculations are carried out using different internal radii. The channel width is 2.5 mm and the wall thickness 1.0 mm, respectively. The same layout of the 50 nm membrane and of the carrier is used.

The aim is to get a high flow rate together with a high saving of membrane surface. Without the removal of the ring area the membrane area amounts to 13.3 m^2 and the flow through the module is 2256 l/h.

Table V shows the simulation results of the module flow and the membrane area when the internal radius is changed. Increasing the diameter, the permeate flow increases and the membrane area decreases. Up to a radius of 0.010 m the flow increases more than 500 l/h but the reduced membrane area is negligible. At an increased radius the membrane area saving is much higher, whereas the flow reaches a maximum.

Table V. Calculated module flow and membrane area as a function of internal radius

Internal radius in m	Module flow in l/h	Membrane area in m²
0.000	2,256	13.3
0.005	2,589	13.2
0.010	2,771	13.0
0.020	3,087	12.3
0.030	3,349	11.0
0.035	3,383	10.2
0.040	3,411	9.2
0.045	3,311	8.1

The results of the simulation show that a good choice for the radius is 0.04 m. Figure 6 shows the pressure distribution and permeate flow in the module for this configuration.

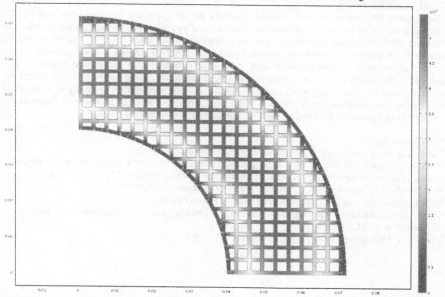

Figure 6. Pressure distribution in the filter module for a reduced cross section with an internal radius of 0.04 m

With this reduced cross section the volumetric permeate flow is double of the flow of the full cylindrical filter module. The corresponding reduction of the membrane area results is more than 30 %. However it has to be emphasized that this saving potential must be compared with the extra costs that result from the production of the new module form.

FUTURE STEPS

The optimization of the module will be expanded for different geometries of the porous carrier. In addition to the 0.144 m module the 0.180 m cylindrical module will be calculated. The optimization will be carried out for the same membrane types and layers.

Due to the long and tortuous path through which the permeate must flow to reach the outside of the monolith, there can be a large pressure drop for the permeate flow. In order to decrease this phenomenon it is necessary to modify the shape of the module. These modifications form permeate conduits within the monolith, and the conduits conduct permeate from within the monolith to a permeate collection zone. A future aim is to simulate the permeate flow rate with the use of conduits and to optimize the number and form of the conduits for different diameters of ceramic modules.

CONCLUSION

The ceramic filter modules with internal channels are compact and resistant units and can be used in a variety of industrial applications. Geometrical optimizations of a complete module can be done with numerical procedures. For an optimization the knowledge of the characteristic quantities of a ceramic cylindrical module is essential. These quantities include porosity of the carrier, dimensions of the channels and membrane parameters as membrane thickness and porosity. Fine tuning of channel shapes should be investigated, using Finite Element Programs. The results of the simulation analysis show that a reduction in cost due to saving of membrane area can be reached removing a ring area inside the carrier. The internal geometrical shape of the carrier can be optimized by varying the wall thickness and the channel width.

REFERENCES

[1]Technical Brief CeraMem's Crossflow Membrane Modules
[2]P. Dolecek, Cakl J. Permeate flow in hexagonal 19-channel inorganic membrane under filtration and backflush operating modes Journal of Membrane Science 149, pp. 171-179 (1998).
[3]R.B. Bird, W.E. Stewart, E.N. Lightfoot Transport Phenomena (2002).
[4]H.S. Carslaw, J.C. Jaeger Conduction of Heat in Solids (1960).
[5]P.C. Carman „Fluid flow through a granular bed – Transactions of the Institution of Chemical Engineers" v. 15., London, (1937).
[6]COMSOL Multiphysics User's guide Version 3.2 (2005)

WEIBULL STATISTICS AND SCALING LAWS FOR RETICULATED CERAMIC FOAM FILTERS

Mark A. Janney
Selee Corporation
700 Shepherd Street
Hendersonville, NC 28792

ABSTRACT

Reticulated ceramic foam is one of the standard filter forms used in cast iron, aluminum, and steel foundry applications. We have studied the flexure strength of these filters as a function of sample size and sample test geometry in both 3-point and 4-point flexure. Good agreement was demonstrated among the various results based on the use of Weibull scaling laws. In particular, we have shown that the results scale according to a volume flaw distribution for as-fired test parts. We have also shown that at last for one foam composition, the properties of as-fired and diamond-sawed test bars are similar.

BACKGROUND

Ceramic reticulated foam is mainly used to filter molten metals, including aluminum, cast iron, steel, and superalloys. Other applications include chemical processing equipment (catalyst supports, radiation shields, fluid diffusers), furnace and kiln furniture, and natural gas burners. Commercial foams are manufactured from a variety of ceramics including alumina, partially-stabilized zirconia, silicon carbide, mullite, and zirconia-toughened alumina. Pore size ranges from as large as 3 pores per inch (ppi) to as fine as 100 ppi. The choice of pore size is dictated by the application. In general, foundry operations use coarser pore sizes while billet operations use finer pore sizes. Foam densities range between about 8 and 25 vol %, depending on the

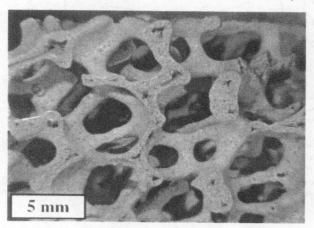

Figure 1. Overview of ceramic reticulated foam. Polished section, epoxy mount, stereomicroscope image.

application. For filtration, lower density is preferred to increase permeability, which leads to lower flow resistance and greater throughput of molten metal. For kiln furniture, higher density is preferred to increase strength. Figure 1 shows a typical, coarse pore size ceramic foam.

Ceramic reticulated foam is made by a templating method. The process starts with open cell, reticulated polyurethane foam. A ceramic slurry is infiltrated into the polymer foam to coat the struts of the foam. Excess slurry is removed from the foam to create a highly porous green body. The green body is fired to burn out the polyurethane foam and to sinter bond the ceramic particles.

Brezny, et al[1,2,3] have reported extensively on the properties of reticulated ceramics and reticulated vitreous carbon (RVC) foams. In particular, they looked at the effect of pore size to part thickness. Figure 2 shows how average strength and standard deviation of strength vary with part thickness for a 10 ppi (pore per inch) RVC foam. Strength increased and standard deviation decreased with thickness. Their results agreed well with the micromechanical model developed by Gibson and Ashby[4] once a correction was made to include an increase in strut strength with decreasing strut size. Brezny, Green and Dam[3] showed that the strength of ceramic foams depend critically on the details of the micro- and macro-structure of the foam and, in particular, on voids and cracks that form in the struts as a result of processing.

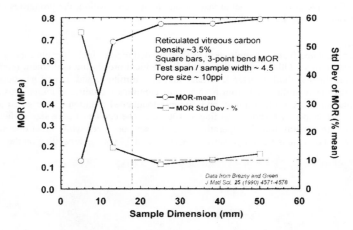

Figure 2. Results from Brezny and Green[1] showing that strength increases and standard deviation decreases as the thickness of a reticulated carbon foam sample increases.

The present work looks at three issues associated with the strength of commercial reticulated ceramic foams:
(1) the effect of pore size and thickness on strength,
(2) the role of green vs fired cutting (edge damage), and
(3) the applicability of Weibull fracture statistics to reticulated ceramic foam.

MATERIALS AND METHODS

Most of this investigation was conducted using a commercial, partially-stabilized zirconia foam. Pore size was either 4 or 8 ppi. Part density was ~12% theoretical. Sample thickness ranged between 0.5 and 1 inch.

Strength was determined using both 4-point MOR (50 over 150mm span) and 3-point MOR (46mm or 75mm span) to provide a wide range of sampled volume during strength testing. All sample tensile surfaces were tested in the as-fired condition. Because the surfaces of the samples had high and low spots (an inherent property of the reticulated foam), 1mm pads of neoprene rubber were placed between the loading rollers and the sample surface to provide a compliant layer that spread the load uniformly across the sample, Figure 3.

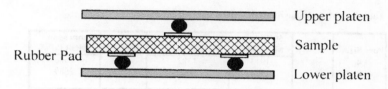

Figure 3. Schematic of MOR test apparatus for reticulated ceramic foam.

RESULTS AND DISCUSSION
Effects Of Pore Size And Thickness On MOR

The initial study examined the effect of pore size and part thickness on strength. Table 1 summarizes the experimental set up, which was a full-factorial 2 x 2 experiment. Twenty (20) MOR bars were broken for each test condition. Test bar dimensions were 1 x 8 inches x part thickness (25 x 200 x 13 or 19 mm).

Table 1. Experimental design for initial strength study.

Experiment #	Pore size (ppi)	Thickness (inch / mm)
2	4	0.50 / 13
4	4	0.75 / 19
6	8	0.50 / 13
8	8	0.75 / 19

Two sets of bars were fabricated for each test. One set of bars was cut in the green state using a water jet cutter. A second set of bars was cut wet, on a diamond saw after firing. All of the bars were first broken in 4-point MOR using the 75mm over 300 mm test span fixture. After breaking, the bars were in two sections. The longer of the two sections was re-broken in 3-point bending using either a 46mm (green cut) or a 75mm (fired cut) span. By using this approach, we were able to obtain fracture data for two different bar lengths from a single bar.

Table 2. Initial Strength Study - MOR Bars - Cut Green

Exp #	Pore Size (ppi)	Thickness (mm)	MOR 50/150 mm span (kPa)	Std Dev (% mean)	46 mm span (kPa)	Std Dev (% mean)
2	4	13	1881	8.5	2671	14.6
4	4	19	1945	18.2	2707	18.2
6	8	13	2010	14.4	2656	14.7
8	8	19	1937	8.9	2682	11.6
All			1943		2679	

Table 2A. Mean Value Analysis - MOR Bars - Cut Green

Pore Size		50/150 mm span MOR (kPa)	Variance (kPa^2)		46 mm span MOR (kPa)
4		1913	1382		2680
8		1974	982		2669
Thickness					
13 mm		1945	958		2664
19 mm		1941	703		2695

Table 3. Initial Strength Study - MOR Bars - Cut After Firing

Exp #	Pore Size (ppi)	Thickness (mm)	MOR 50/150 mm span (kPa)	Std Dev (% mean)	46 mm span (kPa)	Std Dev (% mean)
2	4	13	2150	11.9	2627	9.2
4	4	19	1859	9.8	2379	13.9
6	8	13	2053	23.6	2626	16.0
8	8	19	1980	19.4	2370	21.4
All			2011		2500	

Table 3A. Mean Value Analysis - MOR Bars - Cut Fired

Pore Size		50/150 mm span MOR (kPa)	Variance (kPa^2)		46 mm span MOR (kPa)
4		2004	824		2503
8		2017	3212		2498
Thickness					
13 mm		**2102**	2410		**2627**
19 mm		**1919**	1627		**2374**

Tables 2 and 3 summarize the data for the initial tests. Tables 2A and 3A give the Mean Value Analysis for each test; i.e., an analysis that compares the effects of pore size and part thickness variables.

Consider first the results for the two different pore sizes. There was essentially no effect of pore size on MOR values for either bars cut green or bars cut after firing. This was somewhat unexpected given the earlier work of Brezny and Green,[1] which showed a significant increase in strength as the pore size was decreased. Their interpretation was that as the pore size was reduced (for a constant thickness) the number of struts available to resist fracture increased and hence the strength increased.

The difference between the present results and those of Brezny and Green may be due to the differences in the structure of the struts that make up the two materials. The struts in reticulated carbon foam are solid with little, if any porosity. In contrast, the struts in reticulated ceramic foam contain at least two types of porosity, Figure 4. First, there is the large central void left from the burnout of the polyurethane reticulated foam template used to make the ceramic foam green body. The central void is typically between 1/4 and 1/3 the diameter of a strut. Second, there is residual porosity from the sintering operation - ceramic foam struts are typically not fully dense. The central voids that run down the middle of struts are quite large (typically 50 to 500 μm). One might suspect the central voids constitute the strength limiting flaws in the foam. In some cases, central voids extend to the surface of a strut, in which case, they constitute a significant break in the continuity of the strut structure. Another difference between this work and the work done by Brezny and Green is that they studied foams having much finer pore sizes, typically between 10 and 100 ppi.

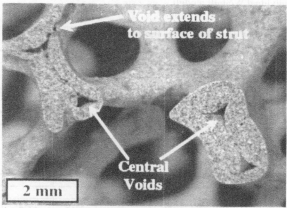

Figure 4. Reticulated ceramic foam struts contain large central pores that are a remnant of the starting polyurethane foam template. Sometimes voids extend to the surface of a strut. Polished section, epoxy mount, stereomicroscope image.

Next, consider the effects of sample thickness on strength. One might predict thicker samples to be stronger than thinner ones, for a constant pore size. Such was not the case for either set of samples. For the bars cut green, there was no effect of sample thickness. For the bars cut after firing, the thicker samples were actually weaker than the thinner bars.

Finally, overall there was little difference in the strength of the parts whether they were cut green or cut after firing. This suggests that the inherent flaws in the foam are large relative to the flaws that might be introduced during diamond sawing operation. This would be consistent with the central voids or other processing-induced flaws being the strength-limiting features in these parts.

Weibull Behavior of Reticulated Ceramic Foam
A series of MOR samples were tested specifically to investigate the applicability of Weibull statistics to the fracture behavior of sintered ceramic reticulated foam. The same commercial, partially-stabilized zirconia foam was used for this study as was used in the work presented previously in Tables 2 and 3. A much larger specimen size was used to provide as much difference in effective volume and surface area as possible. The starting test bars were cut in the green state and sintered to overall dimensions of 25 x 50 x 300mm and 25 x 100 x 100mm. The pore size was 4 ppi for these tests. Fifty (50) MOR bars were broken for each test condition. The test conditions used and the summary results are given in Table 4. The strength ratios given are for values associated with the largest test volume (Test 10) divided by those associated with the other two tests (Tests 11 and 12). The Weibull probability plots for these tests are shown in Figure 5. Clearly, the data exhibit a straight line in the Weibull plot and thus Weibull statistics can be used to analyze these data. For the comparison analysis shown in Table 4, an average Weibull modulus, m, of 10.5 was used. This is a reasonable approach given the uncertainty of the estimate of the individual Weibull moduli.

Table 4. Summary of Weibull analysis for large reticulated ceramic foam bars								
Test	Test Condition	m	Mean Strength (kPa)	Se Ratio*	Ve Ratio**	Actual Strength Ratio	Predicted Strength Ratio (Surface)	Predicted Strength Ratio (Volume)
10	25x50x300mm 4-point bend 75 over 225 mm	11.5	1965					
11	25x50x300mm 3-point bend 75 mm span	9	2650	11000	14	0.74	0.41	0.77
12	25x100x100mm 3-point bend 75 mm span	12	2570	5700	7	0.76	0.44	0.83

*Ratio of effective surface areas of larger bar to smaller bar.
**Ratio of effective volumes of larger bar to smaller bar.
Analysis from GD Quinn, "Weibull Strength Scaling for Standardized Rectangular Flexure Specimens" J.Am.Ceram.Soc. , **86** [3] 508–10 (2003).

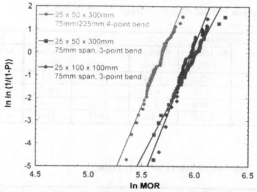

Figure 5. Weibull plots for reticulated ceramic foam demonstrating the applicability of Weibull statistics to describe the MOR of different size samples.

If one assumes that Weibull statistics are valid the next logical question is: "Do surface flaws or volume flaws control fracture?" In dense ceramic systems, determination of fracture origins can be used to help answer this question. In highly porous systems, however, finding fracture origins is virtually impossible. Instead, one can look at scaling phenomena to determine whether surface or volume flaws are controlling behavior; i.e., how does the strength vary with specimen size.

Table 4 summarizes the following parameter for the three tests: (1) the effective surface (Se) ratios, (2) the effective volume (Ve) ratios, (3) the actual strength ratios, (4) the predicted strength ratio if surface flaws are controlling, and (5) the predicted strength ratio if volume flaws are controlling. The actual ratios of strengths are 0.74 (Test 10/Test 11) and 0.76 (Test 10/Test 12). For surface flaw control, the predicted ratios are 0.41 and 0.44; for volume flaw control, the ratios are 0.77 and 0.83, respectively. From these results, one would conclude that volume flaws are controlling in this system.

Further analysis was done using the test results from the preliminary strength study - Tests 2, 4, 6, and 8 (Tables 2 and 3). We re-analyzed the data from those test using Weibull statistics, Tables 5 and 6. Comparing the ratios of large bar strength (75/150mm, 4-point bending) to small bar strength (either 46mm or 75mm, 3-point bending) showed that the actual strength ratios compare favorably with those computed for volume flaw distributions. The actual ratio of large-bar to small-bar strengths is typically between 0.7 and 0.8. Surface flaw control would have predicted strength ratios between 0.25 and 0.45.

SUMMARY AND CONCLUSIONS
The present work demonstrated that the fracture behavior of ceramic reticulated foam can be accurately described using Weibull statistics. In particular, it was shown that Weibull volume scaling laws and not Weibull surface scaling laws apply.

Table 5. Weibull Analysis - MOR Bars Cut Green

Exp #	Pore Size (ppi)	Thickness (mm)	MOR 50/150mm (kPa)	Weibull Slope	46mm span (kPa)	Weibull Slope	MOR Ratio Actual	Predicted (volume)
2	4	13	1881	13.8	2671	8.1	0.704	0.770
4	4	19	1945	6.6	2707	6.7	0.719	0.700
6	8	13	2010	8.4	2656	8.1	0.757	0.730
8	8	19	1937	13.6	2682	10.4	0.722	0.790

Table 6. Weibull Analysis - MOR Bars Cut After Firing

Exp #	Pore Size	Thickness (mm)	MOR 50/150mm (psi)	Weibull Slope	75mm span (psi)	Weibull Slope	MOR Ratio Actual	Predicted (volume)
2	4	13	2150	9.9	2627	13.1	0.818	0.820
4	4	19	1859	11.9	2379	7.9	0.781	0.810
6	8	13	2053	4.5	2626	7.6	0.782	0.740
8	8	19	1980	5.9	2370	5.5	0.836	0.730

Further, this work showed that for the coarse pore size (4 and 8 ppi), partially-stabilized zirconia foams studied here, there was no effect of pore size on strength. This suggests that the effects of voids and other defects caused by processing are more important than the size of pores created by the template urethane foam.

Finally, no significant differences were observed between samples cut in the green state and those cut on a diamond saw after firing. This suggests that the flaws inherent in the foam (such as the central voids left by the urethane template) are large relative to the flaws that might be introduced during diamond sawing operation.

REFERENCES

[1]R. Brezny, D.J. Green, "Characterization of Edge Effects in Cellular Materials," *J.Matls.Sci.*, 25, 4571-78 (1990).

[2]R. Brezny and D.J. Green, "The Effect Of Cell Size On The Mechanical Behavior Of Cellular Materials," *Act Metall.Mater.*, **38**, 2517-26, 1990.

[3]R. Brezny, D.J. Green, and C.Q. Dam, "Evaluation of Strut Strength in Open-Cell Ceramics," *J.Am.Ceram.Soc.*, **86**, 508–10 (2003).

[4]L. J. Gibson and M. F. Ashby, "The Mechanics of Three-Dimensional Cellular Materials ." *Proc.* R. Soc. Ser. A, **382**, 43-59 (1982).

EFFECT OF TEST SPAN ON FLEXURAL STRENGTH TESTING OF CORDIERITE HONEYCOMB CERAMIC

Randall J. Stafford, Rentong Wang
Cummins Incorporated
Research and Technology
Columbus, IN 47202

ABSTRACT

A standard test method for porous honeycomb ceramics is required since suppliers and end users conduct flexural tests of these complex geometry materials in different specimen configurations and test spans resulting in conflicting data. Generally these methods used are variations of the current ASTM standard test method on flexural strength, C1161, or its predecessor MIL-STD-1942A. The ASTM C28.04 subcommittee (Advanced Ceramic Applications) has been developing a new test standard for flexural strength testing of porous honeycomb structure materials. This method has recommended specimen dimensions and test spans.

To prepare for the anticipated use of the new test method, Cummins conducted a series of comparisons for the test methods used by Cummins and its suppliers to determine flexure strength of cordierite honeycomb ceramics. The objective of this comparison was to determine if the historical data can be related to data generated using the 2006 draft version of the new ASTM test method. The Weibull parameters (characteristic strength and modulus) were used as the basis for comparison. Specimen size effect of different test spans was evaluated by a model combining Weibull statistics with Weakest Link principal.

INTRODUCTION

The use of honeycomb ceramics in aftertreatment has a long history. The application of these structures as automotive catalysts was one of the initial uses. Subsequent applications in the diesel and natural gas markets as catalysts supports for emission control on stationary generator sets and then as passive catalyst/filter combinations in retrofit bus and vocational vehicles have broadened the range of operating conditions and the sizes of the individual components.

More demanding applications on the honeycomb ceramic have been in process since the early 1990's with implementation of particulate filters in diesel exhaust which require active regeneration. The diesel industry has implemented this application in 2007 to meet the latest mandated diesel emission regulations[1]. Active regeneration requires addition of energy to the exhaust stream to bring the catalyst and/or filter to the optimal operating temperature for reduction of nitrous oxides (NO_x) and combustion of carbon particulate. This addition of energy also introduces thermal cycling effects on the honeycomb materials with axial and radial temperature gradients and thermal and mechanical stresses which were not as prevalent in the earlier applications.

These more complex stress states of the honeycomb material and the interactions with the exhaust stream coupled with the desire to use computational models to reduce the amount of long term testing leads to the need to have accurate and accepted material property values. The best available test method for measuring strength has been ASTM C1161, Standard Test Method for Flexural Strength of Advanced Ceramics at Ambient Temperature.[2] This test method has

113

been modified from the standard specimen dimensions and test spans by the honeycomb ceramic suppliers and users with several different combinations of specimen size and span ratios currently in use. The influences of test technique, specimen alignment, fixture articulation and specimen preparation on the test result are well known in the ceramic industry for monolithic ceramics. Testing of a honeycomb structure is subject to the same influences. Direct application of ASTM C1161 for the honeycomb structure uses standard monolithic beam formulas which treat the honeycomb structure as an equivalent homogenous continuum. In addition, this calculation does not differentiate between honeycombs of different cell density (e.g. 15.5 cell/cm^2 versus 31 cell/cm^2 versus 62 cell/cm^2) or cell wall thickness, which are not addressed in the results.

Future improvements in life of honeycomb ceramic catalysts and filters through application of engine/aftertreatment controls, material property modifications and structure modifications will be accelerated by detailed understanding and application computer models. The models for thermal stress and life prediction require inputs of Weibull modulus, characteristic strength, thermal expansion, thermal conductivity, elastic modulus, slow crack growth and other parameters. These inputs need to be validated data of high precision to give the best outcome of the computer model.

The proposed new ASTM test method, Flexural Strength of Advanced Ceramics with Engineered Porosity (Honeycomb Cellular Channels) at Ambient Temperature has a recommendation for a preferred test specimen size and load and support spans to standardize the testing between suppliers and users of the honeycomb structures. In addition to the common test parameters, there is a recommended calculation to convert the honeycomb bar strength to a material strength of the wall. This material strength will be calculated in a consistent manner by all users of the test method and the results from suppliers and users can be directly compared and can be utilized in the modeling codes to evaluate performance and predicted life.

This work was conducted to evaluate the test parameters of the 2006 draft version of the new ASTM test method and compare to two of the current variations of ASTM C1161 where historical data was available. The analysis was limited to cordierite honeycomb filter material with the assumption that the results will be generally applicable to cordierite materials. Applicability of this analysis to materials other than cordierite was not part of this work and has not been determined.

EXPERIMENTAL

The test specimens for this work were prepared from one 12 inch diameter by 10 inch long extruded honeycomb particulate filter. This filter was a commercially available cordierite (Corning Duratrap CO®) with an applied catalytic coating for soot combustion. The nominal cell density was 31 cells/cm^2 (200 cells/in^2) and the nominal substrate wall thickness was 0.305 mm (0.012 in). The specimens were cut to dimensions of ~26 mm by ~13 mm by >100 mm with the long axis of the specimen aligned with the open channels in the extrusion direction of the filter. These dimensions corresponded to a specimen with 7 complete cells in thickness and 14 complete cells in width. The specimens were randomly divided into three groups for testing. The specimens in one group were recut to give specimen dimensions of ~20 mm by ~10 mm (12 cells by 6 cells). All specimens were prepared for testing by sanding with 180 grit SiC paper to remove most of the residual wall segments protruding above the surface. The sanding was carefully monitored to prevent contact and damage to the wall surfaces to be tested.

The specimens were tested to fracture in four point bending using a Sintech 20G load frame with a 500 N load cell. The test load was applied at a constant crosshead rate of 0.5 mm/min until fracture or 20% load drop in the specimen. The specimens were supported in a fully articulating four point bending fixture at three different span configurations as shown in Table 1. These configurations correspond to the spans currently used by Cummins, a cordierite supplier and the 2006 draft version of the new ASTM flexural standard for porous honeycomb ceramics.

Table 1. Test Span and Specimen Configurations

Test User	Inner Load Span (l), mm	Outer Support Span (L), mm	Specimen Width (b), mm	Specimen Thickness (h), mm	Specimen Length, mm
Supplier	19.05	88.9	25	13	≥ 100
Cummins	20	40	20	10	≥ 100
ASTM	45	90	25	13	≥ 100

The exact specimen width and thickness were measured at the fracture site on each bar. Two measurements were made for both width and thickness to assess the difference in strength from using dimensions which include the residual wall stubs and dimensions which are limited to the rectangular size specimen shape without the wall stubs. The wall thickness was measured from ceramographic sections taken adjacent to the fracture site on a random selection of approximately 50% of the specimens from each sample group.

RESULTS AND DISCUSSION

The fracture strength was initially calculated assuming the specimen was a homogenous bar resulting in a bar strength, σ_{MBS} as shown in Equation 1.

$$\sigma_{MBS} = \frac{Mc}{I_B} \qquad (1)$$

Where M is the applied moment, c is the distance from the neutral axis to the bottom of the specimen where the maximum tensile stress is generated and I_B is the moment of inertia of the cross section. The moment of inertia is defined in Equation 2 where b is the specimen width and h is the specimen thickness.

$$I_B = \frac{bh^3}{12} \qquad (2)$$

The calculated results of bar strength for the three different configurations are shown in Table 2.

Table 2. Weibull Parameters for Bar Strength Calculations

Inner Span, mm	Outer Span, mm	Weibull Shape (m)	Weibull Scale (σ_{MBS}), MPa
19.05	88.9	26.4	2.92
20	40	29.7	3.11
45	90	25.4	2.88

The bar strength includes both structure and material contributions. To decouple the structure and material contributions, the bar strength was converted to the measure of the material, wall strength (σ_{WS}), by taking into account the actual honeycomb structure of the specimens. Wall strength is calculated using the same general formula (Equation 3) with the same values of M and c but the moment of inertia term, I_{HC}, is determined from Equation 4.

$$\sigma_{WS} = \frac{Mc}{I_{HC}} \tag{3}$$

$$I_{HC} = \frac{bh^3}{12} - \left(n*m* \left(\frac{b_c*h_c^3}{12} \right) + 2*m*b_c*h_c* \sum_{i=1,step2}^{n} \left(\frac{n-i}{2}*(h_c+t_w) \right)^2 \right) \tag{4}$$

Where n and m are the number of complete cells in the thickness and width of the specimen, respectively, and b_c and h_c are the cell open area width and thickness, respectively, calculated in Equations 5 and 6 from the measured total dimensions, number of cells and measured average wall thickness as shown in Figure 1.

$$b_c = \frac{b-(m+1)*t_w}{m} \tag{5}$$

$$h_c = \frac{h-(n+1)*t_w}{n} \tag{6}$$

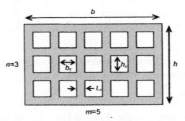

Figure 1. Test specimen dimensions and nomenclature

The wall strength data was analyzed using Minitab 14 software assuming a Weibull distribution and using the maximum likelihood estimation. The wall strength for the three test configurations are compared in Figure 2 as Weibull plots. The Weibull statistics are shown in the key of the plot. The Weibull slopes (shape) were compared using a Chi-squared distribution. The characteristic strength (Weibull scale) were also compared using a Chi-squared distribution. The Cummins specimen span (20/40) was found to be statistically different from the other two spans. The comparative confidence intervals for the three test configurations are shown in Table 3. These confidence intervals are normalized to the basis configuration in column 1, so in order for two configurations to be considered equivalent, the confidence interval must contain the whole number value of 1.0. As can be seen in the comparison, the characteristic strength for the 20/40 configuration is statistically different (higher value) than the other configurations and the 19.05/88.9 and 45/90 configurations are statistically equivalent. The Weibull slopes were also compared and found to be within the same confidence bands leading to the conclusion that the Weibull slopes for all three configurations were equivalent.

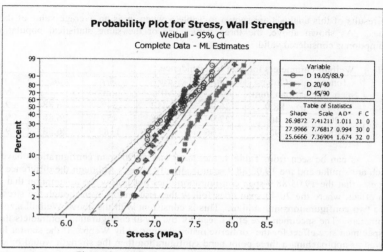

Figure 2. Weibull plot of wall strength, σ_{WS}

Table 3. Statistical Comparison of Weibull Shape and Scale Parameters

Basis	Comparison	Estimate	Lower CI (95%)	Upper CI (95%)
Shape 19.05/88.9	Shape 20/40	1.037	0.614	1.753
	Shape 45/90	0.951	0.584	1.548
Shape 20/40	Shape 45/90	0.917	0.569	1.477
Scale 19.05/88.9	Scale 20/40	1.048	1.021	1.076
	Scale 45/90	0.994	0.968	1.021
Scale 20/40	Scale 45/90	0.949	0.924	0.974

The Weibull[3] model was used to determine the size independent parameters for the different test configurations. The Weakest Link theory[4] was used in determining whether a size effect was present in the materials. These calculations assume the specimens are uniaxially stressed and the $\sigma_{0f}{}^{*} A_{0f}{}^{1/m}$ product is a constant. The model used is shown in Equation 7.

$$F = 1 - \exp\left[-\frac{\int_A \sigma^m(x)dA}{\sigma_0^m} \right] = 1 - \exp\left[-\left(\frac{\sigma_{max}}{\sigma_{0f}{}^{*}} \right)^{m_f{}^{*}} \right] \tag{7}$$

Where $m_f{}^{*}$ and $\sigma_{0f}{}^{*}$ are size dependent Weibull parameters and are converted to the size independent Weibull parameters using Equations 8 through 10.

$$m = m_f{}^{*} \tag{8}$$

$$\sigma_0 = \sigma_{0f}{}^{*} A_{0f}{}^{1/m} \tag{9}$$

$$A_{0f} = \int_A \left(\frac{\sigma(x)}{\sigma_{max}} \right)^m dA \tag{10}$$

The results of this analysis are shown in Table 4 where m is an average value of the three m_f^* values. As shown above, the three values are of the same statistical population so this assumption is considered valid.

Table 4. Weibull Parameters for Size Independence.

Inner Span, mm	Outer Span, mm	m_f^*	σ_{0f}^* MPa	A_{0f} mm^2	m	σ_0 MPa
19.05	88.9	26.99	7.412	547	26.88	9.37
20	40	27.99	7.768	452	26.88	9.75
45	90	25.66	7.369	1183	26.88	9.59

As can be seen from Table 4, the 20/40 and 45/90 span configurations have σ_0 values which are similar and the 19.05/88.9 span is slightly lower. Although the difference is small, it suggests that the 19.05/88.9 span configuration does not follow the assumption that there is an area effect, where the 20/40 span does follow the assumption even though the stress areas for these two configurations are similar. This is consistent with the above Weibull size dependent parameters. The specimen size effect seen for the 20/40 configuration contradicts the report of no specimen size effect for the cordierite material reported by Webb.[5] If the shorter load span is viewed as approaching a three point bend configuration then the strength would be expected to increase for the 19.05 and 20 mm span specimens. In this work, the configuration closest to the three point bend configuration (19.05/88.9) has the lowest strength, so this assumption is not supported by the data. The current analysis does not give a clear explanation for the specimen size effect, so further detailed analysis of the configurations in a finite element model may be necessary to determine if there are stress effects which are not currently accounted for in the calculations.

ERROR ANALYSIS
The specimens for each of the three configurations were measured using two methods that might be expected during testing. All stress calculations include the moment of inertia correction to give the wall strength of the material. For the initial stress calculations, the nominal substrate wall thickness was used, although the catalyst coating may change this wall thickness. The resulting differences in stress were compared along with the difference from using a nominal versus measured wall thickness. Four cases were used in the analysis as defined in Table 5.

Table 5. Case Definitions for Error Analysis

Case	Specimen Dimension Variable	Wall Thickness Variable
A	Measured dimensions include residual wall stubs	Assumed nominal wall thickness from substrate manufacturer
B	Measured dimensions include residual wall stubs	Measured wall thickness from microscopy
C	Measured dimensions do not include residual wall stubs	Assumed nominal wall thickness from substrate manufacturer
D	Measured dimensions do not include residual wall stubs	Measured wall thickness from microscopy

Weibull plots of Cases A, B, C and D are shown in Figures 3 through 6. The Weibull parameters for Cases A through D are shown in Table 6.

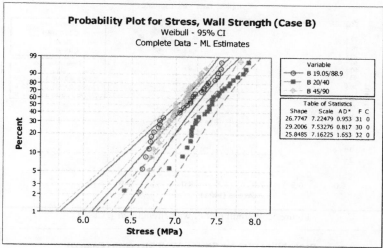

Figure 3. Weibull plot of wall strength, σ_{WS}, Case A

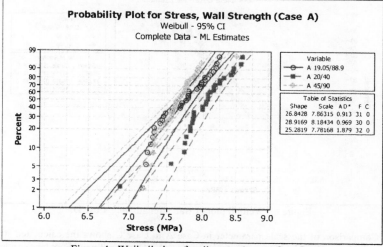

Figure 4. Weibull plot of wall strength, σ_{WS}, Case B.

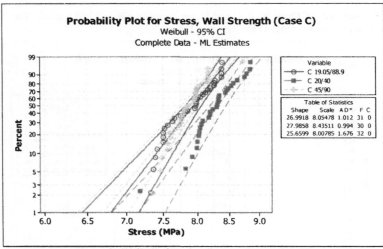

Figure 5. Weibull plot of wall strength, σ_{WS}, Case C.

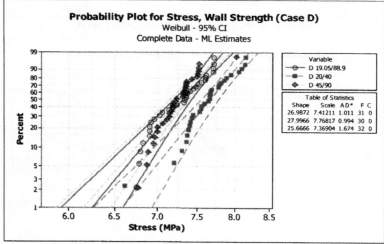

Figure 6. Weibull plot of wall strength, σ_{WS}, Case D.

Comparison of the scale parameter in Case A to Case C shows the calculated stress for Case C to be lower then the calculated stress for Case A. This result would not be expected from the measured dimensions. where Case A width, b, and thickness, h, are greater than Case C. The difference is due to the wall thickness value used in the moment of inertia calculation. The honeycomb moment of inertia, I_{HC}, uses calculated values of cell width, b_c, and cell thickness, h_c, representative of the open honeycomb dimensions. As shown in Equations 5 and 6 and Figure 1, this calculation is dependent on the cell count, overall dimension and wall thickness. Comparing Case A to Case C shows the honeycomb moment of inertia term is on the order of 8.5% lower for Case A, therefore, the calculated stress will be higher for Case A than for Case C.

As noted above in the Case A to Case C comparison, the same effect of wall thickness is found in the Case B to Case D comparison. The honeycomb moment of inertia term is on the order of 8.5% lower for Case B, therefore, the calculated stress will be higher for Case B than Case D.

Table 6. Weibull Parameters for Error Analysis Case Comparisons

Span Configuration	Weibull Parameter	Case			
		A	B	C	D
19.05/88.9	Scale, σ_{WS}	7.863	7.225	8.055	7.412
	Shape, m	26.82	26.78	26.99	26.99
20/40	Scale, σ_{WS}	8.184	7.533	8.435	7.768
	Shape, m	28.92	29.20	27.99	27.99
45/90	Scale, σ_{WS}	7.782	7.162	8.008	7.369
	Shape, m	25.28	25.85	25.66	25.66

Comparison of the shape parameter for the four cases shows that there is no significant difference between the three configurations. Comparison of the scale parameter for the four cases shows large variations in the wall strength dependent on the measurements. The mean measurements for the specimen dimensions and wall thickness are shown in Table 7. The effects of the measurement differences are correlated to the changes in stress value for the four cases in Table 8.

In the comparison of wall thickness change (Case A to Case C and Case B to Case D), the two comparisons were found to be equivalent. The correlation shown in Table 8 is representative of both comparisons. In the comparison of specimen dimension effects (Case A to Case B and Case C to Case D), the two comparisons were found to be equivalent. The correlation shown in Table 8 is representative of both comparisons.

Table 7. Wall Thickness and Specimen Dimension Parameters

Span Configuration	Nominal Wall Thickness, mm	Measured Wall Thickness, mm	Specimen Width (w/ stubs), mm	Specimen Width (w/o stubs), mm	Specimen Thickness (w/ stubs), mm	Specimen Thickness (w/o stubs), mm
19.05/88.9	0.305	0.336	25.54	25.17	12.79	12.65
20/40	0.305	0.336	21.91	21.60	11.05	10.89
45/90	0.305	0.336	25.52	25.15	12.79	12.63

Table 8. Correlation of Wall Thickness and Specimen Dimension Change to Stress Change

Span Configuration	Wall Effect		Dimension Effect		
	Wall Thickness Change, %	Stress Change, %	Specimen Width Change, %	Specimen Thickness Change, %	Stress Change, %
19.05/88.9	+10.2	+2.9	-0.7	-0.5	-8.0
20/40	+10.2	+3.1	-0.7	-0.7	-7.9
45/90	+10.2	+2.6	-0.7	-0.6	-8.0

When the cases are compared and correlated between the change in measurement and change in stress, the greatest effect on the stress is from the specimen measurements. Where the specimen measurement changes by 0.7% (approximately 0.15 mm in width or 0.07 mm in thickness) there is a corresponding change in stress by 8%. A change in the cell wall thickness of 10% results in a corresponding change in stress of less than 3%. Therefore, careful measurement of the specimens, preferably at the fracture site, is necessary for accuracy and high confidence in the data.

CONCLUSIONS

Computer modeling requires inputs of validated data which is representative of the material properties of interest. Since the models typically have terms that include exponents, small changes in the input data can have large effects on the results of the model. For example, stress is an input to the life prediction model as a term $\sigma_{WS}{}^m$, so a higher value of wall strength, σ_{WS}, can increase the predicted life greatly.

The initial analysis of the data shows there is a slight effect for the smallest specimen span ratio (20/40) in relation to the stressed area of the specimen. The longer support spans did not show any area effect on the calculated strength values. More rigorous finite element analysis of the 20/40 specimen may reveal further information on the cause of the size effect. The ultimate result of this work shows that this smaller span gives an overprediction of the wall strength on the order of 5%.

The second result of this work is the analysis of the measurements. When the cases are compared and correlated, the greatest effect on the change in strength is the difference (accuracy) of the specimen dimension measurements. The results show that small changes in the dimensions (~0.7%) cause a change the calculated wall strength by 8%. Changes in wall thickness have much lesser effect on the strength, where a change in wall thickness of ~10% results in a change of the calculated wall strength of <3%. The wall thickness results may be considered a worst case as the sample used for this work was coated with a catalyst and the wall thickness was expected to be greater than the nominal thickness value from the cordierite manufacturer.

These results show the necessity for a standard test method which includes the test procedure and calculation procedure to determine the properties which are of interest for computer modeling. Errors in measurement or in assumptions for dimensional values (wall thickness) can affect the final calculated value and greatly affect the modeling results.

REFERENCES
[1]40 CFR Part 86, Federal Register, Vol. 65, No. 107, 86.007-11 and 86.007-15, June 2, 2000 (proposed 2007 diesel emission regulations).

[2] ASTM C1161, Standard Test Method for Flexural Strength of Advanced Ceramics at Ambient Temperature, Annual Book of ASTM Standards, Vol 15.01, ASTM International, West Conshohocken, PA, 2006.

[3] Weibull, W., "A Statistical Theory of Strength of Materials", 1939, Ing. Vet. Ak. Handl., No. 151, Royal Inst. Of Tech., Stockholm.

[4] Batdorf, S. B., "Fundamentals of the Statistical Theory of Fracture", Fracture Mechanics of Ceramics, R. C. Bradt, D. P. H. Hasselman, and F. F. Lange, Eds., Vol. 3, Plenum Press, New York, NY, 1977, pp 1-30.

[5] Webb, James E.; Widjaja, Sujanto; Helfinstine, John D., "Strength Size Effects in Cellular Ceramic Structures", Proceedings of the 30th International Conference on Advanced Ceramics and Composites, Andrew Wereszczak and Edgar Lara-Curzio, ed., The American Ceramic Society, Westerville, OH, 2006.

Bioceramics

BIOMATERIALS MADE WITH THE AID OF ENZYMES

Hidero Unuma
Department of Chemistry and Chemical Engineering, Faculty of Engineering, Yamagata University
4-3-16 Jonan, Yonezawa, Yamagata 992-8510, Japan

ABSTRACT

Ceramics-based biomaterials have to be precisely designed and processed for their chemical and structural features to achieve their best performance. The use of enzymes in aqueous solution-based processing of biomaterials may make it easy to control the chemical and structural features of the resultant materials. The advantages of the use of enzymes in the processing of biomaterials are: (1) ceramic precipitates can be deposited under mild conditions such as near neutral pH and near room temperature, (2) the rate of the precipitation can easily be controlled by enzyme concentration, and (3) immobilization of enzymes on various templates makes it easy to control the morphology of the precipitate. This article describes the advantages of the use of enzymes in the processing of hydroxyapatite (HA)/polymer composites and hollow/porous spheres of hydroxyapatite, Y_2O_3, and Fe_3O_4.

OVERVIEW OF ENZYMES AND THEIR ROLES IN CERAMIC PROCESSING

Enzyme is protein that can catalyze a specific reaction of a specific group of chemicals. Products of some enzymatic reactions can act as precipitants of metal ions in aqueous solutions. For example, ammonia is produced by enzymatic hydrolysis of urea (Eq. 1) at room temperature while the hydrolysis proceeds only at elevated temperature without the enzyme, urease.

$$H_2N\text{-}CO\text{-}NH_2 + 3H_2O \longrightarrow 2NH_4OH + CO_2 \tag{1}$$

Enzymes have been used in the preparation of inorganic precipitates to generate the precipitates. So far, alkaline phosphatase, carbonic anhydrase, urease, glucose oxidase with peroxidase, acylase, lipase have been used to prepare hydroxyapatite[1-6], basic aluminum sulfate[6-10], tin oxide hydrate[11], nickel dimethylglyoximate[12], manganese oxinate[13], magnetite[14-16], calcium carbonate[17, 18], strontium and barium carbonates[19, 20], calcium oxalate hydrates[21, 22], and silica[23]. The use of enzymes in the preparation of ceramics or ceramic precursors brings us a number of advantages as described below.

Firstly, in most cases, enzymatically assisted chemical reactions proceed under mild conditions, such as in near neutral pH and near room temperature. This feature may be especially beneficial for the processing of biomaterials because it is sometimes necessary to hybridize ceramics with polymers which are not durable to high temperatures or in highly acidic/alkaline conditions.

Secondly, the kinetics of enzymatic reactions follows the Michaelis-Menten equation, which roughly means that the rate of an enzymatic reaction is almost proportional to the enzyme concentration when the substrate exists in a large excess. Therefore, the overall rate of ceramic precipitation can easily be controlled by changing the enzyme concentration in the solution from which ceramics is precipitated.

Thirdly, it is easy to immobilize enzymes on various templates. When an enzyme is immobilized on a template, the precipitant of a metal ion is supplied only in the vicinity of the enzyme. Then, the resultant ceramics are formed preferentially on the surface of the template; thereby the morphology of the resultant ceramics can easily be controlled.

This paper reviews the studies of the biomaterials processed using these advantaged of the use of enzymes.

BIOMATERIALS PREPARED WITH THE AID OF ENZYMES
Hydroxyapatite (HA)/collagen composites

It was Banks and coworkers[1] that first synthesized hydroxyapatite (HA)/collagen composites using an enzyme, in the course of their study of biomineralization. They immobilized alkaline phosphatase onto reconstituted collagen and soaked the collagen in a solution containing calcium β-glycerophosphate. As the enzymatic hydrolysis of β-glycerophosphate proceeded, phosphorus acid was produced which subsequently reacted with calcium ion to precipitate hydroxyapatite. They repeated the immobilization/precipitation process and fabricated HA/collagen composites. It should be noted that the HA crystallites showed preferential orientation similar to that in natural bone.

The aforementioned technique was used years later to fabricate different types of HA/collagen composites. Doi et al.[2] found that their composite showed resorbability in muscle tissue with no evidence of toxicity. Yamauchi et al.[3] fabricated HA/collagen multilayer assembly and showed that the assembly had mechanical flexibility and HA particles adhered strongly to the collagen.

The HA/collagen composites seem to be promising biomaterials because of its structural similarity to natural bone, in vivo resorbability, and mechanical properties. However, as it takes a very long period of time for fabrication, it would be difficult for this technique to be used for industrial production of biomaterials.

HA coatings on silk, poly(L-lactic acid) (PLLA), and collagen

The author's group[4, 5] used immobilized urease to deposit HA on a silk cloth and a porous PLLA matrix. Compared with alkaline phosphatase, urease is active in a wider pH range and more readily available.

As raw silk cloth has a sheath on whose surface a considerable amount of carboxyl group aligns, it is easy to immobilize enzymes by using commercial water-soluble carbodiimide as a carboxyl activator. They immobilized urease (originated from Jack bean, 5000U/g) on silk cloth, and soaked the cloth in a solution containing 0.06 mol/dm^3 of Ca(NO$_3$)$_2$, 0.10 mol/dm^3 of H$_2$NH$_4$PO$_4$, and 0.11 mol/dm^3 of urea at 310K for 1, 3, and 6h. As urea was hydrolyzed to ammonia and carbon dioxide, HA was precipitated according to Eq.2 preferentially on the surface of the silk.

$$10Ca^{2+} + 6PO_4^{3-} + 2NH_4OH \longrightarrow Ca_{10}(PO_4)_6(OH)_2 + 2NH_4^+ \qquad (2)$$

Fig. 1 shows scanning electron micrographs of the surfaces of silk on which HA was deposited for various time periods. Within the first 1h, the surface of silk started to be covered by HA, and in 3h, the silk was totally covered with a thick layer of HA.

Fig. 1 Scanning electron microscope of silk surfaces on which hydroxyapatite was deposited for various time periods, (a) 0h, (b) 1h, (c) 3h, and (d) 6h.

The HA layer deposited with the aid of urease was characterized by FT-IR spectroscopy and X-ray diffraction analysis to be bone-like apatite because carbonate ion was incorporated in the lattice, its composition was calcium-deficient with respect to the ideal stoichiometry, and the crystallinity was low.

The deposition of HA took place very quickly. Considering that it took as long as 7 days for HA to cover the silk surface by the biomimetic method using 1.5 SBF (simulated body fluid)[24], the present method can deposit HA layers 50 times more quickly than the conventional biomimetic method.

The same method was able to be used for the coating of inner and outer surfaces of PLLA sponge. PLLA sponge was prepared by freeze-drying a dioxane solution containing 5 mass% of PLLA after cooling the solution at 280K for 2h and 77K for 30 min[25]. The sponge was then immersed in a NaOH solution (water: dioxane = 1:1 in volume) to partially hydrolyze the surface of the sponge. Then the sponge was modified with urease and soaked in $Ca^{2+}/PO_4^{3-}/$urea solutions in the same manner as the case of silk cloth.

Fig. 2 shows scanning electron micrographs of the inner surface of PLLA sponge. The inner surface was almost covered with HA particles within 24 h.

Yan and coworkers[26] evaluated the *in vitro* biological efficacy of the present HA/PLLA sponge composite in terms of osteoblast-like cell proliferation and differentiation. They found that the HA coating on PLLA surfaces enhanced the cell activity especially in the early stage of the cell culture. The HA deposits were not detached from the PLLA during the *in vitro* test.

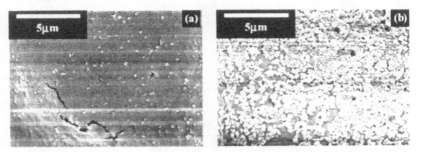

Fig. 2   Scanning electron micrographs of the inner surfaces of poly(1-lactic acid) sponges on which hydroxyapatite was deposited for (a) 3h and (b) 24h.

Ceramic microspheres

The present authors[27] demonstrated that spherical and hollow hydroxyapatite particles can be prepared using alginate gel particles containing urease as template. Ammonium alginate solution (3 mass%) containing 0.01 mass% of urease was dropwise added with a Pasteur pipette to another solution containing 0.05 mol/dm^3 of Ca(NO$_3$)$_2$, 0.01 mol/dm^3 of H$_2$NH$_4$PO$_4$, and 0.11 mol/dm^3 of urea. The droplets of alginate-urease solution transformed into opaque gel spheres about 2mm in diameter within a few seconds by the cross-linking effect of Ca^{2+} ion. While the gel spheres were left immersed in the Ca^{2+}/PO$_4$$^{3-}$/urea solution at 310K, HA was deposited around the gel spheres as urea was enzymatically hydrolyzed. This process is schematically illustrated in Fig. 3. After 48h of immersion, the spheres were taken out, dried, and heat treated at 1073K for 2h in air.

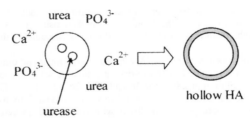

Fig. 3   Schematic illustration of the formation of hollow HA shell around alginate gel template

Fig. 4 shows the optical and scanning electron micrographs of the resultant particles. The shells of the particles were smooth as can be seen in the figure (a). and the inside was hollow. Photo (b) in the figure shows the cross-section of the shell, showing the thickness of the shell being approximately 150μm.

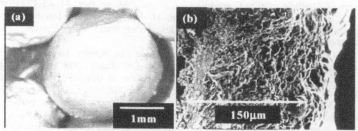

Fig. 4 (a) Optical and (b) scanning electron micrographs of hollow hydroxyapatite particle. Photo (b) shows the cross-section of the shell of the particle shown in photo (a).

The size of the resultant particle is dependent on that of the template. Kawashita and coworkers synthesized hollow/porous Y_2O_3[28] and Fe_3O_4[29] microspheres 20 to 30 μm in diameter using alginate and carboxymethyl cellulose (CMC) gel particles as templates. Both microspheres can be used for radiotherapy and hyperthermia treatment for cancer, and it is better that the densities are as low as possible. They used an atomizer to generate small droplets containing either alginate or CMC with urease which were to be gelled in solutions containing either Y^{3+} or Fe^{3+} ions together with urea. After depositing precursors of respective ceramics with the aid of urease, the gel particles were heat treated under optimum conditions to be converted to porous Y_2O_3 and Fe_3O_4 microspheres with low density. Considering that such porous ceramic microspheres have not been prepared by other means, their success in preparing porous ceramic microspheres is a typical example of the advantage of the use of enzymes.

SUMMARY

The use of enzymes in the processing of biomaterials can be beneficial especially in preparing ceramic/polymer composites, ceramic coatings on polymers, and hollow/porous ceramic microspheres, because it makes easy to control the chemical and structural features of ceramics or ceramic precursors prepared in aqueous solutions. Composites of HA/collagen and HA/PLLA have been proved to show excellent biomedical performance. HA coatings can be prepared in much shorter time than by the conventional biomimetic method. Hollow or porous ceramic microspheres for cancer remedy can easily be prepared. The use of enzymes is certainly an option for us to choose in the processing of ceramics-based biomaterials with unique structures which would be difficult or inconvenient to make by other means.

REFERENCES

[1]E. Banks, S. Nakajima, L. C. Shapiro, O. Tilevitz, J. R. Alonzo, and R. R. Chianelli, "Fibrous Apatite Grown on Modified Collagen," *Science*, **198**, 1164-66 (1977).

[2]Y. Doi, T. Horiguchi, Y. Moriwaki, H. Kitago, T. Kajimoto, and Y. Iwayama, "Formation of Apatite-Collagen Complexes," *J. Biomed. Mater. Res.*, **31**, 43-49 (1996).

[3]K. Yamauchi, T. Goda, N. Takeuchi, H. Einaga, and T. Tanabe, "Preparation of Collagen/Calcium Phosphate Multilayer Sheet Using Enzymatic Mineralization," *Biomater.*, **25**, 5481-89 (2004).

[4]H. Unuma, and A. Ito, "Deposition of Bone-like Apatite Layers on the Surface of Poly(L-Lactic Acid) Using Immobilized Urease," *Key Eng. Mater.*, **309-311**, 667-70 (2006).

[5]H. Unuma, M. Hiroya, and A. Ito, "Deposition of Bone-like Hydroxyapatite on the Surface of Silk Cloth with the Aid of Immobilized Urease," *J. Mater. Sci.: Mater. Med.*, in press.

[6]D. McConnell, W. J. Frajola, and D. W. Deamer, "Relation between Inorganic Chemistry and Biochemistry of Bone Mineralization," *Science*, **133**, 281-82 (1961).

[7]H. Unuma, S. Kato, T. Ota, and M. Takahashi, "Homogeneous Precipitation of Alumina Precursors via Enzymatic Decomposition of Urea," *Adv. Powder Technol.*, **9**, 181-90 (1998).

[8]F. Kera, and G. Sahin, "Hydrated Aluminium Sulfate Precipitation by Enzyme-Catalyzed Urea Decomposition," *J. Eur. Ceram. Soc.*, **20**, 689-94 (2000).

[9]S. Kato, T. Makino, H. Unuma, and M. Takahashi, "Enzyme Catalyzed Preparation of Hollow Alumina Precursor Using Emulsion Template," *J. Ceram. Soc. Japan*, **109**, 369-71 (2001).

[10]R. E. Simpson II, C. Hageber, A. Rabinovich, and J. H. Adair, "Enzyme-Catalyzed Inorganic Precipitation of Aluminium Basic Sulfate," J. Am. Ceram. Soc., 81, 1377-79 (1998).

[11]S. Kato, H. Unuma, T. Ota, and M. Takahashi, "Homogeneous Precipitation of Hydrous Tin Oxide Powders at Room Temperature Using Enzymatically Induced Gluconic Acid as a Precipitant," *J. Am. Ceram. Soc.*, **83**, 986-88 (2000).

[12]S. Hikime, H. Yoshida, and M. Taga, "Application of Urease for the Precipitation of Nickel Dimethylglyoximate by Urea from Homogeneous Solution," *Talanta*, **14**, 1417-22 (1967).

[13]S. Hikime, H. Yoshida, M. Taga, and S. Taguchi, *Talanta*, **19**, 569-72 (1972).

[14]T. Hamaya, and K. Horikoshi, "*In Vitro* Magnetite Formation by Urease Dependent Reaction," *Agric. Biol. Chem.*, **53**, 851-52 (1989).

[15]T. Hamaya, and K. Horikoshi, "Variation in the Iron Substances Produced from Ferrous Iron through a Urease Dependent Reaction," *Agric. Biol. Chem.*, **53**, 1989-90 (1989).

[16]T. Hamaya, T. Takizawa, H. Hidaka, and K. Horikoshi, "A New Method for the Preparation of Magnetic Alginate Beads," *J. Chem. Eng. Japan*, **26**, 223-24 (1993).

[17]I. Sondi, and E. Matijevic, "Homogeneous Precipitation of Calcium Carbonates by Enzyme Catalyzed Reaction," *J. Colloid Interfa. Sci.*, **238**, 208-14 (2001).

[18]K. L. Bachmeier, A. E. Williams, J. R. Warmington, and S. S. Bang, "Urease Activity in Microbiologically-Induced Calcite Precipitation," *J. Biotechnol.*, **93**, 171-81 (2002).

[19]I. Sondi, and E. Matijevic, "Homogeneous Precipitation by Enzyme-Catalyzed Reactions. 2. Strontium and Barium Carbonates," *Chem. Mater.*, **15**, 1322-26 (2003).

[20]S. D. Skapin, and I. Sondi, "Homogeneous Precipitation of Mixed Anhydrous Ca-Mg and Ba-Sr Carbonates by Enzyme-Catalyzed Reaction," *Cryst. Growth Design*, **5**, 1933-38 (2005).

[21]S. Kato, H. Unuma and M. Takahashi, "Enzyme-Catalyzed Synthesis of Hydrated Calcium Oxalate," *Adv. Powder Technol.*, **12**, 493-505 (2001).

[22]S. Kato, T. Ota, H. Unuma, and M. Takahashi, "Enzymatically Assisted Precipitation of Hydrous Calcium Oxalate," *Ceram. Trans.*, **112**, 53-58 (2001).

[23]P. Buisson, H. El Rassy, S. Maury, and A. C. Pierre, "Biocatalytic Gelation of Silica in the Presence of a Lipase," *J. Sol-Gel Sci. Technol.*, **27**, 373-79 (2003).

[24]A. Takeuchi, C. Ohtsuki, T. Miyazaki, H. Tanaka, M. Tanaka, and M. Tanihara, "Deposition of Bone-Like Apatite on Silk Fiber in a Solution That Mimics Extracellular Fluid," *J. Biomed. Mater. Res.*, **65A**, 283-89 (2003).

[25]R. Zhang and P. X. Ma, "Porous Poly)l-Lactic Acid)/Apatite Composites Created by Biomimetic Process," *J. Biomed. Mater. Res.*, **45**, 285-93 (1999).

[26]Z. Pan, X. Huang, D. Yang, H. Unuma, and W. Yan, "Biological Efficacy of Modified Structure of Polymer Surfaces for Bone Substitute," *Key. Eng. Mater.*, **330-332**, 707-10 (2007).

[27]H. Unuma, "The Use of Enzymes for the Processing of Biomaterials", *Int. J. Adv. Ceram. Technol.*, **4**, 14-21 (2007)

[28]M. Kawashita, Y. Takayama, T. Kokubo, G. H. Takaoka, N. Araki, and M. Hiraoka, "Enzymatic Preparation of Hollow Yttrium Oxide Microspheres for *In Situ* Radiotherapy of Deep-Seated Cancer," *J. Am. Ceram. Soc.*, **89**, 1347-51 (2006).

[29]M. Kawashita, K. Sadaoka, T. Kokubo, T. Saito, M. Takano, N. Araki, and M. Hiraoka, "Enzymatic Preparation of Hollow Magnetite Microspheres for Hyperthermic Treatment of Cancer," *J. Mater. Sci.: Mater. Med.*, **17**, 605-10 (2006).

USE OF VATERITE AND CALCITE IN FORMING CALCIUM PHOSPHATE CEMENT SCAFFOLDS

A. Cuneyt Tas
Department of Biomedical Engineering
Yeditepe University
Istanbul 34755, Turkey

ABSTRACT

A series of novel orthopedic calcium phosphate (CaP+CaCO$_3$) cements have been developed. The common point in these cements was that they all utilized single-phase CaCO$_3$ (calcite or vaterite) in their powder components. The major phase in the end-product of these cements was carbonated, Ca-deficient, apatitic calcium phosphate, together with some varying amounts of unreacted CaCO$_3$. Calcite powders used were needle-like or acicular in shape, whereas the vaterite powders were monodisperse, and spherical or ellipsoidal in shape. A new method for synthesizing spheroidal vaterite powders has also been developed. Setting solutions for calcite-based cement scaffolds were prepared by acid-base neutralization of concentrated H$_3$PO$_4$ and NaOH. 0.5 M phosphate buffer solution was used to transform precipitated vaterite powders into CaP at 37° to 70°C. Setting solutions possessed pH values ranging from 3 to 7.4 at room temperature. Resultant cements were micro- (5 to 50 micron pores) and macroporous (200 to 700 micron pores). In the macroporous cements, total porosity was variable from 20 to 45%. Setting times of those cements were adjustable over the range of 12 to 25 minutes. Compressive strengths of these cements varied from 2 to 3 MPa, depending on their porosity. CaP+CaCO$_3$ cements thus obtained had relatively high surface areas (30 to 85 m^2/g) whose surfaces were covered with nanocrystallites similar in size to the nanoplatelets found in biological collagen-calcium phosphate composites. Cement scaffolds were characterized by XRD, FTIR, SEM, ICP-AES, surface area and compressive strength measurements.

INTRODUCTION

Calcium carbonate (CaCO$_3$), being an important material of marine and geological biomineralization processes, has three anhydrous polymorphs; vaterite, aragonite and calcite. Amorphous calcium carbonate may also be added to this list as the fourth form.[1] At the ambient temperature and one atmosphere, calcite is the most stable and common polymorph of calcium carbonate, while vaterite[2] is known to be the least stable. Owing to its instability, vaterite is rare in nature as it would readily convert into one of the more stable calcium carbonate phases, aragonite or calcite, and possibly to monohydrocalcite (CaCO$_3$·H$_2$O). However, Lowenstam and Abbott[3] reported that a special type of sea squirts, i.e., *Herdmania momus*, ascidiacea, exhibited vaterite as the mineralization product of the spicules of their skeletons. The remainder of marine organisms (excluding those yet to be discovered) uses aragonite or calcite as the biomineralization phases in forming their hard tissues, due to the scarcity of dissolved phosphorus (< 0.1 ppm) in the sea water.[4] The thermodynamic instability of vaterite makes it quite suitable for the rapid formation of carbonated and apatitic calcium phosphate scaffolds as synthetic bone substitutes [1]. To the best of our knowledge, the conversion of vaterite to calcium phosphates (CaP) has not been thoroughly studied.

Synthetic, implantable bone-like materials are useful in many different biomedical applications. For example, bone-like scaffolding material can be implanted to fill large bone defects caused by trauma or removal of cancerous or otherwise diseased bone or cysts.[5] The ideal biomaterial should have a chemical and structural design so as to induce a response similar to that of natural fracture healing when placed in an osseous defect, when invaded by mesenchymal cells, fibroblasts and osteoblasts before new trabeculae of bone infiltrate into the porous structure of the implant from the walls of the defect.[6-10] Bone mineral consists of small crystals (3 to 5 nm in thickness and 40-100 nm in length, aligned parallel to the collagen fibrils) of CO_3-containing (5.5 wt%), alkali and alkaline earth ion (Na, K, Mg)-doped, Ca-deficient, non-stoichiometric, apatitic calcium phosphate (Ap-CaP) with a large and reactive surface area of 100-200 m^2/g.[11, 12] Bone mineral is similar to, but far from being identical with, the mineral Ca-hydroxyapatite (HA: $Ca_{10}(PO_4)_6(OH)_2$). A synthetic bone substitute should have high porosity dominated by macropores similar to the structure of trabecular bones, with pore sizes ranging from 100 to 700 μm.[13] Macroporosity in bones allows for vascularization and provides a habitat for the cells, whereas microporosity mainly acts as fluid channels for nutrient supply and transmission.

The development of calcium phosphate cements (CPC) started almost three decades ago with the formulation of in situ-setting calcium phosphate pastes for dental repair and restoration applications.[14-17] Cements provide the surgeon with the unique ability of manufacturing, shaping, and implanting the bioactive bone substitute material on a patient-specific base, in real time, in the surgery room. Moreover, during their preparation in the surgical theatre, such cement pastes can be impregnated with the patient's own bone marrow cells or platelet-rich plasma to enhance their *in vivo* osteogenetic properties.[18] Early CPCs commonly employed either α-tricalcium phosphate (α-TCP: α-$Ca_3(PO_4)_2$) or tetracalcium phosphate (TTCP: $Ca_4(PO_4)_2O$), or both, in their powder components. Driskell *et al.*[14] described in 1974 the very first calcium phosphate cement paste, which used α-TCP powders. α-TCP and TTCP do share a useful property through which both undergo in situ hydrolysis into nanocrystalline, non-stoichiometric Ap-CaP or carbonated CDHA (calcium-deficient hydroxyapatite: $Ca_9(HPO_4)(PO_4)_5OH$), upon contact with aqueous solutions at 20-37°C.[19-22] For a comprehensive assessment of commercialized CPC formulations, classified with respect to their end-products (i.e., after setting), the reader is referred to a recent study of Bohner *et al.*[23]

$CaCO_3$ is not yet seen as a major starting material in commercialized CPCs. The only exception to this was seen in the formulations of *Calcibon*® and *Norian*® [23], which both employed about 10 to 12% of $CaCO_3$ in their powders for the purpose of obtaining carbonated apatite as the end-product. Very recently, Combes *et al.*[24] reported the very first experimental $CaCO_3$ cement, which had a biphasic starting material of amorphous calcium carbonate (doped with either strontium or magnesium) and crystalline vaterite. The end-products of these low-compressive strength cements were again calcium carbonate (vaterite + aragonite) because of the use of 0.9% (w/w) NaCl isotonic solution as their cement liquid. This solution was not able to transform $CaCO_3$ into Ap-CaP. The *in vitro* resorption of $CaCO_3$ (of calcite form) by the human or rat osteoclasts, in simultaneous comparison with HA and β-TCP bioceramics, has been previously studied by Monchau *et al.*[7]

The purpose of the present study was to develop a low-compressive strength but macroporous and inexpensive $CaCO_3$+Ap-CaP biphasic scaffold for skeletal repair by using single-phase $CaCO_3$ (calcite or vaterite) powders. Commercially available calcite or chemically precipitated (*in house*) vaterite powders were used, and the powder components of the cements did not contain any other additive whatsoever. Cell culture studies on these macroporous cements will be published later.

EXPERIMENTAL PROCEDURE
Vaterite powder

Vaterite powders were synthesized as follows: 12.0 g of $CaCl_2 \cdot H_2O$ (Catalog No: C79-500, 99.9%, Fisher Scientific, Fairlawn, NJ) was first dissolved in 1200 mL of deionized water at room temperature. 12.0 g of gelatin (Catalog No: G7-500, 100 Bloom, FisherSci) was then added to the above solution and dissolved in it by heating the solution to 35°C. Upon dissolving gelatin, the solution was cooled back to room temperature (i.e., 21±1°C). 12.0 g of $NaHCO_3$ (Catalog No: S233-500, 99.9%, FisherSci) was separately dissolved in 300 mL of water and placed into a glass burette, which was positioned over the Ca-gelatin solution. Ca-gelatin solution was stirred at 500 rpm. $NaHCO_3$ solution was then fed into the Ca-gelatin solution at the rate of 13 mL/min in a dropwise manner. At the end of $NaHCO_3$ addition, the solution containing the precipitates was immediately filtered by using a vacuum-assisted Buechner funnel and filter paper (Whatman, No: 41, FisherSci). Precipitates were washed with 2 L of water and then dried overnight at 65°C in air.

Preparation of the 0.5 M phosphate buffer solution for vaterite conversion

800 mL of deionized water was placed in a 1 L-capacity beaker, followed by adding 39.749 g of Na_2HPO_4 (Merck, Catalog No: 1.06585) at room temperature. Upon stirring, one obtained a transparent solution. To this solution 16.559 g of $NaH_2PO_4 \cdot H_2O$ (Merck, Catalog No: 1.06346) was added to raise the pH to about 7.4. Solution was then transferred into a 1 L-capacity media bottle and stored in a refrigerator when not in use.

This solution was used to convert the vaterite powders into Ap-CaP. 800 mg of vaterite powders were first placed into a 250 mL-capacity glass media bottle, followed by adding 200 mL of 0.5 M phosphate buffer solution and sealing the bottle. The bottle was kept undisturbed in a 70°C oven for 17 h. The solids were filtered and then washed with deionized water, followed by drying overnight at 70°C.

Calcite powder

The powder component of the macroporous cement of this study consisted of single-phase $CaCO_3$ (calcite). These cements were initially developed by using the reagent-grade precipitated $CaCO_3$ powders manufactured by Merck KGaA (Catalog No: 1.02076, Darmstadt, Germany). The optimized cement recipe was also tested by using the precipitated $CaCO_3$ (calcite) powders supplied by Fisher Scientific (Catalog No: C63-3, 99.8%, Fairlawn, NJ), as well. The cement recipe successfully worked for both reagent-grade calcite powders.

Preparation of the cement liquid for calcite powders

The setting solution was prepared as follows; first a 100 mL aliquot of concentrated H_3PO_4 (Merck, Catalog No: 1.00573, 85%, 14.8323 M) was placed into an unused erlenmeyer flask of 300 mL capacity together with a Teflon®-coated magnetic stir bar. This erlenmeyer flask was then placed on a magnetic stirrer plate at room temperature (21 ± 1°C). A 138 mL portion of concentrated NaOH (Merck, Catalog No: 1.05590, 32%, 10.8 M) was taken into a burette and placed directly on top of the open mouth of the erlenmeyer containing H_3PO_4. The first 50 mL portion of the NaOH solution in the burette was slowly (in 20 minutes) added in a drop wise manner, to prevent excessive heating and boiling, with a magnetic stirring rate of about 300 rpm. Upon slowly (ca. 10 min) adding another 25 mL aliquot of NaOH, the pH of the solution in the erlenmeyer rose to 0.4 (at 66-67°C). It should be noted that until now only 75 mL of NaOH was added into the erlenmeyer. The slow (ca. 10 min) addition of another 25 mL portion of NaOH increased the pH to 1.60 (at 75°C). With the following addition of another 25 mL aliquot of NaOH, pH became 2.60 (at 75°C). Finally, after adding (again slowly in about 10 min) the last 13 mL portion of NaOH in the burette to the erlenmeyer, the pH rose to 3.2 (at 45-50°C). The total volume of the viscous liquid thus formed in the erlenmeyer was completed to 250 mL by adding deionized water. The solution was cooled to room temperature under constant stirring, and transferred into a 250 mL-capacity Pyrex® media bottle (with a cap) for long-term storage. This solution has been confirmed to preserve its constant pH at 3.2 (at room temperature) for more than two years of storage on the laboratory bench. 1 mL of this solution (called afterwards as "SS" as an abbreviation of "setting solution") thus contained 0.0059 mol of phosphor.

Cement preparation

A plastic cup of cylindrical geometry (approx. 8 cm tall with a diameter of 6.5 cm) was used to mix the calcite powder with the SS (setting solution) to form the cement paste. $CaCO_3$ powder of appropriate quantity was first weighed and then placed into this cup. SS of proper volume was measured with a pipette and then placed into a clean glass bottle. At the moment of mixing the $CaCO_3$ powder and the setting solution, SS in the glass bottle was immediately poured into the plastic cup containing $CaCO_3$. Mixing was manually performed by using a plastic spatula. Mixing was continued for 80 to 90 seconds in all trials, and at the end of mixing the formed paste was cast into a polyethylene, square-shaped weighing boat. Therefore, upon setting (which took place in 4 minutes), porous cement blocks readily took the shape of those weighing boats. Cement blocks were then soaked in deionized water at room temperature for 3 hours, followed by drying at 24°C in an air-ventilated oven.

Sample characterization

Samples were characterized by powder X-ray diffraction (XRD; Model D5000, Siemens GmbH, Karlsruhe), scanning electron microscopy (SEM; Model 630, Jeol Corp., Tokyo and Model S-3500, Hitachi, Tokyo), Fourier-transform infrared spectroscopy (FTIR; Nicolet 550, Thermo-Nicolet, Woburn, MA), water absorption [63], density measurements (Pycnometer; AccuPyc 1330, Micromeritics, Norcross, GA), inductively-coupled plasma atomic emission spectroscopy, (ICP-AES; Model 61E, Thermo Jarrell Ash, Woburn, MA) and compressive strength measurements (Model 4500, Instron Deutschland GmbH). Percentage water absorption values of the porous cement scaffolds (((wet weight—dry weight)/dry weight)*100) were measured in accord with the ASTM standard C20-92.[25] Dry weights of the samples were recorded after drying at 24°C for 72 hours, whereas the wet weights were recorded immediately after soaking the cement samples in deionized water (at 21°C) for 3 hours. Pore sizes of the porous scaffolds were determined by using the linear intercept method on the photomicrographs.

To give more details on the characterization runs; samples for XRD analyses were first ground in an agate mortar using an agate pestle and then sprinkled onto ethanol-damped single crystal quartz sample holders to form a thin layer, followed by tapping to remove the excess of powder. The X-ray diffractometer was operated at 40 kV and 30 mA with monochromated Cu K_α radiation. XRD data (over the range of 20 to 50° 2θ) were collected with a step size of 0.03°

and a preset time of 1 sec at each step. FTIR samples were first ground in a mortar, in a manner similar to that used in the preparation of XRD samples, then mixed with KBr powder in a ratio of 1:100, followed by forming a pellet by using a uniaxial cold press. 128 scans were performed at a resolution of 3 cm^{-1}. Powder samples examined with the scanning electron microscope (SEM) were sputter-coated with a thin Au layer, to impart surface conductivity to the samples. The BET surface area of powder samples was determined by applying the standard Brunnauer–Emmet–Teller method to the nitrogen adsorption isotherms obtained at -196°C using an ASAP 2020 instrument (Micromeritics Corp., Norcross, GA). Powder samples used in the ICP-AES analyses were first dissolved in nitric acid prior to the measurements. Cylindrical cement samples (1 cm diameter, 2 cm height) were loaded under compression until failure, at a crosshead speed of 1 mm/min, with a 1.5 kN load cell, to determine compressive strength. The ASTM standard C1424-04 was observed in strength measurements.

RESULTS AND DISCUSSION
Vaterite synthesis and conversion
The phase purity of vaterite powders was deduced by using the XRD technique defined by Rao.[26] As shown in the XRD trace of Fig. 1a, the formed powders contained around 10% calcite, the rest being vaterite. The IR spectra of the powders also revealed the vaterite structure, with a characteristic band at 744 cm^{-1} (Fig. 1b).[27] When precipitated in gelatin solutions as described above, vaterite was always formed with a spheroidal morphology (Figs. 1c and 1d).

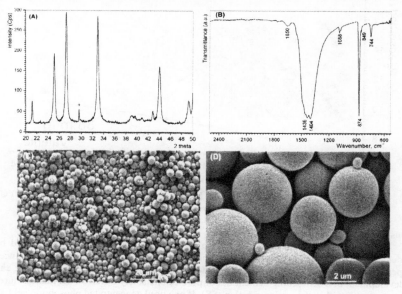

Fig. 1   (a) XRD, (b) FTIR traces, (c) & (d) SEM pictures of precipitated vaterite powders

The only peak denoted in Fig. 1a with + corresponded to calcite, all the other peaks belong to vaterite. ICP-AES analyses performed on these precipitates showed that they had about 700 ppm Na, which originated from the use of $NaHCO_3$ in their preparation. Although a tremendous amount of literature is available on the synthesis of vaterite particles in the presence of exotic polymers or bizarre organic substances,[28, 29] gelatin was not a typical choice. If one intends to use vaterite as a biomaterial for skeletal repair, gelatin is logical to use in synthesis since it is simply the denatured collagen. Collagen is the only biopolymer of human bones.

Vaterite powders aged in 0.5 M phosphate buffer solutions at 70°C for 17 h were totally converted into Ap-CaP. The resultant Ap-CaP powders had again spheroidal particles but with an increased surface area. The combined SEM photomicrographs and the BET surface area data presented in Fig. 2 below summarized the conversion of vaterite spheres.

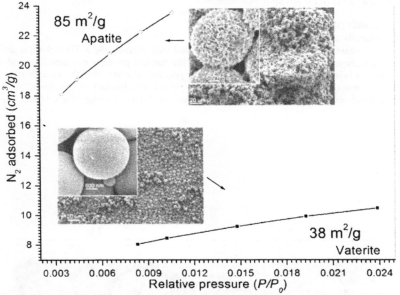

Fig.2   Conversion of ppt[d] vaterite into Ap-CaP by ageing in 0.5 M phosphate buffer at 70°C; surface area and SEM morphology before and after the conversion

Porous cement preparation from calcite powder

This cement was designed to have a unique property; calcium was supplied by its powder component, whereas phosphor was only received from its setting solution. Since this inexpensive, low-strength cement[30] had a powder component comprised of single-phase calcite, the development of a self-setting cement out of pure calcite solely relied on the manufacture of a proper cement liquid or the setting solution (SS). If one used concentrated phosphoric acid alone, the reaction of the acid with calcite would be too severe and instantaneous (resulting in the formation of a mixture of acidic phosphates, such as $Ca(H_2PO_4)_2$ and $Ca(H_2PO_4)_2 \cdot H_2O$), and the handling or shaping of the material into a specific shape would be almost impossible. Partial neutralization of phosphoric acid with concentrated sodium hydroxide solution, in forming the SS, thus seemed to be a viable alternative. Pore formation in this biphasic cement proceeded with the conversion of only a certain fraction of $CaCO_3$ into Ap-CaP upon coming into contact with the mildly acidic SS, and the by-product of this limited conversion reaction was CO_2 gas. The pore cavities were the direct result of physically entrapped bubbles of the evolved CO_2 gas in the setting body.

SEM photomicrographs of Figs. 3a and 3b portrayed the starting morphology of the powder component (i.e., $CaCO_3$) of this cement. Calcite powders used in this study were commercial powders (supplied by Merck KGaA and Fisher Scientific Corp.) and both of these precipitated chalk powders were morphologically similar to one another. Both calcite powders were found to have a surface area of about 6.15 ± 0.25 m^2/g (BET runs in triplicate on each), and similarly consisted of 0.5 μm-thick and 1.5 μm-long spindle-shaped agglomerated particles.

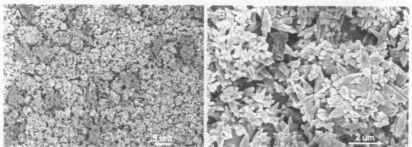

Fig. 3　SEM photomicrographs of starting calcite powders, (*a*) low, (*b*) high mag image

Table I listed three optimal recipes developed for the preparation of these cements. The nominal, starting Ca/P molar ratio in these cements could be adjusted either by simply varying the amount of calcite powder or the volume of the SS to be used. However, since the SS developed for calcite powders was so mild that the conversion of $CaCO_3$ to Ap-CaP would never reach to completion.

Table I. Optimal recipes for the preparation of biphasic ($CaCO_3$+Ap-CaP) porous cements by using pure calcite powders

Powder (g)	SS (mL)	SS/P (mL/g)	Mixing time (sec)	Nominal Ca/P (molar)	%CaCO₃ left[*] (rest Ap-CaP)
29.75	30	1.0084	80	1.67	82 ± 3
26.90	30	1.1152	85	1.51	79 ± 3
22.30	30	1.3453	90	1.25	77 ± 4

determined by using the ratio of XRD peak intensities of calcite and Ap-CaP phases

It should be apparent from Table I that when one keeps the mixing time limited to between 80 and 90 seconds, there would not be enough time for a complete reaction between calcite and the SS. If one increased the mixing (or agitation) time to about 4 or 7 minutes, more calcite would convert to Ap-CaP, and total porosity and pore sizes would significantly decrease (data not shown). Figures 4a and 4b showed the optical micrographs of the porous cement scaffolds produced at the nominal Ca/P molar ratio of 1.51. Figures 5a through 5d, on the other hand, depicted the SEM photomicrographs of the same. Pores of these scaffolds are big enough to allow the attachment and proliferation of bone cells, i.e., osteoclasts and osteoblasts.

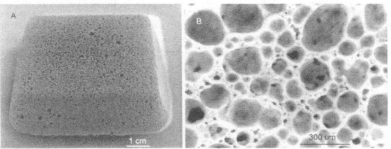

Fig. 4   Optical micrographs of the set cement body (*Table I; Ca/P = 1.51*), (*a*) macro view of the body cast into a large weighing cup, (*b*) optical microscope image of the same

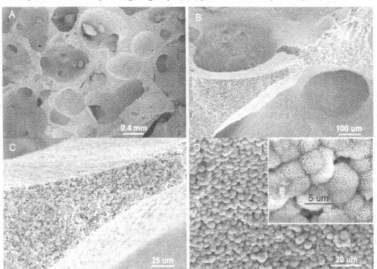

Fig. 5   SEM photomicrographs of set cement body (*Table I; Ca/P = 1.51*); (*a*) to (*c*) pore morphology of the cement body, (*d*) close-up image of the Ap-CaP skin

Upon casting the bubbling paste (after 80-90 seconds of agitation/kneading with a spatula) into a plastic weighing boat (Fig. 4a), the paste completed its setting within the next 4 to 5 minutes. After 4 minutes, the cement body was stable and workable. Figures 5b and 5c revealed the skin of Ap-CaP formed along the entire external surfaces. The micromorphology of the Ap-CaP skin shown in the inset of Fig. 5d resembled that of Ap-CaP biomimetically deposited via SBF (synthetic body fluid) solutions onto metals.[31]

The variation in the SS/P ratios was found to have a small but detectable effect on the degree of conversion of calcite to Ap-CaP. On the other hand, the change in SS/P ratios employed did not produce a significant change in percentage water absorption, density and compressive strength of the resultant cement bodies (as shown in Table II), as well as total apparent porosity. Unfortunately, in this study, the percentage porosity was mainly estimated by visual inspection of the photomicrographs, in combination with the water absorption and density measurements. According to the results of measurements by using the linear intercept method, pore sizes in the samples were varying between 20 and 750 μm. Micropores (20 to 80 μm) were typically located around the macropores (90 to 750 μm), and the struts defining the macropores were porous, as well (Figs. 4b and 5c).

Table II. Variation of the physical properties of biphasic ($CaCO_3$+Ap-CaP) porous cements

Powder (g)	SS (mL)	Setting time (min)	Water absorption (%)	Density[*] (g/cm^3)	Compressive strength[*] (MPa)
29.75	30	4	172 ±5	0.60	0.71
26.90	30	4	169 ±3	0.63	0.67
22.30	30	4	179±6	0.59	0.69

[*] *reported as the averages of measurements in triplicate*

It should be noted that when a whole cement block (as shown in Fig. 4a) was placed into 150 mL of deionized water (in a 400 mL-capacity beaker) at RT, its pH was recorded as 6.4±0.2. However, upon changing that water, after 10 minutes of soaking, with a fresh 150 mL aliquot, the pH of the water rose to 6.8±0.2. After 1 h of soaking, the pH of the water became stabilized at 7.0±0.1. This soaking step was deemed to be necessary to substantially remove the sodium-H_2PO_4 species originating from the mildly acidic SS.

Conversion of cement scaffolds to Ap-CaP

The bottom trace of Fig. 6 showed one of the characteristic XRD spectra for the porous cement blocks soaked in water for 3 hours, following the setting. It shows the presence of about 80% calcite (ICDD PDF 05-586). The top XRD trace of Fig. 6 confirmed the formation of single-phase Ap-CaP of low crystallinity upon soaking these biphasic cement scaffolds in a 0.5 M phosphate buffer solution at 80°C for 36 h. The XRD trace shown in Fig. 6b was obtained from the agate mortar-ground powders of these 0.5 M phosphate buffer-soaked and washed scaffolds.

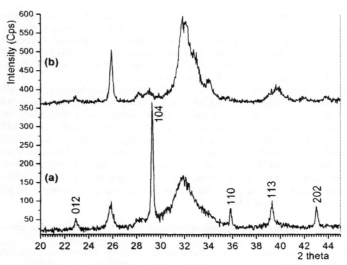

Fig. 6   XRD traces of cement scaffolds (Table I; Ca/P = 1.51 composition); (*a*) cement scaffold after 3 h of soaking in water at RT, peaks of calcite were indicated by their Miller indices and the rest belonged to Ap-CaP of low crystallinity, (*b*) cement scaffold after 36 h of soaking in 0.5 M phosphate buffer at 80°C

FTIR spectra of the same samples were plotted in Figure 7, respectively. In Fig. 7a, the bands observed at 3643, 2513, 1797, 1587, 1450, 871, 848, and 713 cm⁻¹ were characteristic of calcite. On the other hand, in Fig. 7b, the weak stretch for OH⁻, which was observed at 3571 cm⁻¹, indicated the presence of hydroxyl ions in Ap-CaP. Soaking of the porous cement scaffolds in a 0.5 M phosphate buffer solution at 80°C caused the conversion of calcite into Ap-CaP.

Water absorption and density measurements performed on these converted scaffolds showed that the values given in Table II did not change. However, the compressive strength of the converted scaffolds increased to 2.4±0.4 MPa from the starting value of ca. 0.7 MPa. It should be noted that the compressive strength of trabecular bones is between 2 and 10 MPa, whereas that of cortical bones is in the vicinity of 100 MPa. The weak cement scaffolds of this study can only match the compressive strength of trabecular bones at its lower limit. Many commercial cements could reach compressive strengths in excess of 50 MPa. [16, 23] However, those high-strength cements, which employed α-TCP or TTCP powders in their formulations (with lesser amounts of $CaCO_3$), were not able to show any significant overall *in vivo* resorbability even after one year of implantation.

ICP-AES analyses of the agate mortar-ground powders of the converted cement scaffolds (from the samples of Figs. 4(b) and 5(b)) gave a Ca/P molar ratio of 1.56±0.3, together with a Na concentration of 7700±250 ppm. The reader is referred to the works of Yubao *et al.*[32], Ivanova *et al.*[33], and Kasten *et al.*[34] for more detailed treatises (including *in vitro* and *in vivo* experiments) on non-stoichiometric Ap-CaP.

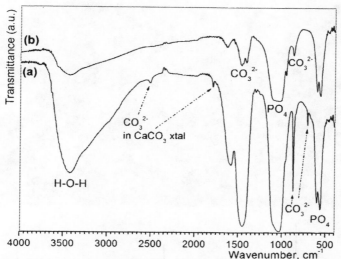

Fig. 7 FTIR traces of cement scaffolds (Table I; Ca/P = 1.51 composition); (*a*) cement scaffold after 3 h of soaking in water at RT, (*b*) cement scaffold after 36 h of soaking in 0.5 M phosphate buffer at 80°C

Pure calcite powders (Fisher Sci. Corp., Catalog No: C63-3) could also be converted to Ap-CaP by simple soaking in a dilute phosphate solution at 60 to 80°C. We have previously shown that even the surfaces of calcitic marble blocks would transform to Ap-CaP by using such a soaking procedure at 60°C.[35] To transform the calcite powders, this time we used the Soerensen's buffer solution (5.26 g KH_2PO_4 and 8.65 g of Na_2HPO_4 dissolved in 1 L water). Soaking experiments were performed in 250 mL-capacity Pyrex® media bottles, which contained 3.1 g calcite powder (whose morphology was given in Fig. 3) and 225 mL of Soerensen's buffer in each. Sealed bottles were first kept in a microprocessor-controlled oven at 60°C for 24 h, and then the solutions were replenished with an unused solution at every 24 h for the next 48 hours period. For the last 48 hours of the total soaking time of 72 h, the temperature of the oven was increased to 80°C. Powders in the media bottles were filtered out, washed with deionized water and dried overnight at 80°C. Resultant powders had the morphology illustrated in Fig. 8a. Characteristic XRD spectra of these Ap-CaP powders were given in Fig. 8b. ICP-AES analyses of these powders returned a Ca/P molar ratio of 1.58±0.2. BET surface area of these powders were measured as 40±2 m^2/g. This procedure was presented here as a robust and economical way of synthesizing Ap-CaP powders of high surface area in bulk form. It also served to confirm the phase nature of the "Ap-CaP skin" observed on our porous cement scaffolds as depicted in Fig. 5d.

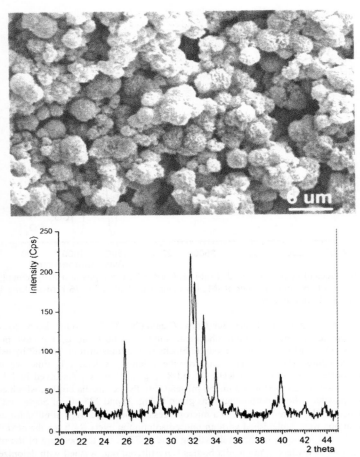

Fig. 8 (a) SEM photomicrograph of Ap-CaP powders synthesized by soaking calcite powders in Soerensen's buffer at 60-80°C for 72 h, (b) XRD trace of the same powder

The interest in the conversion of $CaCO_3$ into Ap-CaP, to obtain porous biomaterials, started with the pioneering work of Prof. Della M. Roy in the early 70's.[36-38] In these extremely competent studies, Roy et al.[37, 38] investigated the conversion of natural coral skeletal aragonite (Porites) by using high temperatures (140 to 260°C) and high autoclave pressures (550 to 1055 kg/cm^2) in the presence of acidic (such as, $CaHPO_4$ and $Ca(H_2PO_4)_2 \cdot H_2O$) and basic (such as,

$(NH_4)_2HPO_4)$ phosphates, as well as of sodium or potassium orthophosphates and acetic acid. Hydrothermal reactions were carried out from 12 to 48 h under the above-mentioned experimental conditions. Over the last three decades, the early studies of Roy et al.[36, 38] were duplicated, improved, or inspired by and the scientific literature related to these is immense, but just to name a few proficient examples, the works of Zaremba et al.[39], Su et al.[40], Ni et al.[41], and Jinawath et al.[42] must be cited.

Nowadays, converted coral and other marine skeletal species are also available as commercial porous implant biomaterials.[43, 44] However, it should be noted that in most coral or marine species the major phase was biological aragonite, and its conversion to Ap-CaP was more difficult than that of calcite or vaterite of this study. This was apparent from the need for higher temperatures and higher autoclave pressures in converting biological aragonite to Ap-CaP. Experimental studies which started only with calcite or vaterite powders and which produced macroporous Ap-CaP biomedical scaffolds from those were quite rare, and this study was conceived to fill this gap.

From now on, vaterite shall be expected to be seen in the powder components of new self-setting, orthopedic CaP cement formulations.

CONCLUSIONS

(1) Vaterite ($CaCO_3$) powders were synthesized in mixed solutions of Ca-chloride and gelatin by slowly adding a $NaHCO_3$ solution to those. Formed vaterite particles had a spheroidal morphology and surface area greater than 35 m^2/g.

(2) Soaking of vaterite powders in 0.5 M phosphate buffer solution at 70°C caused them to completely transform into Ap-CaP. Upon this transformation, the surface area of the samples increased to 85 m^2/g.

(3) Single-phase calcite ($CaCO_3$) powders were used to form a macroporous, biphasic (calcite+Ap-Cap) and weak cement scaffold.

(4) Concentrated orthophosphoric acid, partially neutralized by NaOH to pH 3.2, was used as the setting solution of the above-mentioned cement.

(5) 0.5 M phosphate buffer solutions were able to fully convert the calcite+Ap-CaP biphasic scaffolds into pure Ap-CaP, upon soaking these weak scaffolds in those solutions at 80°C.

(6) Soerensen's buffer was used to synthesize single-phase Ap-CaP powders, with a Ca/P molar ratio of 1.58 and BET surface area ≥40 m^2/g, by soaking pure calcite powders in these solutions at 60 to 80°C.

ACKNOWLEDGEMENTS

The author is cordially grateful to Mr. Sahil Jalota and Dr. Sarit B. Bhaduri for their generous help with some of the characterization runs. The author was a research scientist at Merck Biomaterials GmbH located in Darmstadt, Germany from September 2001 to April 2003. The author was then a research associate-professor at Clemson University (South Carolina) between May 2003 and April 2006.

REFERENCES

[1]Y. Kojima, A. Kawanobe, T. Yasue, and Y. Arai, "Synthesis of Amorphous Calcium Carbonate and Its Crystallization," J. Ceram. Soc. Jpn., 101, 1145-52 (1993).

[2]H. Vater, "Ueber den Einfluss der Loesungsgenossen auf die Krystallisation des Calciumcarbonates," Z. Kristallogr. Mineral., 27, 477-512 (1897).

[3]H. A. Lowenstam and D. P. Abbott, "Vaterite: A Mineralization Product of the Hard Tissue of a Marine Organism (Ascidiacea)," Science, 188, 363-65 (1975).

[4]A. S. Posner, "Crystal Chemistry of Bone Mineral," Physiol. Rev., 49, 760-92 (1969).

[5]J. A. Jansen, J. W. M. Vehof, P. Q. Ruhe, H. Kroeze-Deutman, Y. Kuboki, H. Takita, E. L. Hedberg, and A. G. Mikos, "Growth Factor-loaded Scaffolds for Bone Engineering," J. Control. Release, 101, 127-36 (2005).

[6]R. A. Robinson and M. L. Watson, "Collagen-Crystal Relationships in Bone as seen in the Electron Microscope. III. Crystals and Collagen Morphology as a Function of Age," Ann. NY. Acad. Sci., 60, 596-628 (1955).

[7]F. Monchau, A. Lefevre, M. Descamps, A. Belquin-Myrdycz, P. Laffargue, and H. F. Hildebrand, "In Vitro Studies of Human and Rat Osteoclast Activity on Hydroxyapatite, β-Tricalcium Phosphate, Calcium Carbonate," Biomol. Eng., 19, 143-52 (2002).

[8]H. Mutsuzaki, M. Sakane, A. Ito, H. Nakajima, S. Hattori, Y. Miyanaga, J. Tanaka, and N. Ochiai, "The Interaction between Osteoclast-like Cells and Osteoblasts Mediated by Nanophase Calcium Phosphate-hybridized Tendons," Biomaterials, 26, 1027-34 (2005).

[9]A. F. Schilling, W. Linhart, S. Filke, M. Gebauer, T. Schinke, J. M. Rueger, and M. Amling, "Resorbability of Bone Substitute Biomaterials by Human Osteoclasts," Biomaterials, 25, 3963-72 (2004).

[10]O. Zinger, G. Zhao, Z. Schwartz, J. Simpson, M. Wieland, D. Landolt, and B. D. Boyan, "Differential Regulation of Osteoblasts by Substrate Microstructural Features," Biomaterials, 26, 1837-47 (2005).

[11]Z. Molnar, "Additional Observations on Bone Crystal Dimensions," Clin. Orthop., 17, 38-42 (1960).

[12]W. E. Brown, "Crystal Growth of Bone Mineral," Clin. Orthop., 44, 205-220 (1966).

[13]R. Hodgskinson, C. F. Njeh, M. A. Whitehead, and C. M. Langton, "The Non-linear Relationship between BUA and Porosity in Cancellous Bone," Phys. Med. Biol., 41, 2411-20 (1996).

[14]T. D. Driskell, A. L. Heller, and J. F. Koenigs, "Dental Treatments," US Patent No: 3,913,229, October 21, 1975.

[15]R. Z. LeGeros, A. Chohayeb, and A. Shulman, "Apatitic Calcium Phosphates: Possible Dental Restorative Materials," J. Dental Res., 61, 343-47 (1982).

[16]W. E. Brown and L. C. Chow, "Dental Restorative Cement Pastes," US Patent No: 4,518,430, May 21, 1985.

[17]W. E. Brown and L. C. Chow, "A New Calcium Phosphate, Water Setting Cement"; pp. 352-79 in Cements Research and Progress 1986. Edited by P. W. Brown. American Ceramic Society, Westerville, OH, 1988.

[18]J. Wiltfang, F. R. Kloss, P. Kessler, E. Nkenke, S. Schultze-Mosgau, R. Zimmermann, and K. A. Schlegel, "Effects of Platelet-rich Plasma on Bone Healing in Combination with Autogenous Bone and Bone Substitutes in Critical Size Defects: An Animal Experiment," Clin. Oral Implants. Res., 15, 187-93 (2004).

[19]M. Tamai, T. Isshiki, K. Nishio, M. Nakamura, A. Nakahira, and H. Endoh, "Transmission Electron Microscopic Studies on an Initial Stage in the Conversion Process from α-Tricalcium Phosphate to Hydroxyapatite," J. Mater. Res., 18, 2633-38 (2003).

[20]H. Monma, M. Goto, H. Nakajima, and H. Hashimoto, "Preparation of Tetracalcium Phosphate," Gypsum Lime, 202, 151–55 (1986).

[21]R. I. Martin and P. W. Brown, "Hydration of Tetracalcium Phosphate," Adv. Cement Res., 5, 119–25 (1993).

[22]S. Jalota, A. C. Tas, and S. B. Bhaduri, "Synthesis of HA-seeded TTCP ($Ca_4(PO_4)_2O$) Powders at 1230°C from $Ca(CH_3COO)_2·H_2O$ and $NH_4H_2PO_4$," J. Am. Ceram. Soc., 88, 3353-60 (2005).

[23]M. Bohner, U. Gbureck, and J. E. Barralet, "Technological Issues for the Development of More Efficient Calcium Phosphate Bone Cements: A Critical Assessment," Biomaterials, 26, 6423-29 (2005).

[24]C. Combes, B. Miao, R. Bareille, and C. Rey, "Preparation, Physical-chemical Characterisation and Cytocompatibility of Calcium Carbonate Cements," Biomaterials, 27, 1945-54 (2006).

[25]"Standard Test Method for Apparent Porosity, Water Absorption, Apparent Specific Gravity and Bulk Density of Burned Refractory Brick and Shapes by Boiling Water," ASTM Designation C20-92. 1995 Annual Book of ASTM Standards, Vol. 15.01; pp. 5–7. American Society for Testing and Materials, Philadelphia, PA.

[26]M.S. Rao, "Kinetics and mechanism of the transformation of vaterite to calcite," Bull. Chem. Soc. Jpn., 46, 1414-17 (1973).

[27]Q.S. Wu, D.M. Sun, H.J. Liu, and Y.P. Ding, "Abnormal Polymorph Conversion of Calcium Carbonate," Cryst. Growth Des., 4, 717-20 (2004).

[28]D. Walsh, B. Lebeau, and S. Mann, "Morphosynthesis of Calcium Carbonate (Vaterite) Microsponges," Adv. Mater., 11, 324-28 (1999).

[29]K. Naka, Y. Tanaka, and Y. Chujo, "Effect of Anionic Starburst Dendrimers on the Crystallization of $CaCO_3$ in Aqueous Solution: Size Control of Spherical Vaterite Particles," Langmuir, 18, 3655-58 (2002).

[30]A. C. Tas and S. B. Bhaduri, "Conversion of Calcite ($CaCO_3$) Powders into Macro- and Microporous Calcium Phosphate Scaffolds for Medical Applications," US Patent Appl. Pub. No: US 2006/0110422 A1, May 25, 2006.

[31]S. Jalota, S. B. Bhaduri, and A. C. Tas, "Effect of Carbonate Content and Buffer Type on Calcium Phosphate Formation in SBF Solutions," J. Mater. Sci. Mater. M., 17, 697-707 (2006).

[32]L. Yubao, Z. Xingdong, and K. de Groot, "Hydrolysis and Phase Transition of Alpha-Tricalcium Phosphate," Biomaterials, 18, 737-41 (1997).

[33]T. I. Ivanova, O. V. Frank-Kamenetskaya, A. B. Koltsov, and V. L. Ugolkov, "Crystal Structure of Calcium-deficient Carbonated Hydroxyapatite. Thermal Decomposition," J. Sol. State Chem., 160, 340-49 (2001).

[34]P. Kasten, J. Vogel, R. Luginbuhl, P. Niemeyer, M. Tonak, H. Lorenz, L. Helbig, S. Weiss, J. Fellenberg, A. Leo, H. G. Simank, and W. Richter, "Ectopic Bone Formation Associated with Mesenchymal Stem Cells in a Resorbable Calcium Deficient Hydroxyapatite Carrier," Biomaterials, 26, 5879-89 (2005).

[35]A. C. Tas and F. Aldinger, "Formation of Apatitic Calcium Phosphates in a Na-K-phosphate Solution of pH 7.4," J. Mater. Sci. Mater. M., 16, 167-74 (2005).

[36]D. M. Roy and S. K. Linnehan, "Hydroxyapatite Formed from Coral Skeletal Carbonate by Hydrothermal Exchange," Nature, 246, 220-22 (1974).

[37]D. M. Roy, W. Eysel, and D. Dinger, "Hydrothermal Synthesis of Various Carbonate-containing Calcium Hydroxyapatite," Mater. Res. Bull., 9, 35-40 (1974).

[38]D. M. Roy, "Porous Biomaterials and Method of Making Same," US Patent No: 3,929,971, Dec. 30, 1975.

[39]C. M. Zaremba, D. E. Morse, S. Mann, P. K. Hansma, and G. D. Stucky, "Aragonite-Hydroxyapatite Conversion in Gastropod (Abalone) Nacre," Chem. Mater., 10, 3813-24 (1998).

[40]X.-W. Su, D.-M. Zhang, and A. H. Heuer, "Tissue Regeneration in the Shell of the Giant Queen Conch, Strombus Gigas," Chem. Mater., 16, 581-93 (2004).

[41]M. Ni and B. D. Ratner, "Nacre Surface Transformation to Hydroxyapatite in a Phosphate Buffer Solution," Biomaterials, 24, 4323-31 (2003).

[42]S. Jinawath, D. Polchai, and M. Yoshimura, "Low-temperature Hydrothermal Transformation of Aragonite to Hydroxyapatite," Mater. Sci. Eng. C, 22, 35-39 (2002).

[43]J. Vuola, H. Goeransson, T. Boehling, and S. Asko-Seljavaara, "Bone Marrow Induced Osteogenesis in Hydroxyapatite and Calcium Carbonate Implants," Biomaterials, 17, 1761-66 (1996).

[44]J. Vuola, R. Taurio, H. Goeransson, and S. Asko-Seljavaara, "Compressive Strength of Calcium Carbonate and Hydroxyapatite Implants after Bone-Marrow-Induced Osteogenesis," Biomaterials, 19, 223-27 (1998).

DEPOSITION OF BONE-LIKE APATITE ON POLYGLUTAMIC ACID GELS IN BIOMIMETIC SOLUTION

Toshiki Miyazaki[1], Atsushi Sugino[2,3] and Chikara Ohtsuki[2]

[1]Graduate School of Life Science and Systems Engineering, Kyushu Institute of Technology, 2-4 Hibikino, Wakamatsu-ku, Kitakyushu 808-0196, Japan

[2]Graduate School of Engineering, Nagoya University, Furo-cho, Chikusa-ku, Nagoya 464-8603, Japan

[3]Nakashima Propeller Co., Ltd. 688-1, Jodo-Kitagata, Okayama 700-8691, Japan

ABSTRACT

Apatite-polymer hybrids that mimics natural bone structure are expected as novel bone substitutes having excellent mehcanical performances and high bone-bonding ability, i.e. bioactivity. In this study, we attempted preparation of apatite-polyglutamic acid hybrids through biomimetic process that mimics the principle of biomineralization. Simple chemical modification of the polyglutamic acid gel with calcium chloride solution with various concentrations was attempted to enhance the apatite formation. Apatite-forming ability of the gels was examined in simulated body fluid (SBF, Kokubo solution). Even the untreated gel formed a calcium phosphate in SBF, although it was not identified with the apatite. In contrast, the gels treated with $CaCl_2$ solutions formed the apatite within 7 days. Induction period for the apatite formation decreased with increasing $CaCl_2$ concentration. This type of hybrid is useful for novel bioactive bone substitutes.

INTRODUCTION

So-called bioactive ceramics such as Bioglass®, sintered hydroxyapatite and glass-ceramics A-W have been studied extensively for clinical applications including bone repair, because they have the ability to bond directly to bone [1-6]. The problem is, however, that bioactive ceramics show higher Young's modulus and lower fracture toughness than natural bone. Natural bone is a kind of organic-inorganic hybrid made of organic collagen fibers and inorganic carbonate hydroxyapatite crystals with a characteristic structure that leads to specific mechanical properties, such as a high fracture toughness and high flexibility [7]. The development of organic-inorganic hybrids composed of apatite and organic polymers is attractive because of their novelty in being materials that show bone-bonding ability, i.e. bioactivity, and mechanical properties similar to those of natural bone. In addition, bioactive and machinable materials which can be fabricated into desired shapes during operation have been desired by medical doctors.

The biomimetic process has received much attention for fabricating such a hybrid, where bone-like apatite is deposited under ambient conditions on polymer substrates in simulated body fluid (SBF, Kokubo solution) having ion concentrations nearly equal to those of human extracellular fluid or related solutions. Ion concentrations of SBF are shown in Fig. 1 [8-11]. The heterogeneous apatite deposition in the biomimetic solutions is triggered by the specific kinds of functional groups, as schematically shown in Fig. 2. As such functional groups, Si-OH [10,12],

Ti-OH [12-13], Zr-OH [14], Ta-OH [15], Nb-OH [16], COOH [17], phosphate [17] and sulfonic groups [18-19] are known. However Al-OH group and methyl group are not effective for the apatite formation [12,17]. In addition, release of Ca^{2+} from the materials significantly enhance the apatite nucleation, since the released Ca^{2+} increase the degree of supersaturation in the surrounding fluid with respect to the hydroxyapatite [20].

In this study, we chose polyglutamic acid $((NHCH(COOH)(CH_2)_2CO)_n)$ abundant in carboxyl group as an organic polymer. The polyglutamic acid is water-soluble biodegradable polymer produced by the *Bacillus subtilis* [21-22]. This is also known as a main component of *Natto*, a Japanese traditional food. We synthesized the polyglutamic acid gel through covalent cross-linking using divalent amine and water-soluble carbodiimide. Ability of the apatite deposition on the gels was examined in SBF.

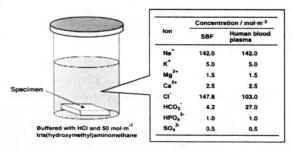

Fig. 1 Ion concentrations of SBF.

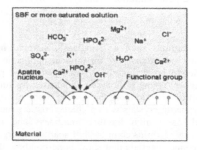

Fig. 2 Schematic representation of biomimetic process.

MATERIALS AND METHODS

Powder of polyglutamic acid produced by *Bacillus strain* was dissolved in distilled water to form aqueous solutions with concentrations of 5 mass%. Then 0.2 g of Ethylenediamine-2(N-hydroxysuccinimide) and 0.96 g of 1-Ethyl-3-(3-dimethylaminopropyl)-carbodiimide hydrochloride (EDC·HCl) was added into 30 cm^{-3} of the solution to progress cross-linking reaction. The solution was poured into a Teflon dish and kept at room temperature for 24 h. The gels obtained were treated with calcium chloride ($CaCl_2$) aqueous solutions with various concentrations ranging 0.01 to 1 M (=kmol·m^{-3}) at 36.5°C for 24 h. The apatite deposition on the obtained gels were examined by soaking in SBF with pH=7.40 at 36.5°C for various periods up to 7 days. After the given period, the gels were removed from the fluid, gently washed with distilled water. Changes in surface structure of the gels before and after soaking in SBF were characterized by thin-film X-ray diffraction (TF-XRD), Fourier-transform infrared (FT-IR) spectroscopy and scanning electron microscope (SEM) equipped with an energy dispersive X-ray microanalyzer (EDX).

RESULTS

Figure 3 shows SEM photographs of the surfaces of PGA gels with and without $CaCl_2$ treatment after soaking in SBF for 7 days. Spherical particles about 2-3 µm in size were observed for the treated gels. Figure 4 shows TF-XRD patterns of the PGA gels treated with and without $CaCl_2$ treatment after soaking in SBF for 7 days. Tiny broad peaks at 26° and 32° in 2θ assigned to low-crystalline apatite were detected on the treated PGA gels. In contrast, these peaks could not be detected on the untreated PGA gels even after soaking in SBF. Figure 5 shows EDX spectra of the surfaces of PGA gels with and without $CaCl_2$ treatment after soaking in SBF for 7 days. Peaks assigned to Ca and P were observed for all the gels. Intensity of the peaks was increased with increasing concentration of $CaCl_2$ solution. Figure 6 shows FT-IR spectra of the PGA gels modified with or without 1M-$CaCl_2$ solution before and after soaking in SBF for 7 days. The peaks assigned to carboxyl groups were detected on the both PGA gels before soaking in SBF at about 1400 and 1600 cm^{-1}. After soaking in SBF for 7 days, the peaks assigned to P-O stretching were newly observed for both the gels at about 1000 cm^{-1}. In addition, the peaks assigned to P-O bending were observed for the gel modified with $CaCl_2$ at about 550 and 600 cm^{-1}. Table 1 summarizes apatite formation of the examined PGA gels in SBF. We can see that induction period for the apatite formation decreases with increasing $CaCl_2$ concentration.

DISCUSSION

We can see from the results described above that even the untreated PGA gels form a calcium phosphate in SBF, although it was not identified with the apatite. This would be amorphous calcium phosphate. These results suggest that PGA gels themselves have an ability of calcification in body environment. In contrast, the PGA gels treated with $CaCl_2$ solution of 0.01 M or more can form the apatite in SBF within 7 days. In addition, intensity of Ca peaks in EDX spectra significantly increases with increasing $CaCl_2$ concentration (See Fig. 5). These results mean that $CaCl_2$ treatment significantly accelerate the apatite formation on the PGA gels. Polyglutamic acid is known as a hydrophilic biodegradable polymer that contains abundant carboxyl groups. Carboxyl groups would act as the heterogeneous nucleation site of the apatite. It has been revealed that negatively charged functional groups are likely to deposit the apatite through selective binding with Ca^{2+} ions and subsequent binding with phosphate ions to form the apatite. Carboxyl groups are also negatively charged in pH region around neutrality [23-24].

Therefore apatite formation on the PGA gels would progress similar to the above process. In addition, the apatite nucleation and growth are enhanced by released Ca^{2+} ions from the treated gels that increase degree of supersaturation with respect to the apatite.

It has been already reported that the apatite can be formed on the surfaces of the polyamide films [25] and silk fibers [26] that contain carboxyl groups. These polymers formed the apatite on their surfaces only in 1.5SBF that has inorganic ion concentrations 1.5 times those of SBF. The PGA gels treated with $CaCl_2$ solutions would have higher apatite-forming ability than these polymers. PGA is a highly hydrophilic polymer with swelling ratio more than 20% [27]. Such a swelling property would contribute to high apatite-forming ability, since penetration of surrounding fluid into the gels supports to form the apatite inside the gels. Finally, this type of apatite-polyglutamic acid hybrids is useful for osteoconductive scaffolds for bone tissue regeneration.

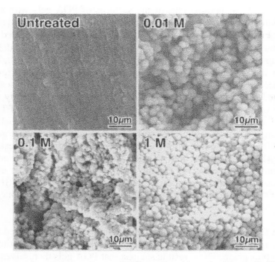

Fig. 3 SEM photographs of the surfaces of PGA gels with and without $CaCl_2$ treatment after soaking in SBF for 7 days.

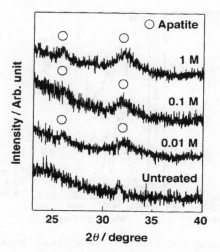

Fig. 4 TF-XRD patterns of the surfaces of PGA gels with and without CaCl₂ treatment after soaking in SBF for 7 days.

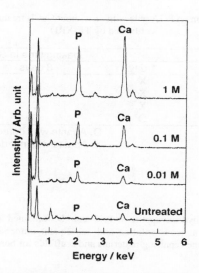

Fig. 5 EDX spectra of the surfaces of PGA gels with and without CaCl₂ treatment after soaking in SBF for 7 days.

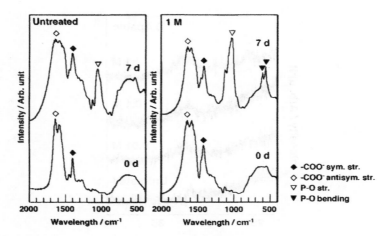

Fig. 6 FT-IR spectra of the surfaces of PGA gels with or without 1M-CaCl₂ treatment before (0 d) and after soaking in SBF for 7 days.

Table 1 Apatite formation on PGA gels with or without CaCl₂ treatment in SBF, which was confirmed by TF-XRD

CaCl₂ concentration	Soaking time in SBF		
	1 day	3 days	7 days
Untreated	X	X	X
0.01 M	X	O	O
0.1 M	X	O	O
1 M	O	O	O

O: Apatite was detected by TF-XRD
X: Apatite was not detected by TF-XRD

CONCLUSION

Polyglutamic acid-hydroxyapatite hybrid can be obtained through simple chemical modification using CaCl₂ aqueous solution and subsequent immersion in SBF. This type of hybrid is useful as novel bone-repairing materials and scaffolds for bone tissue regeneration.

ACKNOWLEDGMENTS
The authors greatly appreciate financial support for this study by Okayama Prefecture Industrial Promotion Foundation (Okayama Challenge Project). One of the authors (T.M) also acknowledges the support by Grant-in-Aid for Encouragement of Young Scientists ((B)16700365), Japan Society for the Promotion of Science.

REFERENCES
[1]L.L. Hench, R.J. Splinger, W.C. Allen, and T.K. Greenlee, "Bonding Mechanisms at the Interface of Ceramic Prosthetic Materials," *J. Biomed. Mater. Res. Symp.*, **2**, 117-141(1972).

[2]M. Jarcho, "Hydroxyapatite Synthesis and Characterization in Dense Polycrystalline Forms," *J. Mater. Sci.*, **11**, 2027-2035 (1976).

[3]T. Kokubo, M. Shigematsu, Y. Nagashima, M. Tashiro, T. Nakamura, T. Yamamuro, and S. Higashi, "Apatite- and Wollastonite-containing Glass-ceramics for Prosthetic Application," *Bull. Inst. Chem. Res., Kyoto Univ.*, **60**, 260-268 (1982).

[4]L.L. Hench, "Bioceramics; From Concept to Clinic," *J. Am. Ceram. Soc.*, **74**, 1487-1510 (1991).

[5]L.L. Hench, "Bioceramics," *J. Am. Ceram. Soc.*, **81**, 1705-1728 (1998).

[6]T. Kokubo, H.M. Kim, and M. Kawashita, "Novel Bioactive Materials with Different Mechanical Properties," *Biomaterials*, **24**, 2161-2175 (2003).

[7]C.K. Loong, C. Rey, L.T. Kuhn, C. Combes, Y. Wu, S. Chen and M.J. Glimcher, "Evidence of Hydroxyl-ion Deficiency in Bone Apatites: an Inelastic Neutron-scattering Study," *Bone*, **26**, 599-602 (2000).

[8]T. Kokubo, H. Kushitani, S. Sakka, T. Kitsugi, and T. Yamamuro, "Solutions Able to Reproduce *in vivo* Surface-structure Changes in Bioactive Glass-ceramic A-W," *J. Biomed. Mater. Res.*, **24**, 721-734 (1990).

[9]C. Ohtsuki, Y. Aoki, T. Kokubo, Y. Bando, M. Neo, and T. Nakamura, "Transmission Electron Microscopic Observation of Glass-Ceramic A-W and Apatite Layer Formed on Its Surface in a Simulated Body Fluid," *J. Ceram. Soc. Japan*, **103**, 449-454 (1995).

[10]S.B. Cho, T. Kokubo, K. Nakanishi, N. Soga, C. Ohtsuki, T. Nakamura, T. Kitsugi, and T. Yamamuro, "Dependence of Apatite Formation on Silica Gel on Its Structure: Effect of Heat Treatment," *J. Am. Ceram. Soc.*, **78**, 1769-1774 (1995).

[11]T. Kokubo, and H. Takadama, "How Useful Is SBF in Predicting *in vivo* Bone Bioactivity? ," *Biomaterials*, **27**, 2907-2915 (2006).

[12]P. Li, C. Ohtsuki, T. Kokubo, K. Nakanishi, N. Soga, and K. de Groot, "The Role of Hydrated Silica, Titania and Alumina in Inducing Apatite on Implants," *J. Biomed. Mater. Res.*, **28**, 7-15 (1994).

[13]M. Uchida, H.M. Kim, T. Kokubo, S. Fujibayashi, and T. Nakamura, "Structural Dependence of Apatite Formation on Titania Gels in a Simulated Body Fluid," *J. Biomed. Mater. Res. A*, **64**, 164-170 (2003).

[14]M. Uchida, H.M. Kim, T. Kokubo, F. Miyaji, and T. Nakamura, "Bonelike Apatite Formation Induced on Zirconia Gel in a Simulated Body Fluid and Its Modified Solutions", *J. Am. Ceram. Soc.*, **84**, 2041-2044 (2001).

[15]T. Miyazaki, H.M. Kim, T. Kokubo, H. Kato, and T. Nakamura, "Induction and Acceleration of Bonelike Apatite Formation on Tantalum Oxide Gel in Simulated Body Fluid," *J. Sol-gel. Sci. Tech.*, **21**, 83-88 (2001).

[16]T. Miyazaki, H.M. Kim, T. Kokubo, C. Ohtsuki, and T. Nakamura, "Bonelike Apatite Formation Induced on Niobium Oxide Gels in Simulated Body Fluid," *J. Ceram. Soc. Japan*, **109**, 934-938 (2001).

[17]M. Tanahashi, and T. Matsuda, "Surface Functional Group Dependence on Apatite Formation on Self-assembled Monolayers in a Simulated Body Fluid," *J. Biomed. Mater. Res.*, **34**, 305-315 (1997).

[18]T. Kawai, C. Ohtsuki, M. Kamitakahara, T. Miyazaki, M. Tanihara, Y. Sakaguchi, and S. Konagaya, "Coating of Apatite Layer on Polyamide Films Containing Sulfonic Groups by Biomimetic Process," *Biomaterials*, **25**, 4529-4534 (2004).

[19]I.B. Leonor, H.M. Kim, F. Balas, M. Kawashita, R.L. Reis, T. Kokubo, and T. Nakamura, "Surface charge of bioactive polyethylene modified with -SO$_3$H groups and its apatite inducing capability in simulated body fluid," pp. 453-456 in *Bioceramics Vol. 17* edited by P. Li, K. Zhang, and C.W. Colwell Jr, Trans Tech Publications Ltd., Switzerland (2005).

[20]C. Ohtsuki, T. Kokubo, and T. Yamamuro, "Mechanism of apatite formation on CaO-SiO$_2$-P$_2$O$_5$ glasses in a simulated body fluid," *J. Non-Cryst. Solids*, **143**, 84-92 (1992).

[21]M. Kunioka, and K. Furusawa, "Poly (γ-glutamic acid) Hydrogel Prepared from Microbial Poly (g-glutamic acid) and Alkanediamine with Water-Soluble Carbodiimide," *J. Appl. Polym. Sci.*, **65**, 1889–1896 (1997).

[22]S.H. Choi, K.S. Whang, J.S. Park, W.Y. Choi, and M.H. Yoon, "Preparation and Swelling Characteristics of Hydrogel from Microbial Poly(γ-glutamic acid) by γ-Irradiation," *Macromol. Res.*, **13**, 339-343 (2005).

[23]H. Takadama, H-M. Kim, F. Miyaji, T. Kokubo and T. Nakamura, "Mechanism of Apatite Formation Induced by Silanol Groups-TEM observation," *J. Ceram. Soc. Japan.*, **108**, 118–121 (2000).

[24]H. Takadama, H-M. Kim, T. Kokubo and T. Nakamura, "TEM–EDX Study of Mechanism of Bonelike Apatite Formation on Bioactive Titanium Metal in Simulated Body Fluid," *J. Biomed. Mater. Res.*, **57**, 441–448 (2001).

[25]T. Miyazaki, C. Ohtsuki, Y. Akioka, M. Tanihara, J. Nakao, Y. Sakaguchi, and S. Konagaya, "Apatite Deposition on Polyamide Films Containing Carboxyl Group in a Biomimetic Solution," *J. Mater. Sci. Mater. Med.*, **14**, 569-574 (2003).

[26]A. Takeuchi, C. Ohtsuki, T. Miyazaki, M. Yamazaki, and M. Tanihara, "Deposition of Bone-like Apatite on Silk Fibre in a Solution that Mimics Extracellular Fluid," *J. Biomed. Mater. Res.*, **65A**, 283-289 (2003).

[27]A. Sugino, T. Miyazaki and C. Ohtsuki, "Apatite-forming Ability of Polyglutamic Acid Gel in Simulated Body Fluid: Effect of Cross-Linking Agent," pp. 683-686 in *Bioceramics, Vol. 19*, ed. by X. Zhang, X. Li, H. Fan and X. Liu, Trans Tech Publications, 2007.

SURFACE STRUCTURE AND APATITE-FORMING ABILITY OF POLYETHYLENE SUBSTRATES IRRADIATED BY THE SIMULTANEOUS USE OF OXYGEN CLUSTER AND MONOMER ION BEAMS

M. Kawashita, S. Itoh, R. Araki, K. Miyamoto and G. H. Takaoka
Ion Beam Engineering Experimental Laboratory,
Graduate School of Engineering,
Kyoto University
Nishikyo-ku, Kyoto 615-8510, Japan

ABSTRACT

Polyethylene (PE) substrates as one example of polymers were irradiated at 10^{15} ions/cm^2 by the simultaneous use of oxygen (O_2) cluster and monomer ion beams, and their apatite-forming ability was examined using a metastable calcium phosphate solution that had 1.5 times the ion concentrations of a normal simulated body fluid (1.5SBF). The acceleration voltage for the ion beams was changed from 3 to 9 kV. Some irradiated substrates were soaked in $1M$-CaCl$_2$ solution for 1 day. Thus obtained substrates were soaked in 1.5SBF at 36.5 °C for 7 days. The hydrophilic functional groups such as COOH groups were formed at the PE surfaces by the irradiation. The amount of the functional groups increased with increasing acceleration voltage, but the hydrophilicity of the substrates decreased slightly at 9 kV. This might be attributed to the carbonization of the PE substrate caused by the irradiation at high acceleration voltage of 9 kV. The irradiated PE substrates formed apatite in 1.5SBF, whereas unirradiated ones did not form it. This indicates that the functional groups such as COOH groups induced apatite nucleation in 1.5SBF. The apatite-forming ability increased with increasing acceleration voltage up to 6 kV, but slightly decreased at 9 kV. The apatite formation was promoted by the subsequent CaCl$_2$ solution treatment. When the substrate is soaked in 1.5SBF, the calcium ions retained at the PE surface by the CaCl$_2$ solution treatment are released into 1.5SBF to increase the ionic activity product of the surrounding fluid with respect to apatite.

INTRODUCTION

Artificial ligaments which are two-dimensional polyethylene terephthalate (PET) fabrics have been used for reconstruction of damaged ligaments. In the reconstruction, the artificial ligaments are fixed to bone by using bone plug and staples. However, a long-term is often required for complete cure, because PET does not bond to living bone directly but bond to the bone through fibrous tissues and hence the fixation of the artificial ligaments becomes sometimes unstable. If the surface of PET is modified to show bone-bonding ability (i.e. bioactivity), it is believed that the term for complete cure can be shortened remarkably. In order to obtain bioactive polymers, we should form functional groups such as Si-OH [1], Ti-OH [2] and -COOH [3] groups effective for apatite nucleation, on the surface of polymers. Various attempts to modify the polymer surface with the bioactive phase have been made [4-9], but it is difficult to form the bioactive phase stably on the surfaces of hydrophobic polymers such as PE.

The ion beam process has a high potential for surface modification of polymers. The predominant properties of the ion beam process are based on the ability to control the kinetic energy precisely by adjusting the acceleration energy. In addition, atomic, molecular and cluster ions are available, and the interaction of these ions with the substrate surface can be greatly varied depending on the ion species used. The cluster is an aggregate of a few tens to several thousands of atoms or molecules, and constitutes a new phase of matter, with significantly different properties than solids, liquids and gases [10]. The effect of cluster ion impact on a solid surface is much different from the single ion effect, and shows the high-density irradiation and the multiple collision. Another feature of the cluster ion impact is the low-energy irradiation, and the damage-free irradiation can be performed by adjusting the incident energy of the cluster ion beam [11-16]. On the other hand, the monomer ion beams with high incident energy produce free radicals and bond scission on the irradiated polymer surfaces that are effective for cross-linking and/or carbonization [17]. Recently, we found that apatite was successfully formed on PE and poly(ethylene terephthalate) (PET) substrates, when the substrates were irradiated with the simultaneous use of oxygen (O_2) cluster and monomer ions and subsequently treated with a bioactive calcium silicate (CaO-SiO_2) sol solution [18]. In that paper, we reported that hydrophilic functional groups such as -COOH and ≡C-OH were formed on the PE surface by the ion beams. On the other hand, it was reported that some specific functional groups, including the -COOH group, were effective for apatite nucleation in the body environment [3,19,20]. Therefore, we can expect that polymers irradiated with the simultaneous use of O_2 cluster and monomer ion beams would show apatite-forming ability in the body environment even without the CaO-SiO_2 sol solution treatment. In this study, we chose the PE substrate as an example of a hydrophobic polymeric material and investigated *in vitro* the apatite-forming ability of the irradiated PE substrates in a metastable calcium phosphate solution (1.5SBF) that had 1.5 times the ion concentrations of a normal simulated body fluid.

EXPERIMENTAL
Simultaneous irradiation of oxygen cluster and monomer ion beams

Details of the experimental apparatus for cluster ion beams have been given elsewhere [12-16]. Briefly, neutral beams of O_2 clusters were produced by adiabatic expansion of a high pressure gas through a glass nozzle into a high vacuum chamber. The O_2 cluster and monomer beams formed were introduced into the ionization and irradiation chamber under the inlet gas pressure of 4.0×10^4 Pa and ionized by electron bombardment with an ionization voltage of 300 V and an electron current of 300 mA. PE substrates $20 \times 20 \times 1$ mm^3 in size (Density: 0.95, Kobe Polysheet EL, Shin-Kobe Electric Machinery, Co., Ltd., Tokyo, Japan) were irradiated at an ion dose of 1×10^{15} ions/cm^2 by the simultaneous use of O_2 cluster and monomer ion beams. Here, the O_2 monomer ion beams mean the beams composed of oxygen atomic and/or molecular ions such as O^+ and/or O_2^+ ions. The acceleration voltage (V_a) for the ion beams was varied from 3 to 9 kV. As a reference, PE substrates were subjected to O_2 plasma treatment at room temperature for 5 min at a frequency of 13.56 MHz and a power of 50 W (BP-2, Samco Inc., Kyoto, Japan).

Soaking in 1.5SBF

The substrates were immersed for 7 days at 36.5°C in 30 mL of a metastable calcium phosphate solution (1.5SBF) [4,5,21,22] that had 1.5 times the ion concentrations of a normal simulated body fluid [23-25]. The 1.5SBF was prepared by dissolving NaCl, NaHCO$_3$, KCl, K$_2$HPO$_4$·3H$_2$O, MgCl$_2$·6H$_2$O, CaCl$_2$ and Na$_2$SO$_4$ (Nacalai Tesque Inc., Kyoto, Japan) in ultrapure water, and buffered at pH 7.40 at 36.5 °C with tris(hydroxymethyl)aminomethane ((CH$_2$OH)$_3$CNH$_2$) and 1 mol/dm^3 HCl aqueous solution (Nacalai Tesque Inc., Kyoto, Japan). The ion concentrations and pH of human blood plasma [26], SBF and 1.5SBF are listed in Table I. Some substrates were soaked for 24 h at 36.5 °C in 30 mL of 1 mol/dm^3 CaCl$_2$ aqueous solution [21,22] prior to the 1.5SBF soaking.

Table I. Ion concentrations and pH value of human blood plasma, SBF and 1.5SBF.

Ion	Ion concentration (mmol/dm³)		
	Human blood plasma	SBF	1.5SBF
Na^+	142.0	142.0	213
K^+	5.0	5.0	7.5
Mg^{2+}	1.5	1.5	2.25
Ca^{2+}	2.5	2.5	3.75
Cl^-	103.0	148.8	222
HCO_3^-	27.0	4.2	6.3
HPO_4^{2-}	1.0	1.0	1.5
SO_4^{2-}	0.5	0.5	0.75
pH	7.2 - 7.4	7.4	7.4

Surface characterization

The contact angle of the substrates was measured by a contact angle meter (CA-D, Kyowa Interface Science Co. Ltd., Saitama, Japan). Surface structural changes to the substrates following irradiation by the O_2 cluster and monomer ion beams and 1.5SBF soaking were investigated with thin-film X-ray diffractometry (TF-XRD; RINT-2500, Rigaku Corp., Tokyo, Japan), X-ray photoelectron spectroscopy (XPS; MT-5500, ULVAC-PHI, Inc., Kanagawa, Japan), Fourier-transform infrared attenuated total reflection spectroscopy (FT-IR ATR; Nicolet Magna 860, Thermo Electron K.K., Kanagawa, Japan) and field emission scanning electron microscopy (FE-SEM; S-4500, Hitachi Ltd., Tokyo, Japan). For XPS, MgKα (1253.6 eV) X-rays were used as an excitation source at the residual pressure of 10^{-9} Pa, and the photoelectron take-off angle (the angle between the sample surface and the detector axis) was set at 45°. For FT-IR ATR, the incident angle was set at 45° and a Ge crystal was used as an internal reflection element.

RESULTS AND DISCUSSION

Table II summarizes the contact angles of PE substrates unirradiated and irradiated by O_2 cluster ion beams, O_2 monomer ion beams and the simultaneous use of O_2 cluster and monomer ion beams. The acceleration voltage and dose were 7 kV and 1×10^{15} ions/cm², respectively. Unirradiated PE substrate gave a high contact angle of 100°, but the contact angle was decreased to around 80° and 45° by irradiation with O_2 cluster ion beams and O_2 monomer ion beams, respectively. On the other hand, with the simultaneous irradiation of O_2 cluster and monomer ion beams, the contact angle was decreased drastically to around 10°. These results indicate that the simultaneous use of O_2 cluster and monomer ion beams is more effective for inducing hydrophilicity on the PE surface than the use of O_2 cluster ion beams or O_2 monomer ion beams.

Figure 1 shows C_{1s} XPS spectra of PE substrates unirradiated (a) and irradiated with O_2 cluster ion beams (b), O_2 monomer ion beams (c) and the simultaneous use of O_2 cluster and monomer ion beams (d). The acceleration voltage and dose were 7 kV and 1×10^{15} ions/cm², respectively. The unirradiated PE substrates gave peak assigned to the ethylene chain (-$(CH_2)_n$-) at 284.6 eV as shown in Fig. 1 (a), and the irradiated ones gave a new peak assigned to ≡C-OH around 286.5 eV [5,27,28] as shown in Figs. 1 (b)-(d). The PE substrate irradiated with the simultaneous use of O_2 cluster and monomer ion beams gave a higher intensity of the ≡C-OH peak than did those irradiated by the O_2 cluster ion beams or O_2 monomer ion beams. Also, the full width at half maximum of the ≡C-OH peak became larger by the simultaneous use of O_2

cluster and monomer ion beams than the O_2 cluster ion beam or O_2 monomer ion beam irradiations. This suggests that many chemical states besides $\equiv$C-OH group were formed on the PE surface by the simultaneous use of O_2 cluster and monomer ion beams, although the detailed surface structural change due to the ion beam irradiation should be investigated in future. In addition, a small peak assigned to -COOH (around 290 eV) [16,27,28] was newly observed after the simultaneous irradiation with O_2 cluster and monomer ion beams. The surface structural change produced by the simultaneous irradiation of O_2 cluster and monomer ion beams can be interpreted as follows. The chemical bonds of PE were partially broken by O_2 monomer ions to form dangling bonds and, at the same time, a large number of oxygen atoms were supplied to the dangling bonds by O_2 cluster ions to form hydrophilic groups. The formation of these hydrophilic groups might be responsible for the low contact angle of the irradiated PE substrate shown in Table II. Therefore, the simultaneous use of O_2 cluster and monomer ion beams was employed in the following experiments.

Figure 2 shows C_{1s} XPS spectra of PE substrates irradiated by the simultaneous use of O_2 cluster and monomer ion beams at different acceleration voltages of 3 kV (a) and 7 kV (b) and a dose of 1×10^{15} ions/cm^2. The intensity of the $\equiv$C-OH peak was larger at the acceleration voltage of 7 kV than at 3 kV. In addition, the -COOH peak was observed only for the acceleration voltage of 7 kV. This indicates that the number of hydrophilic functional groups increased with increasing acceleration voltage, and the surface oxidation was enhanced by irradiation at a higher acceleration voltage.

Figure 3 shows TF-XRD patterns (a) and FT-IR ATR spectra (b) of PE substrates irradiated with the simultaneous use of O_2 cluster and monomer ion beams at acceleration voltages of 3, 6 and 9 kV and a dose of 1×10^{15} ions/cm^2, and then soaked in 1.5SBF for 7 days. In the TF-XRD patterns, no significant change was found after soaking in 1.5SBF, irrespective of the acceleration voltage. However, in the FT-IR ATR spectra, we found some peaks assigned to PO_4^{3-} ions in apatite [29-32] at 550, 650, 1050 and 1120 cm^{-1} beside the peak assigned to C-H bond in PE at 1460 cm^{-1} for the irradiated substrates, whereas no change was observed for the unirradiated substrate. The intensity of the apatite peak was highest at 6 kV. These results indicate that the simultaneous use of O_2 cluster and monomer ion beams has the potential to induce apatite-forming ability on the PE surface, and that an optimum acceleration voltage is present for apatite formation. The difference in the apatite-forming ability between the substrates irradiated at different acceleration voltages can be interpreted as follows. The irradiation at a low acceleration voltage of 3 kV is not able to form sufficient functional groups to induce apatite nucleation on the PE surface, as shown in Fig. 2 (a). On the other hand, irradiation at a high

Table II. Contact angles of PE substrates unirradiated and irradiated with O_2 cluster ion beams, O_2 monomer ion beams, and the simultaneous use of O_2 cluster and monomer ion beams at an acceleration voltage of 7 kV and a dose of 1×10^{15} ions/cm^2.

Substrates	Contact angle / degree
Unirradiated	100
Irradiated with O_2 cluster ion beams	82
Irradiated with O_2 monomer ion beams	46
Irradiated with simultaneous use of O_2 cluster and monomer ion beams	11

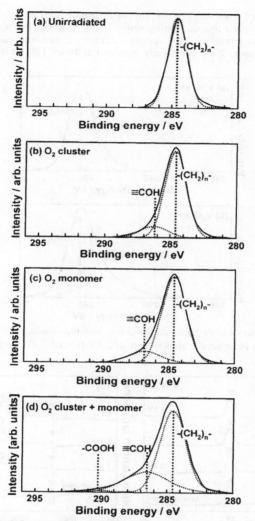

Figure 1. C_{1s} XPS spectra of PE substrates unirradiated (a) and irradiated with O_2 cluster ion beams (b), O_2 monomer ion beams (c), and the simultaneous use of O_2 cluster and monomer ion beams (d) at an acceleration voltage of 7 kV and a dose of 1×10^{15} ions/cm^2.

acceleration voltage of 9 kV might cause structural damage such as carbonization on the PE surface, and hence a good apatite-forming ability is not obtained. In fact, we found that the contact angle of the irradiated PE substrates decreased monotonically with increasing acceleration voltage up to 7 kV but slightly increased at 10 kV [18]. It is considered that the

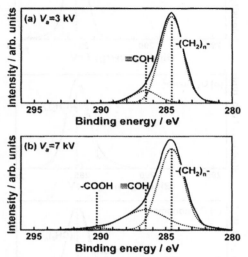

Figure 2. C$_{1s}$ XPS spectra of PE substrates irradiated with the simultaneous use of O$_2$ cluster and monomer ion beams at an acceleration voltage of 3 kV (a) and 7 kV (b) and a dose of 1×10^{15} ions/cm^2.

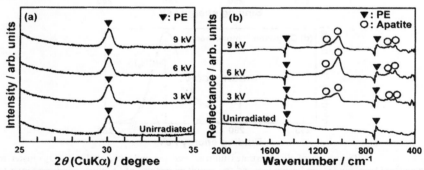

Figure 3. TF-XRD patterns (a) and FT-IR ATR spectra (b) of PE substrates irradiated with the simultaneous use of O$_2$ cluster and monomer ion beams at acceleration voltages of 3, 6 and 9 kV and a dose of 1×10^{15} ions/cm^2, and then soaked in 1.5SBF for 7 days.

surface structural damage caused by irradiation at high acceleration voltage might be responsible for the slight increase in the contact angle.

Figure 4 shows FE-SEM photographs of PE substrates irradiated with the simultaneous use of O_2 cluster and monomer ion beams at acceleration voltages of 3, 6 and 9 kV and a dose of 1×10^{15} ions/cm^2, and then soaked in 1.5SBF for 7 days. Apatite was formed on all the investigated substrates, but the apatite-forming ability was highest at the acceleration voltage of 6 kV. This result coincides well with the result of the FT-IR ATR measurement in Fig. 3 (b).

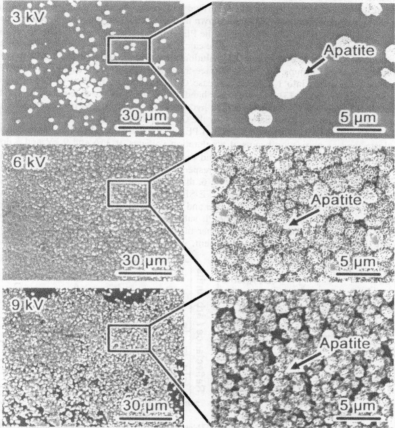

Figure 4. FE-SEM photographs of PE substrates irradiated with the simultaneous use of O_2 cluster and monomer ion beams at acceleration voltages of 3, 6 and 9 kV and a dose of 1×10^{15} ions/cm^2, and then soaked in 1.5SBF for 7 days.

Figure 5 shows TF-XRD patterns (a) and FT-IR ATR spectra (b) of PE substrates irradiated with the simultaneous use of O_2 cluster and monomer ion beams at acceleration voltages of 3, 6 and 9 kV and a dose of 1×10^{15} ions/cm^2, subjected to $CaCl_2$ solution treatment and then soaked in 1.5SBF for 7 days. The $CaCl_2$ solution treatment greatly increased the intensity of the apatite peak in the FT-IR ATR spectra as shown in Fig. 5 (b), and apatite peaks at 26 and 32° in 2θ were observed clearly in the TF-XRD measurement as shown in Fig. 5 (a), irrespective of the acceleration voltage. On the other hand, the unirradiated substrate did not form apatite in 1.5SBF even with the $CaCl_2$ solution treatment. It can be seen from these results that apatite formation on the irradiated PE surface was greatly promoted by the $CaCl_2$ solution treatment. This is explained as follows. As shown in Figs. 1 and 2, hydrophilic functional groups such as $\equiv$C-OH and -COOH were formed on the PE surface by irradiation with the simultaneous use of O_2 cluster and monomer ion beams. The calcium (Ca^{2+}) ions were introduced successfully at the irradiated PE surface by the $CaCl_2$ solution treatment because of the formation of the hydrophilic functional groups. When the surface-treated substrates were soaked in 1.5SBF, the Ca^{2+} ions were released into 1.5SBF to increase the ionic activity product of the surrounding fluid with respect to apatite [31]. On the other hand, Ca^{2+} ions were not able to be introduced on the unirradiated PE surface, owing to its high hydrophobic character, and hence apatite was not formed on the unirradiated substrate even after the $CaCl_2$ solution treatment.

Figure 6 shows FE-SEM photographs of PE substrates irradiated by the simultaneous use of O_2 cluster and monomer ion beams at acceleration voltages of 3, 6 and 9 kV and a dose of 1×10^{15} ions/cm^2, treated in $CaCl_2$ solution and then soaked in 1.5SBF for 7 days. A uniform apatite layer was formed on the substrates, irrespective of the acceleration voltage. Comparison of Fig. 6 with Fig. 4 also confirms, from Fig. 6, that apatite formation was greatly promoted by the $CaCl_2$ solution treatment. As a reference, FE-SEM photographs of PE substrates subjected to O_2 plasma treatment, treated in $CaCl_2$ solution and then soaked in 1.5SBF for 7 days are shown in Fig. 7. The apatite-forming ability of the PE substrates irradiated by the simultaneous use of O_2 cluster and monomer ion beams was higher than that of substrates subjected to O_2 plasma treatment. In the case of O_2 plasma treatment, scission of the chemical bond results from

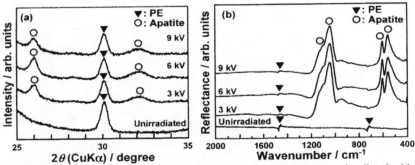

Figure 5. TF-XRD patterns (a) and FT-IR ATR spectra (b) of PE substrates irradiated with the simultaneous use of O_2 cluster and monomer ion beams at acceleration voltages of 3, 6 and 9 kV and a dose of 1×10^{15} ions/cm^2, treated in $CaCl_2$ solution and then soaked in 1.5SBF for 7 days.

collision with the electrons in the plasma. Generally, the electrons in the plasma have a wide range of kinetic energies [33,34]. Therefore, it is difficult to achieve uniform surface modification. In addition, the plasma-modified surface is not stable, and is liable to return to the original state. In contrast, uniform surface modification is easily achieved by ion beam irradiation, because the kinetic energy of the ions can be controlled precisely [11,35]. This might be responsible for the difference in apatite-forming ability between the O_2 ion beam-irradiated PE substrate and the O_2 plasma-treated PE substrate. It is desirable for the clinical application that the bone-like apatite is formed on the substrate surface in normal SBF, but unfortunately, the apatite was not formed on the irradiated surface in SBF in this study. We must further optimize the irradiation condition and obtain the apatite formation in SBF as well as 1.5SBF in future.

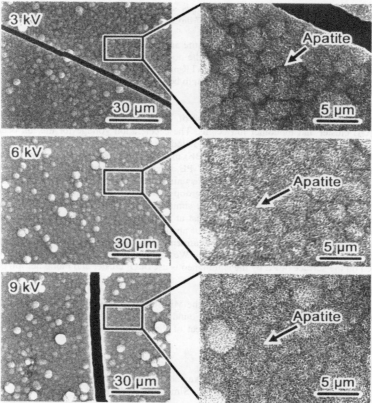

Figure 6. FE-SEM photographs of PE substrates irradiated with the simultaneous use of O_2 cluster and monomer ion beams at acceleration voltages of 3, 6 and 9 kV and a dose of 1×10^{15} ions/cm^2, treated in CaCl$_2$ solution and then soaked in 1.5SBF for 7 days.

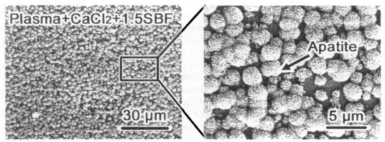

Figure 7. FE-SEM photographs of PE substrates subjected to O_2 plasma treatment, treated in $CaCl_2$ solution and then soaked in 1.5SBF for 7 days.

The adhesive strength between the apatite and the irradiated substrate is not investigated yet, but the apatite was not peeled-off under the normal handling such as placing the substrates on the sample holder in TF-XRD and FT-IR ATR measurements. In future, we will carry out the quantitative measurement of the adhesive strength between the apatite and the substrates.

CONCLUSION

PE substrates were irradiated at a dose of 1×10^{15} ions/cm^2 by the simultaneous use of oxygen (O_2) cluster and monomer ion beams. The acceleration voltage for the ion beams was varied from 3 to 9 kV. Unirradiated and irradiated PE substrates were soaked for 7 days in a metastable calcium phosphate solution (1.5SBF) that had 1.5 times the ion concentrations of a normal simulated body fluid. The irradiated PE substrates formed apatite on their surfaces, irrespective of the acceleration voltage, whereas unirradiated substrates did not form apatite. This is attributed to the formation of functional groups effective for apatite nucleation, such as -COOH groups, on the substrate surface by the simultaneous use of O_2 cluster and monomer ion beams. The apatite-forming ability was highest at an acceleration voltage of 6 kV, suggesting that an optimum acceleration voltage exists for apatite formation. In addition, apatite formation was enhanced greatly by subsequent $CaCl_2$ treatment. This suggests that Ca^{2+} ions present on the substrate surface accelerated apatite nucleation. We can conclude that apatite-forming ability can be induced on the surface of PE by the simultaneous use of O_2 cluster and monomer ion beams.

ACKNOWLEDGEMENTS

This work was partially supported by the Ministry of Education, Culture, Sports, Science and Technology, Japan, Nippon Sheet Glass Foundation for Materials Science and Engineering, Japan and the Nanotechnology Support Project of the Ministry of Education, Culture, Sports, Science and Technology, Japan.

REFERENCES
[1] P. Li, C. Ohtsuki, T. Kokubo, K. Nakanishi, N. Soga, T. Nakamura and T. Yamamuro, Apatite Formation Induced by Silica Gel in a Simulated Body Fluid, *J. Am. Ceram. Soc.*, **75**, 2094-7 (1992).
[2] P. Li, C. Ohtsuki, T. Kokubo, K. Nakanishi, N. Soga, T. Nakamura and T. Yamamuro,

A Role of Hydrated Silica, Titania and Alumina in Forming Biologically Active Apatite on Implant, *Trans. Fourth World Biomat. Cong.*, p. 4.

[3] M. Tanahashi and T. Matsuda, Surface Functional Group Dependence on Apatite Formation on Self-Assembled Monolayers in a Simulated Body Fluid. *J. Biomed. Mater. Res.*, **34**, 305-15 (1997).

[4] M. Tanahashi, T. Yao, T. Kokubo, M. Minoda, T. Miyamoto, T. Nakamura and T. Yamamuro, Apatite Coating on Organic Polymers by a Biomimetic Process, *J. Am. Ceram. Soc.*, **77**, 2805-8 (1994).

[5] M. Tanahashi, T. Yao, T. Kokubo, M. Minoda, T. Miyamoto, T. Nakamura and T. Yamamuro, Apatite Coated on Organic Polymers by a Biomimetic Process - Improvement in Its Adhesion to Substrate by Glow-Discharge Treatment, *J. Biomed. Mater. Res.*, **29**, 349-57 (1995).

[6] A. Oyane, M. Kawashita, T. Kokubo, M. Minoda, T. Miyamoto and T. Nakamura, Bone-Like Apatite Formation on Ethylene-Vinyl Alcohol Copolymer Modified with a Silane Coupling Agent and Titania Solution, *J. Ceram. Soc. Japan*, **110**, 248-54 (2002).

[7] A. Oyane, M. Kawashita, T. Kokubo, M. Minoda, T. Miyamoto and T. Nakamura, Bonelike Apatite Formation on Ethylene-Vinyl Alcohol Copolymer Modified with Silane Coupling Agent and Calcium Silicate Solutions, *Biomaterials*, **24**, 1729-35 (2003).

[8] T. Kokubo, M. Kawai, M. Kawashita, K. Yamamoto and T. Nakamura, Fabrics of Polymer Fibers Modified with Calcium Silicate for Bone Substitute, *Key Engin. Mater.*, **284-6**, 775-8 (2005).

[9] F. Balas, M. Kawashita, T. Nakamura and T. Kokubo, Formation of Bone-Like Apatite on Organic Polymers Treated with a Silane-Coupling Agent and a Titania Solution, *Biomaterials*, **27**, 1704-10 (2006).

[10] M. A. Duncan and D. H. Rouvray, Microclusters, *Sci. Am.*, **261**, 110-5 (1989).

[11] I. Yamada and G. H. Takaoka, Ionized Cluster Beams - Physics and Technology, *Jpn. J. Appl. Phys.*, **32**, 2121-41 (1993).

[12] H. Biederman, D. Slavinska, H. Boldyreva, H. Lehmberg, G. Takaoka, J. Matsuo, H. Kinpara and J. Zemek, Modification of Polycarbonate and Polypropylene Surfaces by Argon Ion Cluster Beams, *J. Vac. Sci. Tech.*, **B19**, 2050-6 (2001).

[13] Takaoka GH, Yamazaki D, Matsuo J. High quality ITO film formation by the simultaneous use of cluster ion beam and laser irradiation. Mater Chem Phys 2002;74:104–8.

[14] G. H. Takaoka, H. Noguchi, T. Yamamoto and T. Seki, Production of Liquid Cluster Ions for Surface Treatment, *Jpn. J. Appl. Phys.*, **42**, L1032-5 (2003).

[15] G. H. Takaoka, M. Kawashita, T. Omoto and T. Terada, Photocatalytic Properties of TiO_2 Films Prepared by O_2 Cluster Ion Beam Assisted Deposition Method, *Nucl. Instr. Meth. Phys. Res. B*, **232**, 200-5 (2005).

[16] G. H. Takaoka, H. Shimatani, H. Noguchi and M. Kawashita, Interactions of Argon Cluster Ion Beams with Silicon Surfaces, *Nucl. Instr. Meth. Phys. Res. B*, **232**, 206-11 (2005).

[17] H. Biederman and D. Slavinska, Ion Bombardment of Organic Materials and Its Potential Application, *Proc. 12th Int. Conf. Ion Imp. Tech. 98*, IEEE, 1999. p. 815-9.

[18] M. Kawashita, H. Shimatani, S. Itoh, R. Araki and G. H. Takaoka, Apatite Deposition on Polymer Substrates Irradiated by Oxygen Cluster Ion Beams, *J. Ceram. Soc. Jpn.*, **114**, 77-81 (2006).

[19] M. Kawashita, M. Nakao, M. Minoda, H.-M. Kim, T. Beppu, T. Miyamoto, T. Kokubo and T. Nakamura, Apatite-Forming Ability of Carboxyl Group-Containing Polymer Gels in a Simulated Body Fluid, *Biomaterials*, **24**, 2477-84 (2003).

[20] T. Kokubo, H.-M. Kim and M. Kawashita, Novel Bioactive Materials with Different Mechanical Properties, *Biomaterials*, **24**, 2161-75 (2003).

[21] A. Takeuchi, C. Ohtsuki, T. Miyazaki, M. Yamazaki and M. Tanihara, Deposition of Bone-Like Apatite on Silk Fibre in a Solution that Mimics Extracellular Fluid, *J. Biomed. Mater. Res.*, **65A**, 283-9 (2003).

[22] K. Hosoya, C. Ohtsuki, T. Kawai, M. Kamitakahara, S. Ogata, T. Miyazaki and M. Tanihara, A Novel Covalently Cross-Linked Gel of Alginate and Silane Having Ability to Form Bone-Like Apatite, *J. Biomed. Mater. Res.*, **71A**, 596-601 (2004).

[23] T. Kokubo, H. Kushitani, S. Sakka, T. Kitsugi and T. Yamamuro, Solutions Able to Reproduce *in vivo* Surface-Structure Change in Bioactive Glass-Ceramic A-W, *J. Biomed. Mater. Res.*, **24**, 721-34 (1990).

[24] S. B. Cho, K. Nakanishi, T. Kokubo, N. Soga, C. Ohtsuki, T. Nakamura, T. Kitsugi and T. Yamamuro, Dependence of Apatite Formation on Silica Gel on Its Structure: Effect of Heat Treatment, *J. Am. Ceram. Soc.*, **78**, 1769-74 (1995).

[25] T. Kokubo and H. Takadama, How Useful is SBF in Predicting *in vivo* Bone Bioactivity? *Biomaterials*, **27**, 2907-15 (2006).

[26] J. E. Gamble, Chemical Anatomy, Physiology and Pathology of Extracellular Fluid, Cambridge, MA: Harvard University Press; 1967. p. 1-17.

[27] D. T. Clark, B. J. Cromarty and A. Dilks, Theoretical Investigation of Molecular Core Binding and Relaxation Energies in a Series of Oxygen-Containing Organic Molecules of Interest in the Study of Surface Oxidation of Polymers, *J. Polym. Sci. Part A: Polym. Chem.*, **16**, 3173-84 (1978).

[28] H. Iwata, A. Kishida, M. Suzuki, Y. Hata and Y. Ikada, Oxidation of Polyethylene Surface by Corona Discharge and the Subsequent Graft-Polymerization, *J. Polym. Sci. Part A: Polym. Chem.*, **26**, 3309-22 (1988).

[29] L. L. Hench, Bioceramics - from Concept to Clinic, *J. Am. Ceram. Soc.*, **74**, 1487-510 (1991).

[30] C. Ohtsuki, T. Kokubo, K. Takatsuka and T. Yamamuro, Compositional Dependence of Bioactivity of Glass in the System $CaO-SiO_2-P_2O_5$: Its *in vitro* Evaluation, *J. Ceram. Soc. Japan*, **25**, 1-6 (1991).

[31] C. Ohtsuki, T. Kokubo and T. Yamamuro, Mechanism of Apatite Formation on $CaO-SiO_2-P_2O_5$ Glasses in a Simulated Body Fluid, *J. Non-Cryst. Solids*, **143**, 84-92 (1992).

[32] H.-M. Kim, F. Miyaji, T. Kokubo, C. Ohtsuki and T. Nakamura, Bioactivity of $Na_2O-CaO-SiO_2$ Glasses, *J. Am. Ceram. Soc.*, **78**, 2405-11 (1995).

[33] J. K. Olthoff, R. J. Van Brunt and S. B. Radovanov, Ion Kinetic-Energy Distributions in Argon Rf Glow Discharges, *J. Appl. Phys.*, **72**, 4566-74 (1992).

[34] J. K. Olthoff, R. J. Van Brunt, S. B. Radovanov, J. A. Rees and R. Surowiec, Kinetic-Energy Distributions of Ions Sampled from Argon Plasmas in a Parallel-Plate, Radio-Frequency Reference Cell, *J. Appl. Phys.*, **75**, 115-25 (1994).

[35] H. Dong and T. Bell, State-of-the-Art Overview: Ion Beam Surface Modification of Polymers towards Improving Tribological Properties, *Surf. Coat. Tech.*, **111**, 29-40 (1999).

CONVERSION OF BORATE GLASS TO BIOCOMPATIBLE PHOSPHATES IN AQUEOUS PHOSPHATE SOLUTION

Mohamed N. Rahaman[1*], Delbert E. Day[1,2], and Wenhai Huang[2,3]
[1]Department of Materials Science and Engineering, University of Missouri-Rolla, Rolla, MO 65409
[2]Graduate Center for Materials Research, University of Missouri-Rolla, Rolla, MO 65409
[3]Institute of Bioengineering and Information Technology Materials, Tongji University, Shanghai, 200092, China

ABSTRACT

Special compositions of CaO-containing borate glass have been shown to react with an aqueous phosphate solution at near room temperature to form hydroxyapatite (HA), $Ca_{10}(PO_4)_6(OH)_2$, the main mineral constituent of bone. In the present work, the formation of other biocompatible phosphates was investigated by converting borate glass containing BaO or both BaO and CaO. The glass compositions were based on previously-studied CaO-containing borate glass by partially or fully replacing the CaO with BaO. The kinetics of the conversion reaction, as well as the composition and structure of the products formed from these BaO-containing glasses were investigated. Borate glass particles (150–300 µm) with the required composition were prepared by conventional glass processing methods and placed in 0.25M K_2HPO_4 solution at 37°C and with a starting pH value of 9.0 or 12.0. At both pH values, the CaO-containing glass particles were converted to HA, with fine needle-like crystals and high surface area (160–170 m^2/g). The BaO-containing glass converted to $BaHPO_4$ at pH = 9.0 and to $Ba_3(PO_4)_2$ at pH = 12.0, with both products having plate-like crystals and much lower surface areas (4–8 m^2/g). The as-formed product from a glass containing both CaO and BaO was amorphous and, on calcination for 8 hours at 700°C, yielded a mixture of two phosphates, a barium phosphate $Ba_3P_4O_{13}$ and tricalcium phosphate $Ca_3(PO_4)_2$.

INTRODUCTION

When placed in an aqueous phosphate solution at near room temperature, special compositions of CaO-containing silicate or borate glass react to form a calcium phosphate material, typically hydroxyapatite (HA), $Ca_{10}(PO_4)_6(OH)_2$, the main mineral constituent of bone [1-6]. A well-recognized example is the silicate-based glass designated 45S5, referred to as Bioglass® (composition 45% SiO_2; 24.5% Na_2O; 24.5% CaO; 6 % P_2O_5; by weight), reported by Hench in 1971 [7], which has received considerable attention for biomedical applications [2,3,8]. However, 45S5 glass and related silicate-based glass compositions convert slowly and incompletely to HA [5,6]. Because of their lower chemical durability, some CaO-containing borate glasses convert faster and completely to HA [5,6,9-11], so they provide more useful starting materials for the production of HA objects. A borate glass, designated 45S5B1, with the same composition as 45S5 but with all the SiO_2 replaced with B_2O_3, was found to be completely converted to HA in less than 4 days when particles (150–300 µm) were placed in dilute K_2HPO_4 solution (0.02–0.25M) with a pH of 7–9 at 37°C [5,6]. In comparison, silicate-based 45S5 glass particles with the same size were only partially converted (<50% by weight) after more than 10 weeks under similar conditions, when the conversion to HA effectively stopped [5,6].

The mechanism of converting silicate-based 45S5 glass to HA in aqueous phosphate solution has been discussed by Hench [2,3]. The conversion of borate-based glass to HA follows a

171

mechanism similar to that for 45S5 glass, but without the formation of an initial SiO_2-rich layer that separates the growing product layer from the shrinking, unconverted glass core [5,6,10]. The conversion mechanism for borate glass in an aqueous phosphate solution is illustrated schematically in Figure 1 for a composition in the system Na_2O-CaO-B_2O_3. Briefly, ions such as Na and B dissolve from the glass into the solution, and the reaction between the Ca ions from the glass and the phosphate ions from the solution lead to the formation of HA. The reaction starts at the surface of the glass and moves inward. An important feature is the pseudomorphic nature of the conversion reaction, with the product retaining the size and shape of the original glass object.

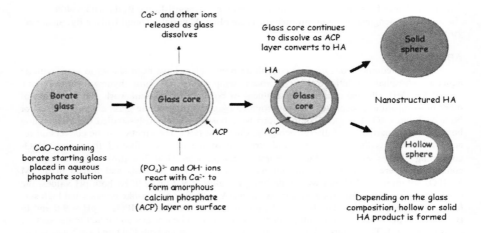

Figure 1. Schematic diagram illustrating the conversion of a CaO-containing borate glass microsphere to hydroxyapatite (HA).

Based on the ability to completely convert CaO-containing borate glass with special compositions to HA in aqueous phosphate solution, the objective of the present work was to investigate whether other alkali-earth phosphates can be prepared by a similar route. The focus of the investigation was on the products formed from borate glass containing BaO or a combinationn of BaO and CaO. These glass compositions were based on previously-studied CaO-containing borate glass by partially or fully replacing the CaO with BaO. The kinetics of the conversion reaction, as well as the chemical composition and structure of the products were studied. The results were compared with those for HA formed from the corresponding CaO-containing borate glass under similar reaction conditions.

EXPERIMENTAL PROCEDURE

Preparation of Glasses

The glass compositions used in the present work are given in Table 1. The first BaO-containing borate glass with the composition $20Na_2O \cdot 20BaO \cdot 60B_2O_3$ (mole percent, mol%), designated 226Ba, was based on a CaO-containing composition: $20Na_2O \cdot 20CaO \cdot 60B_2O_3$ (mol%), designated 226Ca. Previous work [11,12] showed that 226Ca glass particles (90–425 μm) were completely converted to hydroxyapatite (HA) in less than 7 days when reacted in 0.25M K_2HPO_4 solution at 37°C and with a starting pH value of 9.0. The second BaO-containing borate glass, with the composition $10Li_2O \cdot 10CaO \cdot 10BaO \cdot 70B_2O_3$ (mol%), designated 1117Ba, was based on compositions in the system $Li_2O \cdot CaO \cdot B_2O_3$, designated CaLB3 in previous work [13,14]. Depending on the composition, microspheres of the CaLB3 glass 90–106 μm) were completely converted to HA in less than 2–4 days when reacted in 0.25M K_2HPO_4 solution at 37°C and with a starting pH value of 9.0.

Table I. Compositions of Borate Glasses and Characteristics of the As-formed Products Obtained by Conversion in an Aqueous Phosphate Solution (0.25M at 37°C)

Composition (mol%)	226Ca	226Ba	1117Ba
Li_2O	0	0	10.0
Na_2O	20.0	20.0	0
CaO	20.0	0	10.0
BaO	0	20.0	10.0
B_2O_3	60.0	60.0	70.0
As-formed product[1]	HA (pH = 9.0)	$BaHPO_4$ (pH = 9.0)	Amorphous (pH = 9.0)[2]
	HA (pH = 12.0)	$Ba_3(PO_4)_2$ (pH = 12.0)	Amorphous (pH = 12.0)[2]
Measured weight loss (wt%)	73 (pH = 9.0)	59 (pH = 9.0)	65 (pH = 9.0)
	70 (pH = 12.0)	58 (pH = 12.0)	57 (pH = 12.0)
Theoretical weight loss (wt%)[3]	69.3	45.0 (pH = 9.0)	68.4[4]
		52.7 (pH = 12.0)	
Surface area (m^2/g)	170 (pH = 9.0)	8 (pH = 9.0)	18 (pH = 9.0)
	160 (pH = 12.0)	4 (pH = 12.0)	9 (pH = 12.0)

[1]Identified by X-ray diffraction (XRD); [2]Calcination (1 h at 700°C) yields $Ca_3(PO_4)_2$ and $Ba_3P_4O_{13}$; [3]Based on as-formed products identified by XRD; [4]Based on composition of calcined products identified by XRD

The glasses were prepared by melting the required quantities of H_3BO_3, $CaCO_3$, Na_2CO_3, Li_2CO_3, and $BaCO_3$ (Reagent grade; Fisher Scientific; St. Louis, MO) in a platinum/rhodium crucible in air for 1–2 h at 1100°C. The melt was stirred twice at the isothermal temperature and quenched between two steel plates. Each glass was crushed in a hardened steel mortar and pestle to form particles, which were sieved through stainless steel sieves to give particle sizes in the range of 150–300 μm.

Glass Conversion Reaction

The solution used in the experiments to study the conversion of the glass particles to the alkali-earth phosphates was prepared by dissolving K_2HPO_4 (Reagent grade; Fisher Scientific, St. Louis MO) in deionized water (pH = 5.5 ± 0.1) to give a concentration of 0.25M. This concentration was used in previous studies with other glasses [4-6,9-11]. The starting pH value of the solution was adjusted to the required values of 9.0 ± 0.1 or 12.0 ± 0.1 by adding NH_4OH or KOH solution. These basic pH values were chosen primarily to provide stable reaction conditions for studying the conversion of the glasses to the alkali-earth metal phosphates. The mass of glass particles and the volume of the K_2HPO_4 solution (1 g glass in 100 cm^3 solution) in the conversion reaction, used in previous work [5,6], were chosen to ensure that sufficient PO_4^{3-} ions would be available in the solution to react with all the Ca^{2+} or Ba^{2+} ions from the glass to form the phosphate compounds.

The conversion of borate glass to alkali-earth metal phosphates in dilute phosphate solution is accompanied by a weight loss of the particles and a change in the pH value of the solutions, so these parameters were used to monitor the kinetics of the conversion reaction. A system consisting of 1 g of glass in 100 cm^3 solution was used in all the conversion experiments. The procedures for measuring the weight loss and pH as functions of time are described in detail elsewhere [5,6].

Structural and Chemical Characterization of the Reaction Products

Structural and compositional changes in the glasses resulting from the reaction with the phosphate solution were monitored using several techniques. Scanning electron microscopy (SEM, Hitachi S-4700) was used to observe the structure of the reaction products. The composition of the crystalline reaction products was determined using X-ray diffraction (XRD; Scintag 2000) using Cu K_α radiation (λ = 0.15406 nm) in a step-scan mode (0.05° per step) in the range of 10–80° 2θ. X-ray fluorescence (XRF; XEPOS, Spectro Analytic, Marbough, MA) was used to determine the chemical composition of the glass and the as-formed reaction products. The surface areas of the final reaction products were measured using N_2 adsorption at the boiling point of liquid N_2, by the Brunauer-Emmett-Teller (BET) method (Nova–1000 BET; Quantachrome, Boynton Beach, FL).

RESULTS AND DISCUSSION

Conversion of 226Ba Glass

Figure 2 shows data for the fractional weight loss, ΔW, of 226Ba glass particles as a function of the reaction time, t, in aqueous phosphate solution with a starting pH value of 9.0. ($\Delta W = (W_o - W)/W_o$, where W_o is the initial mass and W is the mass at time t.) For comparison, data for 226Ca glass particles are also shown. For each glass, ΔW increased with time, eventually reaching a final limiting value, ΔW_{max}. The measured value of ΔW_{max} was 73 wt% for the 226Ca glass and 59 wt% for the 226Ba glass. Data for the pH value of the phosphate solution during the conversion reaction are shown in Figure 2(b). The pH increased with time, eventually reaching a nearly constant value. The final limiting pH value was slightly higher for the solution reacted with the 226Ba glass (final pH = 9.8) than for the solution reacted with the 226Ca glass (final pH

= 9.4). In general, the pH changes in the solution were small, so they were not expected to have a significant effect on the formation of the phosphate products. The conversion reaction of the 226Ca and 226Ba glasses in a 0.25 M K_2HPO_4 solution with a starting pH of 12.0 was also studied, but the results are omitted for the sake of brevity. In general, the ΔW data showed similar trends to those in Figure 2(a). The values for ΔW_{max} are summarized in Table 1. The pH value of the solution decreased slightly from the starting value, to 11.8 for the 226Ba glass and 11.2 for the 226Ca glass.

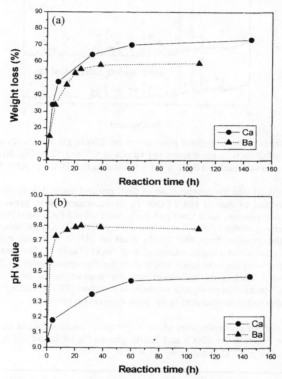

Figure 2. (a) Weight loss of the glass particles and (b) pH of the aqueous phosphate solution vs. time, for conversion of 226Ca and 226Ba glasses in a solution with a starting pH = 9.0.

While the data in Figure 2 showed no drastic difference in the conversion kinetics for the 226Ca and 226Ba glasses, the final products showed interesting structural and compositional differences. At both pH values, the 226Ca glass was converted to HA, as determined by XRD, in accordance with previous work [11,12]. On the other hand, according to XRD (Figure 3), at pH = 9.0 the 226Ba glass was converted to a barium phosphate product that was isostructural with

BaHPO₄ (JCPDS 44-0743), and at pH = 12.0, to a barium phosphate that was isostructural with $Ba_3(PO_4)_2$ (JCPDS 25-0028).

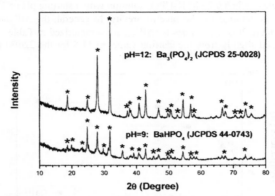

Figure 3. XRD patterns of the as-formed products of the 226Ba glass particles after reaction for 108 h in phosphate solution with for starting pH values of 9.0 and 12.0. The diffraction patterns correspond to those of a standard BaHPO₄ at pH = 9.0 and to $Ba_3(PO_4)_2$ at pH = 12.0.

XRF data showed that the composition of final product formed from the 226Ca glass at pH = 12.0 corresponded to that of HA (Table 2), in accordance with previous observations [11,12]. The major components were CaO and P_2O_5, with a CaO/P_2O_5 ratio (by weight) of 1.3, which is almost identical to the value for stoichiometric HA (1.32). On the other hand, the XRF data showed that the product from the 226Ba glass at pH = 12.0 was not stoichiometric $Ba_3(PO_4)_2$, but that it contained a high concentration of Na_2O (Table 2). The measured BaO/P_2O_5 ratio, equal to 2.4 (by weight), was much lower than the theoretical value (3.24) for tribarium phosphate, $Ba_3(PO_4)_2$. This is an indication that the product might be a Na_2O-substituted $Ba_3(PO_4)_2$. While substitution in phosphate minerals is common [15,16], further work is required to understand the substitution mechanism in the glass conversion process.

Table 2. XRF data for the concentrations of the major oxide components in the as-formed final products obtained by reacting the 226Ca and 226Ba glasses for 145h and 108 h, respectively, in 0.25M K_2HPO_4 solution at 37°C and starting pH = 12.0.

Glass	Product (wt%)			
	Na_2O	CaO	BaO	P_2O_5
226Ca	-	54	-	41
226Ba	10	-	61	25

If it is assumed that all the alkali-earth oxide in the starting glass is converted completely to the stoichiometric phosphate compound, then a theoretical weight loss for the conversion reaction can be determined. For the 226Ca glass, both the XRD and XRF data indicated the formation of HA. The theoretical weight loss for converting the 226Ca glass to HA is 69.3%, which is close to the measured values of 73% at pH = 9.0 and 70% at pH = 12.0 (Table 1). In the case of the 226Ba glass, the XRD data indicated the formation of a compound that is isostructural with $Ba_3(PO_4)_2$ at pH = 12.0. Assuming the formation of stoichiometric $Ba_3(PO_4)_2$, the theoretical weight loss is 52.5%, which is somewhat lower than the measured value of 58%. A reason for the difference between the measured weight loss and the theoretical weight loss might be the formation of an Na_2O-substituted $Ba_3(PO_4)_2$, indicated by the XRF data.

Figure 4(a) and Figure 4(b) show SEM micrographs of the surfaces of the as-formed final products of the 226Ca glass and the 226Ba glass after reaction for 145 h and 108 h, respectively, in the phosphate solution with a starting pH value of 12.0. The product (HA) formed from the 226Ca glass consisted of fine, needle-like crystals, whereas the product from the 226Ba glass consisted of much coarser plate-like crystals. In general, the products formed at pH = 9.0 had a similar morphology to those formed at pH = 12.0, but the crystals were finer. The difference in morphology between the products formed from the 226Ca glass and the 226Ba glass is reflected in the measured surface areas (Table 1). At both pH values (9.0 and 12.0), the surface areas of the HA formed from the 226Ca glass (160–170 m^2/g) were much larger than those for the phosphate products formed from the 226Ba glass (4–8 m^2/g).

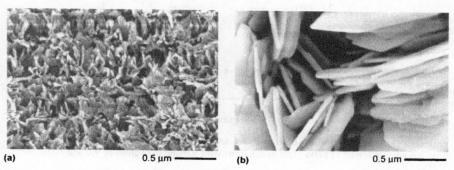

(a)   0.5 µm ———   **(b)**   0.5 µm ———

Figure 4. SEM of the surfaces of the as-formed phosphate products of (a) the 226Ca glass reacted for 145 h at pH = 12.0, and (b) the 226Ba glass reacted for 108 h at pH = 12.0.

Conversion of 1117Ba Glass

Figure 5(a) shows data for the fractional weight loss, ΔW, of 1117Ba glass particles as a function of reaction time, t, in aqueous phosphate solution with starting pH values of 9.0 and 12.0. The pH value of the solution had little effect on the kinetics of the conversion reaction, but the maximum observed weight loss ΔW_{max} at pH = 9.0 (65%) is higher than that at pH = 12.0 (57%). The pH value of the phosphate solution as a function of reaction time is shown in Figure 5(b) for the two starting pH values. The data show trends for similar to those described earlier for the 226Ca and 226Ba glasses, increasing slightly for the starting pH of 9.0 but decreasing slightly for the starting pH of 12.0.

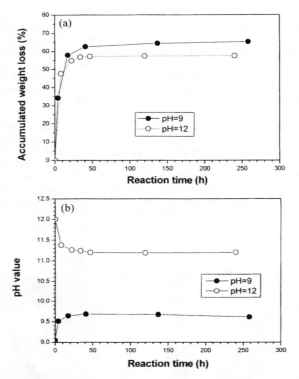

Figure 5. (a) Weight loss of the glass particles and (b) pH of the aqueous phosphate solution vs. time, for conversion of 1117Ba glass in solutions with starting pH values of 9.0 and 12.0.

XRD of the as-formed products, after reaction for ~250 h in the phosphate solution at pH = 9.0 and at pH = 12.0 showed that the products were amorphous (Figure 6). This is in contrast to the crystalline HA and barium phosphate products formed from the 226Ca and 226Ba glasses at even shorter reaction times. It appears that the presence of more than one alkali-earth metal oxide in the glass hinders the formation of a crystalline product during the conversion process. After calcination of the as-formed material (8h at 700°C in air), XRD revealed the formation of a crystalline product consisting of a mixture of two phosphates (Figure 6), identified as tricalcium phosphate $Ca_3(PO4)_2$ (JCPDS 70-2065) and $Ba_3P_4O_{13}$ (JCPDS 79-1530).

SEM revealed that the as-formed product, obtained after reaction for 258 h in the phosphate solution at pH = 9.0, consisted of a mass of agglomerated primary particles (Figure 7(a)). Each agglomerate was composed of a few nearly spherical primary particles with a uniform size of ~500 nm. After calcination, the microstructure of the product consisted of a mixture of two distinct particle morphologies (Figure 7(b)). The more dominant microstructural feature consisted of an interconnected network of coarser particles, while the less dominant feature consisted of agglomerates of much finer particles. The surface area of the as-formed amorphous product, obtained after reaction for 216 h in the phosphate solution at pH = 9.0 was 18 m^2/g. The nearly equiaxial (or spherical) particles formed from the 1117Ba glass have a distinctly different morphology from the needle-like HA crystals formed from Li_2O-CaO-B_2O_3 (CaLB3) glass [13,14] and 226Ca glass, and from the plate-like crystals formed from 226Ba glass (Figure 4). Further work is required to understand the structural development of borate glasses containing more than one alkali-earth metal oxide, such as 1117Ba glass.

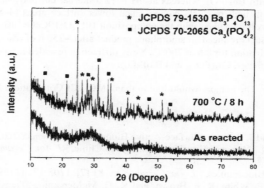

Figure 6. XRD of the as-formed product of the 1117Ba glass after reaction in phosphate solution for 258 h at pH = 9.0, and of the calcined product heated for 8 h at 700°C.

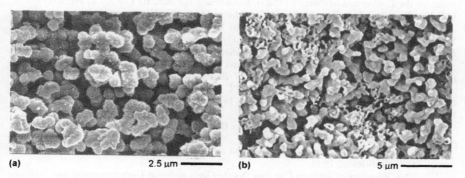

Figure 7. SEM of (a) the as-formed product of the 1117Ba glass after reaction in the phosphate solution for 258 h at pH = 9.0, and (b) the calcined product, heated for 8 h at 700°C.

CONCLUSIONS

A method was investigated for the formation of phosphates of Ca and Ba at near room temperature by the direct conversion of borate glass in 0.25M K_2HPO_4 solution at pH = 9.0 or 12.0. Borate glass particles (150–300 μm), with the composition $20Na_2O \cdot 20AO \cdot 60B_2O_3$ (mole percent), where A is Ca or Ba, were almost completely converted to the alkali-earth phosphate in less than 3 days. The alkali-earth metal in the glass and the starting pH of the solution had little effect on the conversion rate, but they had a significant effect on the structure and composition of the phosphate product. At pH = 9.0 and pH = 12.0, the calcium borate glass yielded hydroxyapatite, $Ca_{10}(PO_4)_6(OH)_2$ with a high surface area (160–170 m^2/g), and nanoscale needle-like crystals. On the other hand, at pH = 9.0, the barium borate glass converted to a barium phosphate that was isostructural with $BaHPO_4$, whereas at pH = 12.0 the barium phosphate was isostructural with $Ba_3(PO_4)_2$, with both products having much lower surface areas (4–8 m^2/g) and coarser plate-like crystals. A borate glass with the composition $10Li_2O \cdot 10CaO \cdot 10BaO \cdot 70B_2O_3$, containing both CaO and BaO produced an amorphous product after ~250 h of reaction in the phosphate solution. Upon calcination for 8 h at 700°C, X-ray diffraction revealed a crystalline product consisting of two phosphates, $Ca_3(PO_4)_2$ and $Ba_3P_4O_{13}$.

Acknowledgement: The authors would like to thank MO-SCI Corporation, Rolla, MO, for XRF analysis.

REFERENCES

[1]L. L. Hench and J. Wilson, "Surface-Active Biomaterials," *Science*, 226, 630-636 (1984).

[2]L. L. Hench LL, "Bioceramics: From Concept to Clinic," *J. Am. Ceram. Soc.*, 74 [7], 1487-1510 (1991).

[3]L. L. Hench, "Bioceramics," *J. Am. Ceram. Soc.*, 81 [7], 1705-1728 (1998).

[4]D. E. Day, J. E. White, R. F. Brown, and K. D. McMenamin, "Transformation of Borate Glasses into Biologically Useful Materials," *Glass Technol.*, 44 [2], 75-81 (2003).

[5]W. Huang, D. E. Day, K. Kittiratanapiboon, and M. N. Rahaman, "Kinetics and Mechanisms of the Conversion of Silicate (45S5), Borate, and Borosilicate Glasses to Hydroxyapatite in Dilute Phosphate Solutions," *J. Mater. Sci: Mater. Med.*, 17, 583-596 (2006).

[6]W. Huang, M. N. Rahaman, D. E. Day, and Y. Li, "Mechanisms for Converting Bioactive Silicate, Borate, and Borosilicate Glasses to Hydroxyapatite in Dilute Phosphate Solutions, *Phys. Chem. Glasses: Europ. J. Glass. Sci. Technol. Part B*, 2006; in press.

[7]L. L. Hench, R. J. Splinter, W. C. Allen, and T. K. Greenlee, Jr., "Bonding Mechanisms at the Interface of Ceramic Prosthetic Materials," *J. Biomed. Mater Res.*, 2, 117-141 (1971).

[8]L. L. Hench, "Bioactive Glasses and Glass Ceramics: A Perspective," in: T. Yamamuro, L. L. Hench, and J. Wilson, Eds., *Handbook of Bioactive Ceramics*, Vol. 1. CRC Press, Boca Raton, FL, 1990, pp. 7-24.

[9]M. N. C. Richard, Bioactive Behavior of a Borate Glass, M.S. Thesis, University of Missouri-Rolla, 2000.

[10]J. A. Wojcik, Hydroxyapatite Formation on a Silicate and Borate Glass, M.S. Thesis, University of Missouri-Rolla, 1999.

[11]X. Han, Reaction of Sodium Calcium Borate Glass to Form Hydroxyapatite and Preliminary Evaluation of Hydroxyapatite Microspheres Used to Absorb and Separate Proteins, M.S. Thesis, University of Missouri-Rolla, 2003.

[12]M. N. Rahaman, W. Liang, D. E. Day, N. Marion, G. Reilly, and J. J. Mao, "Preparation and Bioactive Characteristics of Porous Borate Glass Substrates," *Ceram. Eng. Sci. Proc.*, **26** [6] 3-10 (2005).

[13]K. P. Fears, Formation of Hollow Hydroxyapatite Microspheres," M.S. Thesis, University of Missouri-Rolla, 2001.

[14]S. D. Conzone, "Glass Microspheres for Medical Applications," Ph.D. Thesis, University of Missouri-Rolla, 1999.

[15]A. S. Posner, N.C. Blumenthal, and F. Betts, "Chemistry and Structure of Precipitated Hydroxyapatites," in J. O. Nriagu and P. B. Moore, Eds., *Phosphate Minerals*, Springer-Verlag, Berlin, 1984, pp. 331-350.

[16]R. Z. LeGeros and J. P. LeGeros, "Phosphate Minerals in Human Tissues," in J. O. Nriagu and P. B. Moore, Eds., *Phosphate Minerals*, Springer-Verlag, Berlin, 1984, pp. 351-385.

FABRICATION OF NANO-MACRO POROUS SODA-LIME PHOSPHOSILICATE BIOACTIVE GLASS BY THE MELT-QUENCH METHOD

Hassan M. M. Moawad[1] and Himanshu Jain[2]
Department of Materials Science and Engineering
Lehigh University, Bethlehem, PA 18015

ABSTRACT

We have extended the usefulness of the 24.5CaO - (27.5-x)Na$_2$O - 6P$_2$O$_5$ - (42+x)SiO$_2$ (wt%) (with x=0-10) bioactive glass series by introducing engineered porosity in bulk samples. The glasses are prepared by the melt-quench method with an amorphous phase separated microstructure that varies with SiO$_2$ fraction. For 42-43 and 50-52 wt% SiO$_2$, the phase separation occurs by nucleation and growth, giving two disconnected phases. The glasses containing 44-49 wt%. SiO$_2$ show interconnected structure typical of spinodal decomposition. The degree of interconnectivity of the amorphous phases depends on the thermal history. Heat treatment produces two main crystalline phases of sodium calcium silicates, viz. Na$_2$Ca$_2$Si$_3$O$_9$ and Na$_2$CaSi$_3$O$_8$. Additional calcium silicate (CaSiO$_3$) and calcium phosphate (Ca$_4$P$_6$O$_{19}$) phases are also present. Immersion of samples in a simulated body fluid (SBF) shows that the formation of hydroxyapatite on the surface is significantly higher in the heat treated samples. Controlled leaching in 1N HCl introduces multi-modal nano-macro porosity, which is expected to improve their usefulness as bone scaffolds.

INTRODUCTION

Bioactive glass and glass-ceramics became feasible after Hench et al.[1] reported that some silicate glasses of specific composition within the soda-lime phosphosilicate system could chemically bond with bone tissue. When such materials are implanted in the human body, a bonelike hydroxyapatite (HA) layer is formed on their surface.[2] The biocompatibility is established since the inorganic part of the human bone is a kind of calcium phosphate viz. HA. Hench et al.[3-5] reported that there are some bioactive glasses, which also bond to soft tissues through attachment of collagen to the glass surface. A characteristic of the soft-tissue bonding compositions is a very rapid rate of HA formation.[3] The common characteristic of bioactive glasses and bioactive glass-ceramics is a time-dependent modification of the surface that occurs upon implantation.[4]

Soda-lime phosphosilicates have been used as bioactive glass because of strong bonding to bone tissue and the absence of foreign body response. The bone regeneration properties of soda-lime phosphosilicate bioactive glass are due to the presence of phosphorus, calcium and sodium. For melt-processed soda-lime phosphosilicate bioactive glasses, low silica content and a high calcia and soda content are required for the glass to have a high surface reactivity in an aqueous medium. The primary role of sodium ions in the bioactive glass is to facilitate the glass melting process. The dissolution of Na$^+$, P$^+$ and Ca$^+$ from the glass in solution results in the formation of a silica-rich and a CaO-P$_2$O$_5$ rich layer. Finally, the latter layer promotes the

---

[1] On leave from University of Alexandria, Egypt. Email: hmm205@Lehigh.edu
[2] Author for correspondence. Email: h.jain@lehigh.edu

formation of HA, which has Ca/P ratio compatible with bone structure and is necessary for the tissues to bond.[6]

Besides the chemical composition which determines the bioactivity of soda-lime phosphosilicates, their morphologies and textures are important for their bonding with tissue. So it is useful to design soda-lime phosphosilicates bioactive glasses with properties specific to porous scaffold and articulating surface. An ideal scaffold should be a porous material with two important characteristics: (a) an interconnected network with macroporosity[73] to enable tissue ingrowth and nutrient delivery to the center of the regenerated tissue, and (b) micro or mesoporosity to promote adhesion.[8]

The objectives of this work are: (a) fabricate interconnected multi-modal porosity in the bioactive soda-lime phosphosilicate glass system, and (b) characterize morphology, texture and ability to form HA layer as a function of composition and processing parameters.

EXPERIMENTAL PROCEDURE

The $24.5CaO - (27.5-x)Na_2O - 6P_2O_5 - (42+x)SiO_2$ (wt%) glass series (with x=0-10) was prepared with starting materials: SiO_2 (99.99%), $CaCO_3$ (99%), Na_2CO_3 (99%) and $Ca_5(OH)(PO_4)_3$ (99%). The SiO_2 content was varied within the 42-52 wt% range with 11 different values as listed in Table 1. The P_2O_5 wt%, CaO wt%, and Ca/P ratio were kept constant at 6 wt%, 24.5 wt% and 5.2 respectively. The calculated batches were mixed and ground using an alumina mortar and pestle. The batches were melted in a platinum crucible at 1500°C. The homogenized melts were poured onto a stainless steel mold and were then annealed to relax residual stresses. The result was a glass phase-separated by two quite different mechanisms: The first mechanism produced disconnected, precipitated fine structure. The second mechanism led to interconnected spinodal like texture. To identify glass samples, we have followed the notation XXS proposed by Hench, where the first two letters represent the wt% of SiO_2 e.g. 42S means a glass that contains 42 wt% SiO2. To indicate the transformation of a glass to glass-ceramic, we have modified the notation to XXSGY, where G signifies glass-ceramic and Y indicates a particular heat treatment.

To induce additional multiple phase separation, the samples were further heat treated in two steps: (a) nucleation step at T_n, which included heating to 670°C and holding there for a fixed time, and (b) crystal growth step at T_x, which included heating at the same rate to different two values of T_x (750°C or 1075°C) and holding there for varying times. Table 1 describes the temperature and time of heat treatment schedule for various samples. To create the nano-macro porosity in the prepared glasses, the heat treated glasses were leached in 1N HCl.

To identify the multiple phases and microstructure, the samples were analyzed by X-ray diffraction technique (XRD) and scanning electron microscopy (SEM). A Philips XL30 SEM was used to examine sectioned and polished samples of each glass to elucidate the phase separation and microstructure. The cross-section samples were coated with gold. The elemental distribution in different phases was analyzed with EDAX device equipped in the Hitachi 4300 Field Emission Scanning Electron Microscope. The pore size distribution of the samples was determined by mercury porosimeter (Micromeritics Auto Pore IV).

---

[3] Pores are classified as microporous (diameter<2nm), mesoporous (2<diameter<50nm) and macroporous (diameter>50nm). We have used the term nano to describe both the micro and meso size pores.

The formation of apatite layer is observed in vitro in simulated body fluid (SBF) that contains inorganic ions in concentration corresponding to human blood plasma. The SBF was prepared by dissolving reagent grade NaCl, NaHCO$_3$, KCl, K$_2$HPO$_4$.3H$_2$O, MgCl$_2$.6H$_2$O, CaCl$_2$.2H$_2$O, and Na$_2$SO$_4$ in deionized water. The fluid was buffered at physiological pH 7.4 at 37°C. Ten glass and glass-ceramic (T$_n$ 670°C for 1h and T$_x$ for 750°C for 6h or 9h) samples were used to investigate the reaction of soda-lime phoshosilicate samples with SBF. Each specimen (2mg) was immersed in 1 ml of SBF contained in a polyethylene bottle covered with a tight lid for 20 h.

RESULTS AND DISCUSSION

Figure 1 shows photomicrographs of the various structures and phases observed in the soda-lime phosphosilicate bioactive glass. The morphologies and mechanisms of phase separation vary remarkably in the present samples containing 42-52 wt% SiO$_2$. For SiO$_2$ in the 42-43 and 50-52 wt% ranges, the compositions produce fine structure consisting of disconnected precipitates, which we interpret as a consequence of the composition being in the binodal region. For SiO$_2$ in the 44-49 wt% range, the compositions produce interconnected spinodal texture.

The XRD pattern of the prepared bioactive glass samples, which are subjected to the two step heat treatment, is shown in Figure 2. The crystal phases are identified as Na$_2$Ca$_2$Si$_3$O$_9$, Na$_2$CaSi$_3$O$_8$, Ca$_4$P$_6$O$_{19}$ and CaSiO$_3$ (Wollastonite). The location of diffraction peaks matches the standard powder diffraction file (PDF) card numbers 1-1078, 12-671, 15-177 and 42-550, respectively[9]. The heat treatment of soda-lime phosphosilicate bioactive glasses results in the coarsening of an already phase separated glass structure and crystallization of these phases. In addition, it also induces the formation of new crystalline phase, Wollastonite at high temperature 1075°C (figures 2 (g) and 2 (h)) similar to that found by De Aza et al.[10]

Figure 3 shows microstructure of 43SG1, 45SG1, 49SG1 samples, which were subjected to heat treatment in two steps: (a) nucleation at 670°C, followed by (b) crystal growth at T$_x$, which included heating to 750°C and holding there for 6h. It is clear that the size of micro phases increases with heat treatment. The structure of 43SG1, 45SG1, 49SG1 glass-ceramic has coarsened. The silica-rich phase, instead of forming a continuous network, as in the sample shown in 43S, Figure 1 (a), formed droplets suspended in a phosphorus-rich matrix. The micrograph of heat treated 43SG1 sample in Figure 3 (a) indicates that there is little tendency for these (crystallized) droplets to coalesce to form an interconnecting phase. By comparison, the micrographs of the 45SG1, 49SG1 wt% SiO$_2$ heat treated samples (Figure 3 (b,c)) indicate that there is tendency for the micro phases to form an interconnecting structure. Overall, the effect of heat treatment on interconnected phase can be explained as the formation, growth and coalescence of the second-phase particles.[11]

Figure 4 shows SEM micrographs of 45SG1, 47SG1, and 49SG1 glass-ceramic after the leaching chemical treatment. It is clear that the present melt-quench-heat-etch method has produced a structured porous network composed of interconnected nano-macro porosity. The multi-modal nano-macro interconnected porosity in the heat-and-chemical treated sample 49SG1 is confirmed by mercury porosity data in Figure 5. It shows a broad distribution of macro pores with an average size of ~30 μm, and additional distinct pores within the 0.1-8 μm range. On a much finer scale, there are nano-pores with average size ≈15 nm. The process of producing multi-modal nano-macro interconnected porosity in these samples is described further elsewhere.[9]

Figure 6 shows SEM micrographs of the glass-ceramic samples (44SG1, 45SG1, 47SG1, 49SG1, and 45SG3) after soaking in SBF for 20h. Two layers with different compositions are formed on the surface of the samples. The first layer from the top covers the whole outer surface with HA and is enriched in Ca and P. The HA layer formed on the surface of the 45SG1 sample is identified by XRD pattern, as shown in Figure 7, where the location of marked diffraction peaks matches with the main peaks of HA mentioned in PDF card number 09-0432 and is similar to that found by Peital et al.[12]. It should be noted that deposited particles aggregate in the large pores of samples 45SG1-47SG1, Figure 6 (b, c), while small particle aggregation appears on the surface of 44SG1and 49SG1 samples, see Figure 6 (a, d). Figure 6 (b, e) shows the top layer, HA, covering the whole surface of 45SG1, 45SG3 glass samples that were subjected to the crystal growth treatment for 6h and 9h respectively at 750°C followed by chemical leaching. The thickness of top layer on the surface of 45S heat treated for 9h (Figure 6[e]) is significantly larger than that on the surface of sample heat treated for 6h (Figure 6[b]).

There is an argument in the literature about the effect of crystallization of bioactive glass on the apatite-layer formation. Some authors report that crystallization induced by thermal treatment can turn a bioactive glass into an inert material,[13] whereas others report little effect on the bioactivity.[12, 14] The present work indicates that crystallization enhances HA formation. Comparing the formation of top layer on the surface of 44SG1-49SG1glass-ceramics, we observe the influence of the initial composition, and hence the different phases and the multi-modal porosity, (Fig. 5, 6). It appears that the multi-modal porosity in 44SG1-49SG1 glass-ceramics allows the SBF to infiltrate the material freely and permits a more efficient transport of ions to and from the reactive surface during the 20h soak.

CONCLUSION

The present heat treatment of 45S and 49S glasses exhibits a highly interconnected microstructure. This tendency of interconnected structure of these glasses is due to the growth and coalescence of the second-phase particles. By comparison, the 43S composition exhibits little interconnectivity after the effect of heat treatment.

There is significant effect of microstructure and crystallization-leaching process on apatite-layer formation on the surface of soda-lime phosphosilicate 44S-49S glass-ceramics which was synthesized by the melt-quench method. Testing in SBF shows the formation HA on the surface of 44S-49S glass-ceramics after soaking for 20h, suggesting their good bioactivity. The formation of HA and hence bioactivity is significantly influenced by the initial composition and the presence of different phases that are produced by thermal and chemical treatments.

ACKNOWLEDGEMENT

This work was initiated and continued as an international collaboration with support from National Science Foundation (International Materials Institute for New Functionality in Glass (DMR-0409588) and Materials World Network (DMR-0602975) programs).

REFERENCES
[1]L. L. Hench, R. J. Splinter, W. C. Allen and T. K. Greenle, "Bonding mechanisms at the interface of ceramic prosthetic materials," *J. Biomed. Mater. Res.*, 5 [6] 117-141 (1971).
[2]T. Peltola, M. Joinen, H. Rehiala, E. Levanen, J. B. Rosenholm, I. Kangasniemi and A. Yli-Urpo, "Calcium phosphate formation on porous sol-gel-derived SiO_2 and $CaO-P_2O_5-SiO_2$ substrates in vitro," *J. Biomed. Mater. Res.*, 44 [1] 12-21 (1999).
[3]R. Li, A. E. Clark and L. L. Hench, "An Investigation of bioactive glass powders by sol-gel processing," *J. Appl. Biomater.*, 2 [4] 231-239 (1991).
[4]L. L. Hench and J. Wilson, "Bioceramic," *MRS Bull.*, 16 [9] 62-74 (1991).
[5]L. L. Hench, "Bioceramic: From concept to clinic," *J. Am. Ceram. Soc.*, 74 [7] 1487-1510 (1991).
[6]C.-C. Lin, L.-C. Huang and P. Shen, "$Na_2CaSi_2O_6-P_2O_5$ based bioactive glasses. Part1: Elasticity and structure," *J. Non-Cryst. Solids*, 351 [40-42] 3195-3203 (2005).
[7]S. J. Gregg and K. S. W. Sing, "Adsorption, Surface area and porosity"; 25, 2nd ed., Academic Press, London, New York, 1982.
[8]N. Li, Q. Jie, S. Zhu and R. Wang, "Preparation and characterization of macroporous sol-gel bioglass," *Ceram. Internat.*, 31 [5] 641-646 (2005).
[9]H. M. M. Moawad and H. Jain, "Creation of nano-macro interconnected porosity in bioactive glass-ceramic by the melt-quench-heat-etch method," *J. Am. Ceram. Soc.*, submit to be published (2007).
[10]P. N. De Aza and Z. B. Luklinska, "Effect of glass-ceramic microstructure on its in vitro bioactivity," *J. Mater. Sci.: Mater.Med.*, 14 [10] 891-898 (2003).
[11]T. H. Elemer, M. E. Nordberg, G. B. Carrier and E. J. Korda, "Phase separation in borosilicate glasses as seen by electron microscopy," *J. Am. Ceram. Soc.*, 53 [4] 171-175 (1970).
[12]O. Peitl, E. D. Zanotto and L. L. Hench, "Highly bioactive $P_2O_5-Na_2O-CaO-SiO_2$ glass-ceramics," *J. Non-Cryst. Solids*, 292 [1-3] 115-126 (2001).
[13]P. Li, Q. Yang and F. Zhang, "The effect of residual glassy phase in a bioactive glass-ceramic on the formation of its surface apatite layer in vitro," *J. Mater. Sci.: Mater. Med.*, 3 [6] 452-456 (1992).
[14]O. P. Filho, G. P. LaTorre and L. L. Hench, "Effect of crystallization on apatite-layer formation of bioactive glass 45S5.," *J. Biomed. Mater. Res.*, 30 [4] 509-514 (1996).

Table 1. Temperature and time of heat treatment schedule of various glasses in the 24.5CaO - (27.5-x)Na_2O - $6P_2O_5$ - (42+x)SiO_2 (wt%) series (with x=0-10). All samples were given the same nucleation heat treatment.

SiO_2 (wt.%)	Sample I.D.	Crystal Growth Step (T_x for time t)
42	42SG1	750°C for 6h
43	43SG1	750°C for 6h
44	44SG1	750°C for 6h
45	45SG1	750°C for 6h
46	46SG1	750°C for 6h
47	47SG1	750°C for 6h
48	48SG1	750°C for 6h
49	49SG1	750°C for 6h
50	50S G1	750°C for 6h
52	52SG1	750°C for 6h
45	45SG2	1075°C 6h^{-1}
48	48SG2	1075°C for 6h
49	49SG2	1075°C for 6h
45	45SG3	750°C for 9h

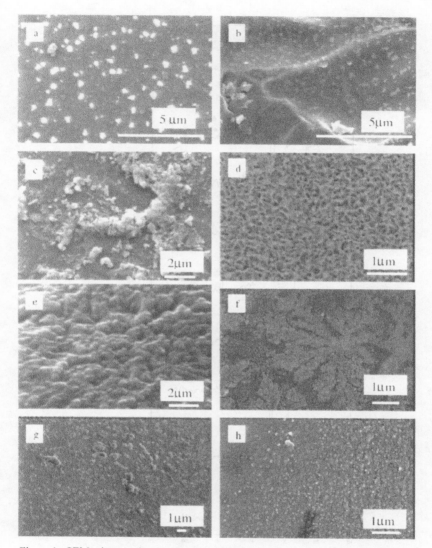

Figure 1. SEM micrographs showing binodal and spinodal phases separation in various glass samples. a) 42S, b) 43S, c) 44S, d) 45S, e) 46S, f) 49S, g) 50S, h) 52S.

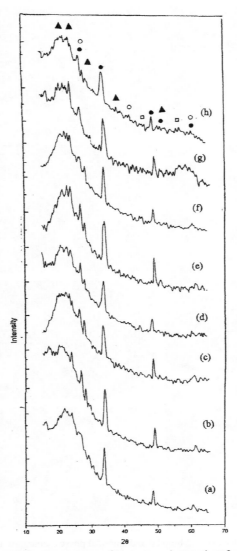

Figure 2. X-ray diffraction patterns of glass-ceramic samples after the nucleation and growth heat treatments: a) 44SG1, b) 45SG1, c) 46SG1, d) 47SG1, e) 49SG1, f) 45SG2, g) 47SG2, h) 49SG2.
(● $Na_2Ca_2Si_3O_9$, ○ $Na_2CaSi_3O_8$, ▲ $Ca_4P_6O_{19}$, $CaSiO_3$)

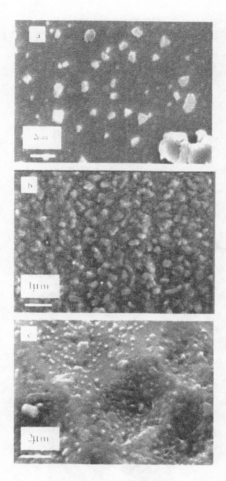

Figure 3. SEM micrographs of glass-ceramic samples after the nucleation and growth heat treatment: a) 43SG1, b)45SG1, c) 49SG1.

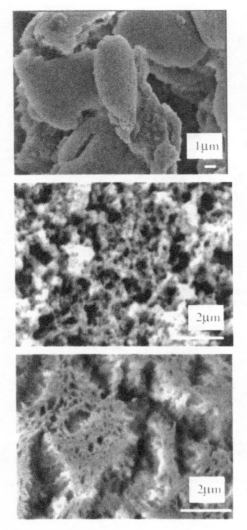

Figure 4. SEM micrographs of the specimens after heat treatment and chemical leaching in 1NHCl: a) 45SG1, b) 47SG1, c) 49SG1.

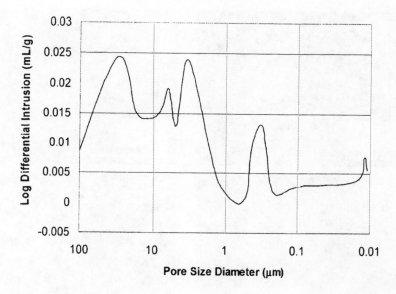

Figure 5. Pore size distribution in 49SG1 glass-ceramic after heat-chemical treatment.

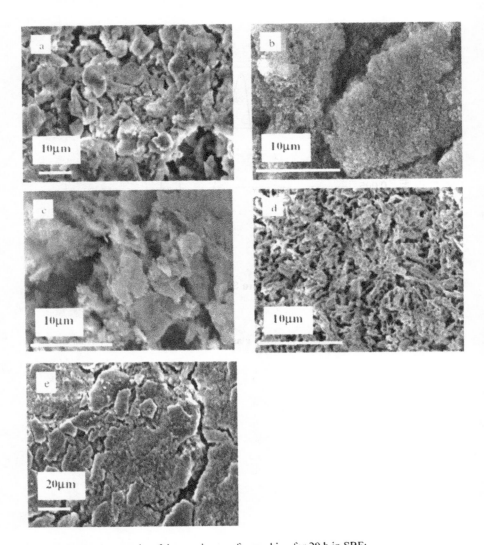

Figure 6. SEM micrographs of the specimens after soaking for 20 h in SBF:
a) 44SG1, b) 45SG1, c) 47SG1, d) 49SG1, e)45SG3.

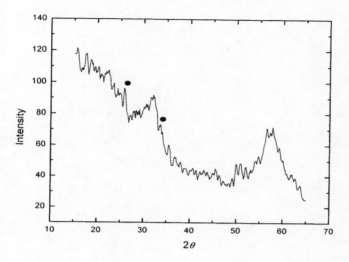

Figure 7. X-ray diffraction pattern of 45SG1 sample after soaking for 20h in SBF. The marked peaks correspond to those of crystalline HA.

PORES NEEDED FOR BIOLOGICAL FUNCTION COULD PARADOXICALLY BOOST FRACTURE ENERGY IN BIOCERAMIC BONE TISSUE SCAFFOLDS

Melissa J. Baumann, I. O. Smith and Eldon D. Case*

Chemical Engineering and Materials Science Department, Room 2527 Engineering Building, Michigan State University, East Lansing, Michigan, 48824-1226, USA

* Corresponding author, e-mail casee@msu.edu, FAX (517)-432-1105.

ABSTRACT
The deleterious effects of unimodal pore size distributions on the mechanical properties of ceramics is well documented, although some unanswered questions remain concerning the modeling of such effects, especially for very high volume fraction porosities, P (where P is greater than roughly 0.35 to 0.50). In contrast, there have been a limited number of studies of the effect of bimodal pore size distributions on mechanical properties of structural ceramics, although some studies show very interesting increases in fracture energy for bimodal pore distributions. Biological function demands bimodal pore distributions for bone tissue engineering scaffold materials and the enhancement of mechanical properties that can result from bimodal porosity in structural ceramics also can be potentially used to optimize the mechanical properties of bioceramics such as hydroxyapatite (HA) bone scaffolds.

INTRODUCTION
Sintering ceramic powders typically generates a unimodal pores size distribution (Figure 1a).[1] As is widely known, as volume fraction porosity P of unimodially distributed porosity increases, the mechanical properties of the ceramic are degraded.

Alternatively, there are processing routes that can produce bimodal pore size distributions (Figure 1b) in sintered ceramics, namely by sintering agglomerated powders[2] or via the volatiles released from either foaming agents[2] or from preceramic precursors[3]. In general, porous ceramics have many potential uses, including as filters[4,5] and substrates for high-temperature catalysts[6]. In particular there are critical applications for porous bioceramics as synthetic bone scaffold materials. In the U.S. alone, there are annually more than half a million bone grafts performed that require bone tissue replacement.[7] Biological function demands bone tissue engineering scaffold materials such as hydroxyapatite (HA), which is similar to the mineral phase of bone, include both large and small pores to allow bone ingrowth and channels for nutrient flow.

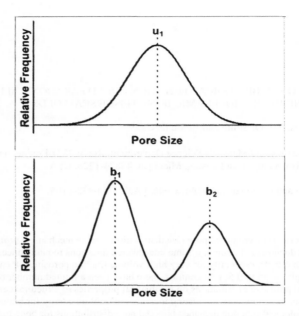

Figure 1 a) Pore size distribution for unimodal pore distribution, where u_1 is the mean pore size and b) pore size distribution for bimodal pore distribution. The parameters b_1 and b_2 are the mean pore sizes for the smaller and larger pore populations, respectively.

This paper presents a paradigm for simultaneously meeting both the biological and the mechanical requirements of bone scaffold materials. A fundamental part of this paradigm is that the researcher need not to resign themselves to the idea that porosity in HA materials will necessarily lead to low mechanical integrity.

While unimodal pore size distributions are in fact detrimental to mechanical integrity, a range of bimodal pore architectures can enhance fracture energy. In the Results and Discussion, we discuss the effect of both unimodal and bimodal pore size distributions on the mechanical properties of ceramics in general and then discuss particular examples of bimodal porosity in HA. A key point is that studies in the non-bioceramics literature demonstrate that there are many differing bimodal pore architectures that can result in increased fracture energy (Figure 2).[8-13] Thus, we are not necessarily constrained to the typical bioceramics approach of attempting to mimic the complicated microstructure of natural bone. However, before discussing the mechanical property implications of unimodal and bimodal pore size distributions, in the next section we discuss the fabrication and characterization of bimodal pore size distributions in HA.

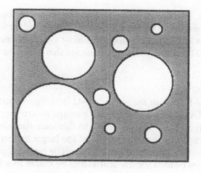

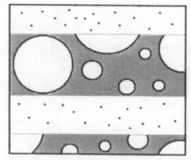

Figure 2 a) Schematic of a microstructure that includes a "random bimodal pore distribution", in which a distribution of large pores is interspersed relatively homogeneous in among a background populations of small pores and b) schematic of a serially bimodal microstructure consisting of a laminate with alternating layers of differing porosity. A relatively dense layer (with a distribution of small pores) is juxtaposed with a layer having a distribution of large pores.

MATERIALS AND METHODS

In this study, HA scaffolds with volume fraction porosities of between 0.15 and 0.50 were produced in the laboratory of one of the authors (MJ Baumann). The HA powders (Hitemco Medical, Old Bethpage, NY) with a mean particle diameter of approximately of 11.0 ± 8.1 μm with an agglomerate size of approximately 34.5 μm ± 19.7 μm (measured using a Microtrac S3000 Pore Size Analyzer), were suspended in a 10^{-3}N KNO_3 electrolyte solution. A hydrogen peroxide (H_2O_2) foaming agent was added to this suspension and the resulting slurry poured into cylindrical cavities in a silicon mold. The samples in the mold were first air dried for one hour at 125 °C in a gravity convection furnace and then gently removed from the mold and immediately sintered in air at 1360 °C for 4 hours using a heating and cooling rate of 10 °C/min. Further details of the processing technique are given elsewhere[2].

The authors have attempted to characterize pore size distributions in porous HA specimens using two different techniques: (1) serial sectioning[2] and by (2) Confocal Laser Scanning

Microscopy[14]. For example, McMullen[2] sampled the pore size distribution of HA specimens using serial sectioning technique, in which the foamed HA cylinders were cut at each end to create parallel sides using a low speed diamond wafering saw. The pores were then observed via SEM, augmented with SPOT Imaging software package that recorded and analyzed pore size distributions.

RESULTS AND DISCUSSION

The foaming technique by one of the author's (MJ Baumann) successfully produced bimodal pore size distributions in HA (Figure 3), where the average macropore size was 256 µm and the average micropore size was 57 µm.[2] The ability (i) to fabricate specimens with a bimodal pore size distribution and (ii) to adjust the relative sizes of the larger and smaller pore size distributions in turn leads to two important advantages. The bimodal pore size distribution provides the pore size ranges needed for the scaffold to be viable as a bone replacement material (in terms of providing the type of porosity required for bone ingrowth and nutrient flow).

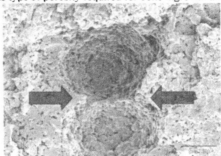

Figure 3 SEM photomicrograph of an 80% dense HA specimen prepared by foaming. Both macropores and micropores are present, as indicated by arrows. The micropores act as channels for metabolic fluid transfer in implanted samples.

Mechanical properties of ceramics with unimodal pore size distributions

The mechanical properties of ceramics such as elastic modulus, E, fracture strength, σ, the hardness, H, and fracture surface energy, γ, are often described in terms of an exponential decrease with increasing volume fraction porosity, P, such that

$$A(P) = A_0 exp(-bP) \tag{1}$$

where the mechanical property A can be any of the properties E, σ, H, or γ.[12,15,16] A_0 is the value of A for P = 0 and b (a dimensionless parameter typically between about 4 and 6) is often assumed to be constant for a material.[12,15] In contrast to E, σ, H, and γ, there have been relatively few studies of the porosity dependence of the fracture toughness, K_{IC}, for either metals or ceramics although in a number of cases, K_{IC} as a function of porosity can be approximated by equation 1.[12]

For unimodal pore size distributions, equation 1 generally provides a good description of the mechanical property-porosity behavior of ceramics in the range from $0 < P < P_C$, where P_C, (the critical value of P), varies with the material and pore morphology although typically, $P_C \approx 0.30$ to 0.40.[12] When $P > P_C$, the material property A, typically decreases much more rapidly than is predicted by equation 1.[12,15,17,18]

As an example of the limited range of validity for equation 1, for the porosity range $0 < P < 0.35$, the Young's modulus, E, as a function of porosity generally is described well by the empirical relationship

$$E(P) = E_o \exp(-bP) \tag{2}$$

where b is a dimensionless constant that ranges from approximately 4 to 6 for a variety of ceramics.[15] However, for $P > 0.35$, E(P) is generally described better by

$$E(P) = E_o \left\{ \frac{1 - P^2}{1 + \left(\frac{1}{\lambda} - 1 \right) P} \right\} \tag{3}$$

where the dimensionless parameter λ may be a function of the average pore shape.[19,20]

Thus, over the volume fraction porosity range from about 0 to 0.35, equations 2 and 3 describe the data equally well (when either b or λ is a free parameter). However, for porosities greater than 0.35, equation 3 fits the Young's modulus-porosity data best.[15,19-21]

A difficulty with equations 1 - 3 is that b and λ are merely fitting parameters, that is, the values of b and λ for a given set of specimens are obtained by a least squares fit of the data to the equation. Another common shortcoming is that the pore "characterization" is vested within a single parameter (b or λ, respectively) over the entire range of porosity. This is counterintuitive since pore morphology changes tremendously during the evolution from a powder compact (where P may be as high as ≈ 0.50 or more) toward a dense body (where $P \rightarrow 0$).[22] Further, as the specimen porosity decreases, the pore shape changes so dramatically that is it difficult to understand how a fixed value of b or λ can be sufficient to characterize data over an extended range of P.[12,15] For example, for tricalcium cement and Portland cement, equation 1 describes the porosity dependence of Young's and shear moduli with $b \approx 2.3$ to 3 for the range $0 < P < 0.35$, but b values ≈ 4.3 to 5.2 are needed to fit equation 1 to the Young's modulus data for $0.35 < P < 0.56$.[13,23]

Thus, porosity-property relationships are complex and not entirely understood, even when the pore size distribution is assumed to be unimodal (Figure 1a). However, if "sub-populations" of small and large pores coexist within a single specimen, such as is the case for natural bone (Figure 4), then such a bimodal pore size distribution (Figure 1b) can have profoundly important implications for optimizing the microstructure of synthetic bone replacement materials.

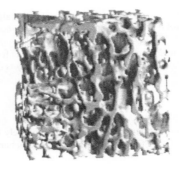

Figure 4   Micro CT (computer tomographic) image of rabbit trabecular bone, courtesy O. Gauthier, National Veterinary School of Nantes, France and R. Müller, ETH, Zurich, Switzerland.

Bimodal porosity in ceramics: Crack-pore interactions

We shall discuss now several instances in the ceramics literature for which bimodal porosity has lead to mechanical property enhancement, including studies of Si_3N_4[12], SiC/SiC laminates[9], and alumina[8] as well as in model crack/pore studies in poly-methymethacrylate (PMMA)[10,11]. For example, crack-pore interactions dominate the fracture surface energy, γ in sintered Si_3N_4 with bimodal porosity where γ has been observed to range from about 6 J/m^2 to about 50 J/m^2 for specimens with volume fraction porosity $P \approx 0.45$ to 0.6.[12] These Si_3N_4 specimens were prepared with a bimodal pore distribution that included artificially induced macropores in the range of 20 to 200 μm in diameter, along with "natural" micropores having diameters in the micron and submicron range. In comparison, Si_3N_4 specimens[12] with unimodal "natural porosity" in the range of about $0.3 < P < 0.4$ displayed fracture energies of about 4 to 6 J/m^2. Therefore, with bimodal porosity, the fracture energy can be more than an order of magnitude higher than similar specimens with a unimodal pore size distribution.

The high values of the experimentally-determined fracture energies for these Si_3N_4 specimens were attributed to crack branching and crack bridging induced by the bimodal nature of the porosity.[12] SEM micrographs supported this interpretation, showing crack branching and/or bridging apparently correlated with large pores.

The type of microstructure produced in these Si_3N_4 specimens[12] could be termed a "random bimodal microstructure", with large pores mixed relatively homogeneously with the background of small pores (Figure 2a). In contrast, "serially bimodal" microstructures were produced by fabricating planar SiC/SiC laminates in which 100 μm thick dense SiC layers where interleaved with 100 μm thick porous SiC layers. Figure 2b shows a laminate with relatively dense layers containing small pores alternating with layers having large pores.[9] The pores were induced in the porous SiC layers by the pyrolysis of a polymeric fugitive phase from particles between 5 and 9 μm in diameter.[9] When macrocracks were driven into the SiC laminate specimens, the porous interlayers successfully deflected the propagating cracks whenever the volume fraction porosity was at least 0.4.[9] Thus, propagating cracks were deflected as they passed from regions of low to high porosity, where the porous region had not only a higher volume fraction porosity, P but also a much larger average pore size than in the denser layers.[9]

In a study of thermal shock damage in porous polycrystalline alumina[8], porosity was introduced by either (i) partially sintering the specimens or (ii) using a fugitive phase (cornstarch). Although the pore sizes were not specified by Clegg et al.[8], the sintered grain size of the alumina specimens was about 3 μm and the nominal particle size for the cornstarch was 10 μm, so that the pore size could easily have been larger than 10 μm if some fraction of the cornstarch particles were agglomerated in the powder compact. The key finding in terms of the pore size distribution was that the fracture energy of the alumina specimens sintered using the fugitive phase was considerably higher than the fracture energy of the alumina sintered without the fugitive phase. In other words, porosity actually increased the fracture energy of the ceramic.[8]

A series of model crack/pore interaction experiments were conducted in which cracks were propagated into an array of holes (model pores) drilled into polymethyl methacrylate specimens[10,11] in an attempt to isolate the crack-pore interaction effects from other crack-microstructure interactions in ceramics based on the grain shape, size distribution, boundary phases, etc. As cracks approached a pair of pores, the crack front was temporarily "pinned" by the pores, bowing outward from the constraints formed by the pores.[10,11]

Bimodal porosity relationships in porous CaP materials

More than 15 years ago, it was recognized that the porosity dependence of the compressive fracture strength of CaP bioceramics could be expressed in terms of the total volume fraction porosity, P, while the tensile strength could be described in terms of the micropore volume.[24] In particular, for porous HA and β-TCP having a bimodal pore size distribution, the compressive fracture strength, σ_c, can be expressed in terms of the total volume fraction of macropores and micropores, P, such that

$$\sigma_c(P) = \sigma_{c0}\exp(-b_cP) \qquad (4)$$

where σ_{c0} is the strength at $P = 0$ and b_c is a dimensionless constant.[24] A least-square fit of the fracture data for porous HA and β-TCP yielded values of 700 MPa and 5 for σ_{c0} and b_c, respectively. The value of $b_c = 5$ is consistent with values of b typically found in the literature for the porosity dependence of the compressive strength of ceramics.[12,15]

However, the relationship between the tensile fracture strength, σ_t, and porosity can be expressed as

$$\sigma_t(P_{mp}) = \sigma_{t0}\exp(-b_tP_{mp}) \qquad (5)$$

where P_{mp} is volume fraction of micropores[25-27] and $\sigma_{t0} = 220$ MPa, $b_t = 20$. Thus, while equation 4 describes the compressive fracture strength in terms of total porosity P, the tensile fracture strength is characterized in terms of only the micropore volume fraction, P_{mp}. In this case, the value of b_t is much larger than the value of b typically observed in equation 1. This example, drawn from the bioceramics literature, points the importance of considering the bimodal pore size distribution for the mechanical properties of HA specimens with a bimodal pore size distribution.

However, the work by de Groot[24] while fundamental in its importance, it does not acknowledge the opportunity that presents itself from the ceramics literature, namely that of enhancing the fracture energy. In general, both the compressive modulus and ultimate compressive strength of HA and β-TCP are highly sensitive to the details of processing and the final microstructure. For example, increases in interconnected porosity resulted in increases in osseointegration[28-32], which in turn alters the mechanical behavior of implanted HA scaffolds. Even for samples having little bone ingrowth[33] the presence of bimodal porosity shifts the fracture mode from the pure brittle fracture behavior typically experienced by ceramics, to an elastic-plastic failure akin to cancellous bone. The degree of bone ingrowth in implanted bone scaffolds in turn is strongly affected by pore morphology and in particular, pore interconnectivity at times as short as 5 weeks in vivo.[33]

Microstructural considerations in addition to the pore size distribution

In addition to porosity, the mechanical properties of ceramics also vary with grain size. In general changes in porosity and grain size are coupled during the densification of ceramic bodies.[2,22] For highly porous ceramics, the grain size of the sintered specimen is expected to be similar to the particle size in the starting powder, since in sinterable powders there tends to be little grain growth until a relative density of roughly 0.95 is obtained.[22] Thus although grain size is in general a very important factor in determining the mechanical properties of a sintered ceramic, in highly porous ceramics one would expect the mechanical properties to be dominated by porosity effects rather than grain size effects, due to the limited grain growth. For example, in studies of HA specimens with volume fraction porosities between about 0.02 to 0.4 and grain sizes ranging from 1.6 μm and 7.4 μm, as porosity increased both the hardness[34] and the dielectric constant[35] decreased exponentially (equation 1) while there was essentially no effect on hardness or dielectric constant due to grain size.[34,35]

CONCLUSIONS

A limited number of studies of structural ceramics (silicon nitride[12,13], SiC[9] and alumina[8,10,11] demonstrate remarkable increases (up to a factor of 10) in fracture energy for specimens having bimodal pore size distributions when compared to specimens of the same material having unimodal pore size distributions. Furthermore, model studies comparing unimodal and bimodal arrays of holes in PMMA sheets indicate an enhancement of fracture energy via bimodal porosity.[10] Although bimodal pore distributions are essential to the biological function of bone tissue, the bioceramics community has not yet attempted to apply the lessons of bimodal pore size/fracture energy enhancement learned by a segment of the structural ceramics community.[8-13] Instead, the bioceramics community often relies on biomimetics to guide them in developing biomaterials. Biomimetics has been a very natural (and fruitful) approach to biomaterial design, but even in nature there can be more than one solution to a given physiological challenge.

In this paper, we propose that while the pore architecture of bone may be the "best answer" to the question of microstructural design for bone, the experience gained from structural ceramics shows that many different bimodal pore architectures may give very good or even comparable answers to the problem of strengthening bone replacement tissue. Moreover, these alternative bimodal pore architectures may be substantially easier to fabricate than a microstructure that duplicates natural bone.

Ongoing and Future Work

To improve our understanding of the design of ideal bone tissue engineering scaffolds and to circumvent the difficulties in fabricating structures that are identical to natural bone, the dual functions of porosity must be recognized and designed into ceramic scaffolds. A powerful tool that we are now beginning to use in these efforts is the software package OOF (Object Oriented Finite Element Code)[36,37], which converts images of either computer-simulated or experimentally determined microstructures into a finite element mesh.

OOF software reads images such as an SEM micrograph (Figure 3) or a micro-CT image (Figure 4) in the ppm (Portable Pixel Map) format and allows the user to analyze selected microstructural features from a given micrograph. Elastic and thermal properties then can be assigned to the features and using Finite Element Analysis, the stress can be calculated based on internal or external loading constraints. OOF has been used to construct finite element mesh based on micrographs of a variety of multiphase microstructures such as SiC particle-reinforced Al matrix composites[37], two-phase aluminum-silicon-based alloys[36], and plasma-sprayed ceramic thermal barrier coatings on metal alloys[38]. Thus, OOF will be used as a guide in optimizing the bimodal porosity for both biological and mechanical function in synthetic bone replacement materials.

ACKNOWLEDGEMENTS

We acknowledge the support of the National Science Foundation (DMR0074439) along with the MSU Foundation.

REFERENCES

1.   Chen PL, Chen IW. Sintering fine oxide powders: I, microstructural evolution. J. Am. Ceram. Soc. 1996;79:3129 - 3141.
2.   McMullen RA. An in vitro investigation of MC3T3-E1 osteoblast proliferation and differentiation on hydroxyapatite based tissue engineered scaffolds [M.S.]. East Lansing: Michigan State University; 2004.
3.   Schmidt H, Koch D, Grathwohl G, Colombo P. Micro/macroporous ceramics from preceramic precursors. J. Am. Ceram Soc. 2001;84:2252 - 2255.
4.   Lee J, Shin K, Shin M, Cho D, Lee H. Properties of catalytic filters for the hot gas cleaning. Mat Sci Forum 2004;449-452(2):1181-1184.
5.   Suzuki Y, Kondo N, Ohji T, Morgan P. Uniformly porous composites with 3-D network structure (UPC-3D) for high-temperature filter applications. Int J Appl Ceram Tech 2004;1(1):76-85.
6.   Rey P, Souto A, Santos C, Guitian F. Development of porous corundum layers on cordierite ceramics. J Eur Ceram Soc 2003;23(15):2983-2986.
7.   Laurencin CT, Khan Y. Bone graft substitute materials. Volume 2004: eMedicine.com, Inc.; 2004.
8.   Clegg WJ, Yuan C, Vandeperre LJ, Wang J. The effect of porosity on the cracking of materials in high temperature applications. Ceramic Eng and Sci Proc 2001;22(4):251 - 257.
9.   Blanks KS, Kristoffersson A, Carlstrom E, Clegg WJ. Crack deflection in ceramic laminates using porous interlayers. J European Ceram Soc 1998;18(13):1945 - 1951.
10. Wang J, Vandeperre LJ, Stearn RJ, Clegg WJ. Pores and cracking in ceramics. J Ceram Process Res 2001;2(1):27 - 30.

11. Wang J, Vandeperre LJ, Stearn RJ, Clegg WJ. The fracture energy of porous ceramics. Key Eng Mat 2002;206-2:2025 - 2028.

12. Rice RW. Porosity of ceramics. New York: Marcel Dekker, Inc.;1998.p 177-181, 285-296.

13. Rice RW. Evaluating porosity parameters for porosity-property relations. J Am Ceram Soc 1993;76:1801 - 1808.

14. Ren F, Smith IO, Baumann MJ, Case ED. Three-dimensional microstructural study of porous hydroxyapatite using laser scanning confocal microscopy (CLSM). Intl. J. Appl. Ceram. Technol. 2005;2(3):200-211.

15. Wachtman JB, Jr. Mechanical properties of ceramics. New York: Wiley;1996.p 409-412.

16. Barsoum MW. Fundamentals of Ceramics. New York: McGraw-Hill;1997.p 411.

17. Green DJ. Fabrication and mechanical properties of lightweight ceramics produced by sintering hollow spheres. J Amer Ceram Soc 1985;68:403-409.

18. Green DJ, Hoagland RG. Mechanical behavior of lightweight ceramics produced by sintering hollow spheres. J Am Ceram Soc 1985;68:395-398.

19. Nielson LF. Elastic properties of two phase materials. Mater Sci Eng 1982;52:39-62.

20. Nielson LF. Elasticity and damping of porous materials and impregnated materials. J Am Ceram Soc 1984;67:93-98.

21. Ashkin D, Haber RA, Wachtman JB. Elastic properties of porous silica derived from colloidal gels. J Am Ceram Soc 1990;73:3376-3381.

22. German RM. Sintering theory and practice. New York: John Wiley and Sons;1996.p 95-121.

23. Helmuth RA, Turk DH. Elastic moduli of hardened cement pastes", Symp. on Structure of Portland Cement Pastes and Concrete. Washington, D. C.: Highway Res. Bd.; 1966. Report nr 90. 135 - 144 p.

24. deGroot K. Effect of porosity and physicochemical properties on the stability, resorption and strength of calcium phosphate ceramics. In: Ducheyne Pea, editor. Bioceramics: Material characteristic versus in vivo behavior. New York; 1988. p 227-234.

25. deGroot K. Effect of porosity and physicochemical properties on the stability, resorption, and strength of calcium phosphate ceramics. Ann N Y Acad Sci 1988;523:227-233.

26. Hench LL. Bioceramics: From concept to clinic. J Am Ceram Soc 1991;74:1487-1510.

27. Bone Engineering. Toronto: em squared incorporated;2000.

28. Klawitter JJ, Bagwell JG, Weinstein AM, Sauer BW, Pruitt JR. An evaluation of bone growth into porous high density polyethylene. J. Biomed. Mater. Res 1976;10:311-323.

29. Hulbert SF, Young FA, Mathews RS, Klawitter JJ, Talbert CD, Stelling FH. Potential of ceramic materials as permanently implantable skeletal prostheses. J. Biomed. Mater. Res 1970;4:433-456.

30. Shimazaki K, Mooney V. Comparative study of porous hydroxyapatite and tricalcium phosphate as bone substitute. J. Orthop. Res. 1985;3(3):301-310.

31. White E, Shors EC. Biomaterial aspects of Interpore-200 porous hydroxyapatite. Dent. Clin. North Am. 1986;30(1):49-67.

32. Lu J, Flautre B, Anselme K, Hardouin P, Gallur A, Descamps M, Thierry B. Role of interconnections in porous bioceramics on bone recolonization in vitro and in vivo. J Mater Sci: Mater Med 1999;10:111-120.

33. Hing K, Best S, Tanner K, Bonfield W, Revell P. Quantification of bone ingrowth within bone-derived porous hydroxyapatite implants of varying density. J Mater Sci: Mater Med 1999;10:663-670.

34. Hoepfner TP, Case ED. The porosity dependence of the dielectric constant for sintered hydroxyapatite. J Biomed Mater Res 2002;60(4):643-650.

35. Hoepfner TP, Case ED. The influence of the microstructure on the hardness of sintered hydroxyapatite. Ceram Intl 2003;29(6):699-706.

36. Saigal A, Fuller ER. Analysis of stresses in aluminum-silicon alloys. Comput. Mater. Sci. 2001;2(1):149 - 158.

37. Chawla N, Patel BV, Koopman M, Chawla KK, Saha R, Patterson BR, Fuller ER, Langer SA. Microstructure-based simulation of thermomechanical behavior of composite materials by object-oriented finite element analysis. Mater. Charact. 2002;49(5):395- 407.

38. Hsueh CH, Haynes JA, Lance MJ, Becher PF, Ferber MK, Fuller ER, Langer SA, Carter WC, Cannon WR. Effects of interface roughness on residual stresses in thermal barrier coatings. J. Am. Ceram. Soc. 1999;82(4):1073 - 1075.

NOVEL ROUTES TO STABLE BIO-TEMPLATED OXIDE REPLICAS

Y. Cheng
Chemistry, School of Biomedical and Molecular Sciences,
University of Surrey,
Guildford, Surrey, GU2 7XH, United Kingdom.

L. Courtney
Chemistry, School of Biomedical and Molecular Sciences,
University of Surrey,
Guildford, Surrey, GU2 7XH, United Kingdom.

P.A. Sermon
Chemistry, School of Biomedical and Molecular Sciences,
University of Surrey,
Guildford, Surrey, GU2 7XH, United Kingdom.

ABSTRACT

Pollen has been treated with zirconium, titanium and silicon alkoxides deposited *via* chemical vapor deposition in order to prepare ceramic nanomaterials that mimic the parent biological structure. The oxides overcoats were equivalent to 3-5 At.% Zr, Si or Ti. The alkoxide-treated samples were gasified in air at 873K and their morphology was then assessed by scanning electron microscopy (SEM) and energy dispersive X-ray analysis (EDX). The oxide overcoats changed the ease of gasification of the pollen.

INTRODUCTION

There is a momentum towards developing inorganic nanostructures based on complex three-dimensional self-assembled bio-templates that are available in a variety of conformations with nm resolution in nature.[1] These bio-templates can be used to cast inorganic phases and produce hollow inorganic replica structures.[2]

Thus, Xu and co-workers have plated Cu in 1μm thick layers onto the surface of pollen, assuming that the pollen was stable in HCl, $CuSO_4$ and NaOH solutions.[3] Mann and co-workers have reported the production of hollow SiO_2 replicas retaining the fine detail of the biotemplate when freeze-dried pollen grains 25μm in size were soaked in $Si(OH)_4$ solutions[4], dried and the biological template thermally removed at 673-873K. One problem, however, was a 50% reduction in size (and an 88% reduction of the volume) compared to the original. Ota and co-workers have reported the vacuum infiltration of wood with a $Ti(OR)_4$ solution, which after drying and calcination produced a titanified structure.[5] The chemical structure of bio-surfaces is clearly important.[6] Because pollen usually contains 5-50% water[7], freezing[8] and dehydration-rehydration may not be fully reversible.[9] Deformation and shrinkage are common problems in green body densification and sintering[10], and the aim here was to achieve nearly zero-shrinkage[11] in biological-ceramic composites (Biocers).

The authors decided to extend previous work undertaken[12] that over-coated a silica sol-gel on a borosilicate substrate with TiO_2 that was introduced as the vapor of titanium isopropoxide.[13] The alkoxide vapor reacted with the surface –OH groups of the silica. This is possible because of the known vapor pressures of alkoxides at moderate temperatures.[14]

EXPERIMENTAL

Air and oxygen-free N_2 (B.O.C.) were used without further purification. *Lilium longiflorum*[15] flowers and pollen samples (obtained from the Royal Botanic Gardens, Kew (London) and from the RHS Gardens (Wisley)) were used. Stamens were relatively large and contained a large amount of pollen. The authors used either the stamen or sections of the stamen that were removed from the flower head and stored in a lidded plastic tub to dry naturally.

Zirconium isopropoxide (70 wt% in 1-propanol), titanium (IV) isopropoxide (97 %) and tetraethoxyorthosilicate (TEOS, 99%) (Aldrich) were used without further purification.

The *Lilium longiflorum* pollen stamens were metal alkoxide coated by a chemical vapor deposition (CVD) method. Rather than remove the pollen[16], here the whole stamen, anther, and pollen from *Lilium longiflorum* was placed in a sealed reactor containing a reservoir of the alkoxide. A stamen was suspended in the N_2 filled apparatus over a reservoir of the appropriate metal alkoxide. An isomantle was used to heat the alkoxide, and a H_2O bubbler was present to absorb unreacted gaseous alkoxide. The CVD reaction was run for 60 min at 493K. The over-coated sample was carefully removed from the system and hydrolyzed by humid air to produce thinly coated pollen. Samples were then air dried and calcined, as appropriate.

Once hydrolyzed and air dried, some treated and untreated samples were subject to temperature programmed oxidation (TPO) analysis in air flowing at 100 cm^3 min^{-1} as the temperature increased at a programmed rate (β). The exhaust stream was analyzed for CO_2 by calibrated IR (Rotork NDIR 401).

RESULTS AND DISCUSSION

ZrO_x Coated Pollen

During the reaction with zirconium isopropoxide the stamen, anthers and pollen became gray, indicating the deposition of the zirconium alkoxide. Figure 1 shows scanning electron micrographs of the pollen untreated (a), untreated but calcined in air at 873K (b) and ZrO_x-coated and then calcined in air at 873K (c). Figure 2 shows the temperature-programmed oxidation profiles of the untreated and ZrO_x-coated pollen. First EDX associated with the SEM suggested that the fresh pollen contained 77.7 At.% C, 17.5 At.% K, 5.67 At.% Al and 0.03 At.% Si, but obviously no Zr. Secondly, SEM suggested that the texture and morphology of the pollen was retained on 873K calcinations in air with the ZrO_x-coated pollen, but not with the uncoated pollen. This is despite the fact that the TPO profiles in Figure 2, CO_2 is released at lower temperature with the ZrO_x overcoat. Thus CO_2 liberation is seen to occur at lower temperatures than those for untreated pollen (*i.e.* the ZrO_x-treated sample exhibited a CO_2 peak at ~650K in Figure 2a, while the untreated pollen only gave a peak at ~700K in Figure 2a). TPO data are summarized in Table 1. The apparent activation energy for CO_2 liberation from ZrO_x-treated pollen at the start of the gasification peak at 623-708K was 119kJ mol^{-1}. Hence the ZrO_x overcoat appears to chemically enhance pollen gasification, rather than hinder it, and so is presumed to be physically porous.

Table 1. Summary of TPO data from Figure 2.

Samples	Start of CO_2 production (K)	Maximum CO_2 rate (K)	Decrease in temp at max CO_2 rate (K)	Max [CO_2] (ppm)
pollen	443	708	-	201
pollen / ZrO_x	463	648	60	141

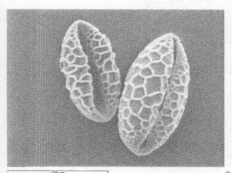

a) Fresh uncoated pollen (scale = 100μm)

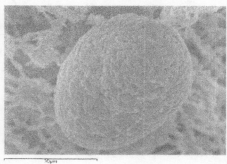

b) uncoated / calcined (873K) pollen (scale = 30μm)

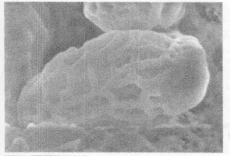

c) ZrO_x-coated / calcined (873K) pollen (scale = 10μm)

Figure 1. Fresh uncoated/uncalcined pollen (a), uncoated/calcined pollen (b) and ZrO_x-coated/calcined pollen (c)

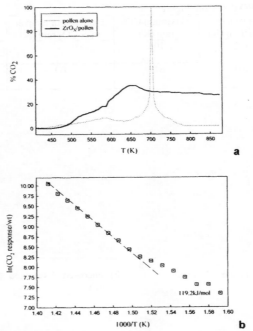

a

b

Figure 2. TPO data showing a) production of CO_2 from pollen with and without a ZrO_x overcoat in air flowing at 100 cm³ min⁻¹ during heating at β = 10K min⁻¹, and b) a pseudo-Arrhenius plot

Figure 1 suggested that the ZrO_x replicas have the same structure and dimensions as the untreated pollen (despite the thermal treatment). Hence it is concluded that the reaction between alkoxide vapor and the hydrated pollen surface at 353K (followed by subsequent gasification of the bio-template at 873K) has produced oxide-coated replicas of the original bio-templating pollen with close to zero-shrinkage. EDX showed that the ZrO_x coated pollen contained 5.39 At.% Zr. Figure 3 shows that the Zr was uniformly distributed over the pollen.

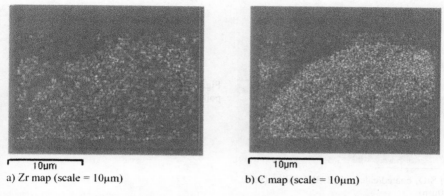

a) Zr map (scale = 10μm)   b) C map (scale = 10μm)

Figure 3. Zr map (a) and C map (b) on the ZrO_2 coated / calcined (873 K) pollen

That the core is resistant to oxidative gasification is a puzzle that remains to be solved. It may be that sporopollenin is relevant. This is a major phenylalanine-rich polymeric component of pollen grains and is known to be very resistant to degradation.[17] Here it must help ensure retention of the pollen shape (if not the fine structure, unless the oxide overcoat is present) under severe oxidation conditions.

SiO$_x$ Coated Pollen

EDX analysis showed that treating the pollen with tetraethoxysilane and calcining at 873K left 6.0 At.% Si on the bio-template. Figure 4a shows that again the pollen structure is retained after heating in air at 873K for 8 h when over-coated with SiO$_x$. The elemental maps in Figure 4 also suggest that Si has been grafted uniformly onto the pollen surface.

TPO in Figure 5 shows that a SiO$_x$-overcoated surface hinders the gasification of the pollen and the release of CO_2 in that this is seen to occur at a higher temperature than those for untreated pollen. Table 2 summarizes relevant TPO data. It seems that the SiO$_x$ hinders pollen gasification and it may be that the SiO$_x$ is unreactive or non-porous.

Table 2. Summary of TPO data from Figure 5.

Samples	Start of CO_2 production (K)	Maximum CO_2 rate (K)	Decrease in temp at max CO_2 rate (K)	Max [CO_2] (ppm)
pollen	443	708	-	201
pollen / SiO$_x$	513	808	-100	201

Figure 4. (a) SEM of SiO$_x$ coated/calcined pollen with b) Si map and c) C map.

a) SiO$_x$ coated/calcined (873K) pollen (scale = 10μm)

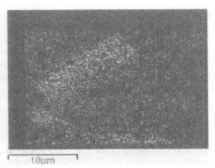

b) Si map (scale = 10μm)

c) C map (scale = 10μm)

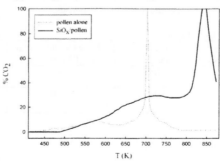

Figure 5. TPO data showing production of CO$_2$ from pollen with and without SiO$_x$. For this process the activation energy appeared to be 183kJ mol^{-1}.

TiO$_x$ Coated Pollen

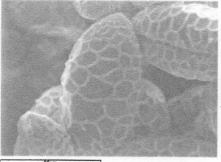

Figure 6. (a) SEM of TiO$_x$ coated/calcined (873K) pollen with b) Ti map and c) C map.

a) TiO$_x$-coated / calcined (873K) pollen (scale = 50μm)

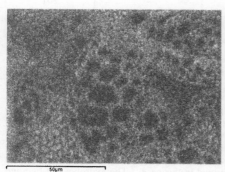

b) Ti map (scale = 50μm)

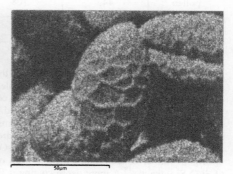

c) C map (scale = 50μm)

Figure 6 shows the structure of the TiO$_x$ coated pollen after heating in air at 873K for 8 h and from this micrograph it is again clear that the TiO$_x$ overcoat (which EDX says is equivalent to 3.4 At.% Ti) is uniform and has allowed the pollen to retain its original structure in this calcinations. In this case TPO indicates (see Figure 7 and Table 3) that CO$_2$ release is greater than on the pollen alone, but occurs at a higher temperature than the untreated pollen. Hence the titania overcoat appears to chemically hinder pollen gasification, rather than enhance it, and so is presumed, also, not to be physically porous.

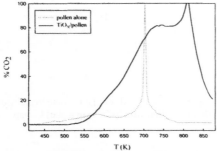

Figure 7. TPO data showing production of CO_2 from pollen with and without TiO_x

Table 3. Summary of TPO data from Figure 7.

Samples	Start of CO_2 production (K)	Maximum CO_2 rate (K)	Decrease in temp at max CO_2 rate (K)	Max [CO_2] (ppm)
pollen	443	708	-	201
pollen / TiO_x	493	838	-130	201

Alkoxide Discussion

It is also interesting to note that the three metal alkoxides used affected gasification temperatures in different ways. For example, the ZrO_x coating lowered the gasification temperature by 60K, whilst the SiO_x and TiO_x coatings raised the temperature by 100 and 130K respectively, when compared to the uncoated pollen samples. This is attributed this to the different nature, reaction rates and mechanisms of hydrolysis and condensation of the metal alkoxide precursors used. For example, the hydrolysis of TEOS at room temperature in moist air results in the creation of silanol species (Si-OH) in a penta-coordinated Si complex, which undergoes further condensation reactions with other silanol groups forming a growing network of Si-O-Si linkages.[18] A more dense coating is formed in this case.[19] The hydrolysis of titanium alkoxides follow the general method outlined for silicon alkoxides, but three of the alkoxides (the ethoxide, isopropoxide and the butoxide) form peroxo complexes that have a deep orange color (e.g. $Ti(O_2)(OH)_{aq}(OC_3H_7)$ is an example of such an orange colored complex). The mechanism leading to their formation is yet unknown.[20] Zirconium alkoxides, on the other hand, are unlike other metal alkoxides in that it is not hydroxo ligands which replace the alkyl groups, but either oxo or aqua species, forming a de-alcolation reaction.[21]

$$Zr(OR)_x(OH)_y \rightarrow ZrO_{x-(x+1)}(OR)_{x+1} + (x-(x+1))ROH$$

This involves four main steps; the first sees a -OH species substitute a -OR group, as in a typical hydrolysis. In the following three steps, this hydroxo ligand separates from the complex together with another alkoxy group, and allows the creation of a Zr=O double bond. These first two steps are repeated for the last two OR groups of the initial alkoxide, so that the final product formed is a pure oxide, ZrO_2.

CONCLUSIONS

The authors have shown that alkoxide vapours allow the grafting of ceramic oxide overcoats of ZrO_x, SiO_x and TiO_x onto pollen (and presumably other biomass) *via* a non-vacuum chemical vapor deposition (CVD) process. These overcoats either catalyze or retard the pollen gasification. This approach allows oxide overcoat to replicate the bio-template and the structure remains even after calcinations to 873K. It may be that these biomaterials will find use as forensic tags that can be followed through extreme conditions, such as combustion or explosions.

Further characterization work will be undertaken on the existing samples, as well as to consider the effect of copper oxide / copper methoxide over-coating on pollen on the replication of these delicate structures. It is contended that biomass conversion can be catalyzed by CuO_x, but whether or not the driving force will be thermal energy or high value chemical intermediates remains to be resolved.

ACKNOWLEDGEMENTS

The authors wish to thank Simon Owen (Royal Botanic Gardens, Kew) and Dawn Edwards (RHS Garden, Wisley) for provision of pollen.

REFERENCES

[1] A. Muller, *J. Chem. Soc. Chem. Commun.*, 803, (2003)
[2] S. Polarz and M. Anmtonietti, *J. Chem. Soc. Chem. Commun.*, 2593, (2002)
[3] L. Xu, K. Zhou, H. Xu, H. Zhang, J. Liao, A. Xu, N. Gu, H. Shen and J. Liu, *Appl. Surf. Sci.*, **183**, 58, (2001)
[4] S.R. Hall, H. Bolger and S. Mann, *J. Chem. Soc. Chem. Commun.*, 2784, (2003)
[5] T. Ota, M. Imaeda, H. Takase, M. Kobayashi, M. Hirashita, H. Miyazaki and Y. Hikichi, *J. Am. Chem. Soc.*, **83**, 1521, (2000)
[6] B. Kellam, P.A. de Bank and K.M. Shakeshaff, *Chem. Soc. Rev.*, **32**, 327, (2003)
[7] E. Pacini, *Plant Systemat. Evolu.*, **222**, 19, (2000)
[8] F.J.M. Bonnier, R.C. Jansen and J.M. Van Tuyl, *J. Plant Physiol.*, **151**, 627, (1997)
[9] J.S. HeslopHarrison and Y. HeslopHarrison, *Biology of the Cell*, **75**, 245, (1992)
[10] A.Z. Shui, S. Tanaka, N. Uchida and K. Uematsu, *Adv. Ceram. Composites*, **247**, 61, (2003), and K. Mori, H. Matsubara, N. Noguchi, M. Shimizu and H. Nomura, *J. Ceram. Soc. Jap.*, **111**, 516, (2003)
[11] U. Soltmann, M. Bottcher, D. Koch and G. Grathwohl, *Mater. Lett.*, **57**, 2861, (2003)
[12] P.A. Sermon and F.P. Getton, *J. Sol-gel Sci. Tech.*, **33**, 113, (2005) and F.P. Getton, V.A. Self, J. M. Ferguson, J.G. Leadley, P.A. Sermon and M. Montes, *Ceram. Trans.*, **81**, 355, (1998)
[13] D.C. Bradley and J.D. Swanwick, *J. Chem. Soc.*, 3207, (1958)
[14] D.C. Bradley and J.D. Swanwick, *J. Chem. Soc.*, 748, (1959)
[15] D.T. Nhut, *Plant Growth Regn.*, **40**, 179, (2003) and M. Griessner and G. Obermeyer, *J. Membrane Biol.*, **193**, 99, (2003)
[16] J.S. Gallo, US Patent No. 5249389 (1993)
[17] G. Eglinton and G.A. Logan, *Philos. Trans. R. Soc. Lon. B. Biol. Sci.*, **333**, (1268), 315 (1991)

[18]C.J. Brinker and G. Scherer, *Sol-Gel Processing: The Physics and Chemistry of the Sol-Gel Process*, Academic Press, London, UK, (1991)

[19]C.G. Swain, M. Esteve and R.H. Jones, *J. Chem. Soc.*, **11**, 965, (1949)

[20]J. Mühlebach, K. Müller and G. Schwarzenbach, *Inorg. Chem.*, **9**, 2381, (1970)

[21]A.C. Pierre, *Introduction to Sol-Gel Processing,* Kluwer Academic Publishers, Norwell, MA, USA. (1998)

Geopolymers

Geopolymers

ROLE OF ALKALI IN FORMATION OF STRUCTURE AND PROPERTIES OF A CERAMIC MATRIX

Pavel Krivenko, DSc (Eng), Professor
V.D.Glukhovsky Scientific Research Institute for Binders and Materials, Kiev National University of Civil Engineering and Architecture
Vozduhoflotsky Prospect 31
Kiev 03037 Ukraine
Email: sribm@mail.vtv.kiev.ua

ABSTRACT

Less than half a century ago just an idea of the presence of free alkalis in a ceramic matrix was considered by the ordinary Portland cement community as an absurd one, and this was a basic postulate accepted in the chemistry of cements. In 1957, a scientist from Ukraine (USSR), Victor Glukhovsky put forward an assumption which was taken as a basis for development and bringing into the practice of construction, a principally new class of cementitious materials which first appeared in the art, under the name of alkaline (now also known under a name of alkali-activated) cements. The validity of these ideas was confirmed by more than 50 years of evolutional development and vast experience was collected on the practical use of new materials in different applications. The paper covers theoretical views on the role played by alkali in the formation of a cement stone structure. Examples of compositional build-up of the alkali-activated cementitious materials vs the quantity of alkali and type of aluminosilicate component are reported as well as a summary of 40-years' experience obtained from observation of concrete structures made with these cements.

INTRODUCTION

Alkali metal compounds were excluded from the traditional hydraulic cement constituents due to their high solubility. At the same time, the studies held in order to reveal the reasons for the excellent durability of ancient cements, together with data collected on the stability and composition of natural mineral formations testified that this postulate was not correct.

According to the data reported in[1-7], a reason for the excellent durability is attributed to a considerably greater content of the alkali metal compounds in ancient cements compared to contemporary Portland cement stone. This was found to result in the formation of alkaline hydroaluminosilicate compounds which are analogous to natural zeolites in a cement stone structure, along with calcium hydrosilicates.

THEORETICAL BACKGROUND OF ALKALINE/ALKALI-ACTIVATED/ CEMENTS

The use of alkali(s) in cement has a long history going back to the beginning of the twentieth century. Earlier, some researchers considered these compounds to be used only for "loosening" in indirect accelerated methods of assessment of slag activity in slag Portland cements[8, 9, 10]. According to the literature,[11, 12] in the case of introduction of sodium hydroxide or its mixture with sodium sulfates, carbonates and chlorides a conclusion was made that the alkalis acted only as catalysts because they remained after hardening in a free/non-bound/ state.

V.D. Glukhovsky was the first who, as long ago as in 1957, showed that a requirement applied to restricted solubility of the mineral binding material and its rate, which, according to theories existing during those times caused their ability to hydration harden, should be laid upon not with regard to starting/constituent materials, but to the resulting products of their hydration and hardening.

These important conclusions were put as a background in creating a principally new class of cementitious materials viz., alkaline or alkali-activated cements which were themselves a mixture of alkali metal compounds of the first Group of the Periodic Table and aluminosilicates of natural and artificial or man-made origin.[13, 14] The idea itself of using these systems as cementitious ones was based, first of all, on geological data that sodium, potassium, sodium-potassium-calcium aluminosilicate compounds, which were known to have higher stability and resistance to atmospheric reagents, were present in the Earth's crust, side by side with the calcium compounds (Figure 1).

Secondly, this idea was based on the results of experimental studies, which proved that the alkali hydroxides and salts of alkali metals come into interaction with clay minerals, aluminosilicate glasses and crystalline substances of natural and artificial origin with the formation of water-resistant alkaline and alkaline-alkali-earth, aluminosilicate, hydration products, analogous to natural minerals of the zeolite and mica types.[15]

Thus, an idea for the creation of cementitious systems was transformed in the following manner:

"Old scheme" (ordinary Portland cement, (high) alumina cement)

$$\left.\begin{array}{l} \text{CaO-SiO}_2 \\ \text{CaO-Al}_2\text{O}_3 \end{array}\right\} +\text{H}_2\text{O} \longrightarrow \begin{array}{l} \text{CaO} - \text{SiO}_2 - \text{H}_2\text{O} \\ \text{CaO} - \text{Al}_2\text{O}_3 - \text{H}_2\text{O} \end{array}$$

"New scheme" (alkali- activated cement)

$$\left.\begin{array}{l} \text{Al}_2\text{O}_3\text{-SiO}_2 \\ \text{MeO-Al}_2\text{O}_3\text{-SiO}_2 \end{array}\right\} \text{MeO} + \text{H}_2\text{O} \longrightarrow \begin{array}{l} \text{Me}_2\text{O} - \text{Al}_2\text{O}_3 - \text{SiO}_2 - \text{H}_2\text{O} \\ \text{MeO} - \text{Me}_2\text{O} - \text{Al}_2\text{O}_3 - \text{SiO}_2 - \text{H}_2\text{O} \end{array}$$

Under conditions of elevated temperature, the formation of the artificial stone takes place according to the following:

$$\begin{array}{l} \text{MeO} - \text{Al}_2\text{O}_3 - \text{SiO}_2 - \text{H}_2\text{O} \\ \text{MeO} - \text{Me}_2\text{O} - \text{Al}_2\text{O}_3 - \text{SiO}_2 - \text{H}_2\text{O} \end{array} \xrightarrow{T^\circ C} \begin{array}{l} \text{MeO} - \text{Al}_2\text{O}_3 - \text{SiO}_2 \\ \text{MeO} - \text{Me}_2\text{O}_3 - \text{Al}_2\text{O}_3 - \text{SiO}_2 \end{array}$$

where MeO = CaO, MgO, SrO

$\text{Me}_2\text{O} = \text{Na}_2\text{O}, \text{K}_2\text{O}, \text{Li}_2\text{O}$

Scheme for Feldspar Weathering
(according to I. J. Ginzburg)[16]

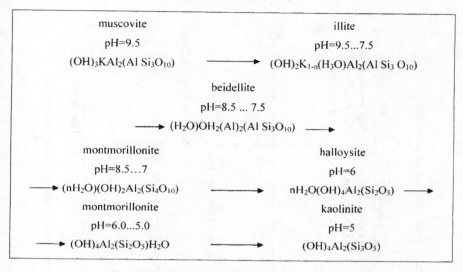

Scheme of Formation of Sedimentary Products

Zeolites of Sedimentary Origin

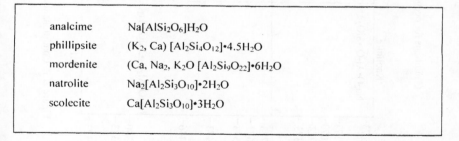

Figure 1. Geological data on weathering and sedimentation processes of natural rocks.

Conditions for occurrence of various zeolite based on clays and sodium carbonates.

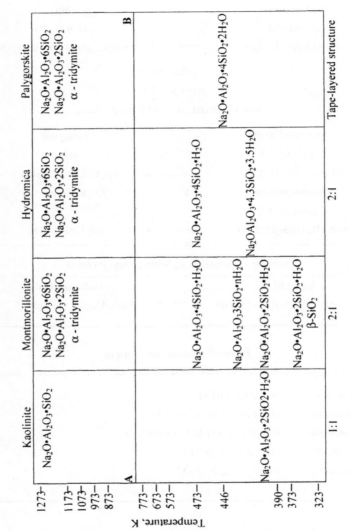

Figure 2(a). Conditions for occurrence of various zeolite based on clays and sodium carbonates.

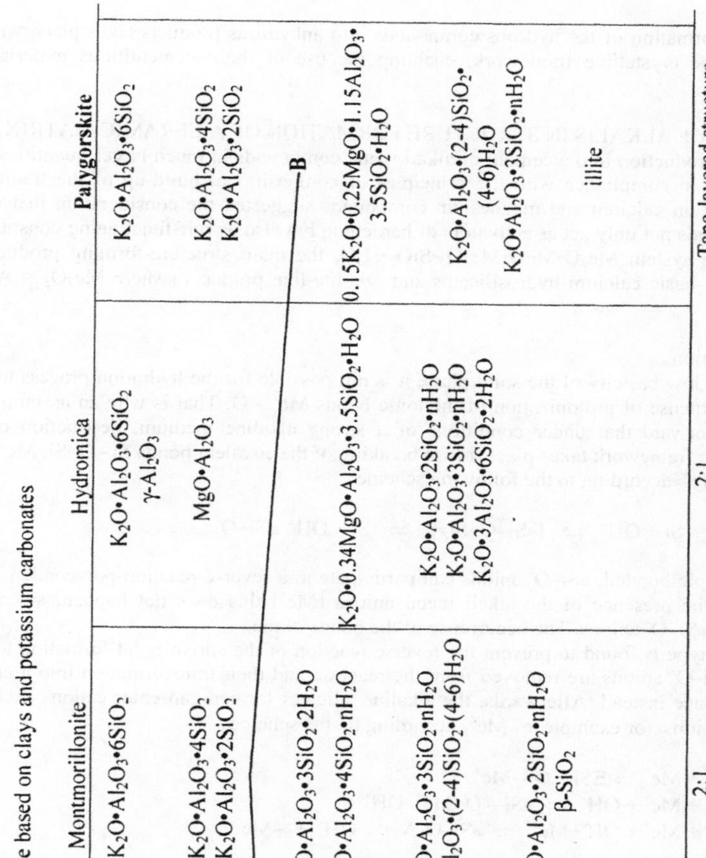

Conditions for occurrence of various zeolite based on clays and potassium carbonates

Temperature, K	Kaolinite 1:1	Montmorillonite 2:1	Hydromica 2:1	Palygorskite illite, Tape-layered structure
1273	$K_2O \cdot Al_2O_3 \cdot 2SiO_2$	$K_2O \cdot Al_2O_3 \cdot 6SiO_2$	$K_2O \cdot Al_2O_3 \cdot 6SiO_2$ $\gamma\text{-}Al_2O_3$	$K_2O \cdot Al_2O_3 \cdot 6SiO_2$
1173				
1073				
973		$K_2O \cdot Al_2O_3 \cdot 4SiO_2$ $K_2O \cdot Al_2O_3 \cdot 2SiO_2$	$MgO \cdot Al_2O_3$	$K_2O \cdot Al_2O_3 \cdot 4SiO_2$ $K_2O \cdot Al_2O_3 \cdot 2SiO_2$
873				
773				
673		$K_2O \cdot Al_2O_3 \cdot 3SiO_2 \cdot 2H_2O$		
573		$K_2O \cdot Al_2O_3 \cdot 4SiO_2 \cdot nH_2O$		
473	$K_2O \cdot Al_2O_3 \cdot 2SiO_2 \cdot nH_2O$ $K_2O \cdot Al_2O_3 \cdot 4SiO_2 \cdot H_2O$			
446	$K_2O \cdot Al_2O_3 \cdot 3SiO_2 \cdot nH_2O$ $K_2O \cdot Al_2O_3 \cdot (2-4)SiO_2 \cdot (4-6)H_2O$	$K_2O \cdot Al_2O_3 \cdot 3SiO_2 \cdot nH_2O$ $K_2O \cdot Al_2O_3 \cdot (2-4)SiO_2 \cdot (4-6)H_2O$	$K_2O \cdot 0.34MgO \cdot Al_2O_3 \cdot 3.5SiO_2 \cdot H_2O$ $K_2O \cdot Al_2O_3 \cdot 2SiO_2 \cdot nH_2O$ $K_2O \cdot Al_2O_3 \cdot 3SiO_2 \cdot nH_2O$ $K_2O \cdot 3Al_2O_3 \cdot 6SiO_2 \cdot 2H_2O$	$0.15K_2O \cdot 0.25MgO \cdot 1.15Al_2O_3 \cdot 3.5SiO_2 \cdot H_2O$ $K_2O \cdot Al_2O_3 \cdot (2-4)SiO_2 \cdot (4-6)H_2O$ $K_2O \cdot Al_2O_3 \cdot 3SiO_2 \cdot nH_2O$
390		$K_2O \cdot Al_2O_3 \cdot 2SiO_2 \cdot nH_2O$ $\beta\text{-}SiO_2$		
373				
323	$K_2O \cdot Al_2O_3 \cdot 2SiO_2 \cdot nH_2O$			

A   B

Figure 2(b). Conditions for occurrence of various zeolite based on clays and potassium carbonates.

Transformation of the hydrous compounds into anhydrous products takes place without breaking of the crystalline framework, enabling the use of these cementitious materials as thermocements.

ROLE OF ALKALIS IN STRUCTURE FORMATION OF A CERAMIC MATRIX

The introduction into a cement of alkali metal compounds in much larger quantities than was permitted in compliance with the principles of compositional build-up of the traditional cements based on calcium and magnesium compounds suggested the consideration that alkali metal compounds not only act as activators of hardening but also as self-functioning constituents of the binding system $Me_2O \cdot MeO \cdot Me_2O_3 \cdot SiO_2 \cdot H_2O$, the main structure-forming products of which are low-basic calcium hydrosilicates and zeolite-like products (where $Me_2O_3 = Al_2O_3$, Fe_2O_3).

Normal Conditions

Due to low basicity of the solid phase it is not possible for the hydration process to take place at the expense of protonization of the ionic bonds Me^{2+}- O. That is why an assumption[16] may be put forward that under conditions of a strong alkaline medium, destruction of the aluminosilicate framework takes place due to breaking of the covalent bonds Si $-O-Si$, $Me^{3+}-O-Me^{3+}$, Si$-O-Me^{3+}$ according to the following scheme:

$$\equiv Si - O - Si + OH^- \Leftrightarrow [\equiv Si-O-Si\equiv] \Leftrightarrow Si - OH^+ \equiv Si-O$$

The triple bonded, $\equiv Si-O^-$ anions can participate in a reverse reaction-polycondensation. However, in the presence of the alkali metal anions (Me^+) this does not happen, since they neutralize the $\equiv Si-O^-$ anions. The occurrence of the bonds of the $\equiv Si-O^-$ - Me^+ type is found to prevent the reverse reaction of the siloxo-bond formation to take place. The $\equiv Si-O^-$ anions are removed from the reaction and their transformation into a colloid phase takes place instead. Afterwards, the alkaline silicates formed can enter cation- exchange reactions with ions, for example, of Me^{2+} according to the scheme:

$$\equiv Si-O^- + Me^+ \Leftrightarrow \Xi Si - O - Me^+$$
$$\equiv Si-O^- + Me^+ + OH^- \Leftrightarrow \equiv Si - O-Me^+-OH^-$$
$$\equiv Si-O^- + Me^+ + OH^- + Me^{2+} \Rightarrow \equiv Si-O-Me^{2+}-OH^- + -Me^+$$

A validity of this assumption is supported by the wave-like character of curve reflecting changes in Na_2O-content of the leached solution as seen in Figure 3.

Thus, at the initial stage of hydration of the binder/cement (stage of dispergation (hydration through dissolving)) an alkali cation was found to act as a catalyst maintaining an ionic force in the liquid phase, which is necessary for destruction of strongly covalent bonds, and for participating in the transformation of the products of destruction into a colloid phase.

The anion constituent of the alkaline activator strongly affects the process of destruction of the solid phase. As has been established, the anion plays two roles. The anions introduced with the alkaline silicates and aluminates are analogous to the primary products of destruction of the aluminosilicate framework, serve as their additional reserve and, naturally, are the most effective. The anions introduced with the alkaline carbonates, sulfates, silicofluorides, or chlorides change the properties of the solution by participating in complex-formation or complexing or processes resulting in the release of the reaction products into a solid phase. This was found to affect the final properties of the cement stone (Table I).

Accumulation of the products of destruction resulting from adsorption and chemical dispergation of the solid phase is found to create the restrained conditions for the formation of a dispergation-coagulation structure in which polycondensation processes evolve or develop.

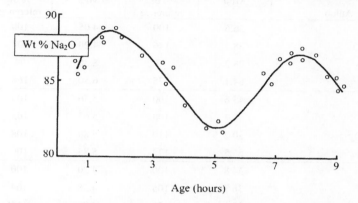

Age (hours)

Figure 3. Na_2O-content of leached solution of alkaline cement (C-A-S glass + NaOH) *vs* age.

A determining role in the condensation microstructure formation is played by Ca^{2+} cations that are capable of interacting with silica- and aluminosols, and participate in solid phase reactions through a cation exchange reaction with alkaline cations of the Me^+ type.

The quantity of the alkaline activator was found to essentially affect this process as well as composition and stability of the resulting hydration phases. Therefore, according to the literature,[18] the following sequence of phase formation is followed in the composition $Ca(OH)_2$ - SiO_2 - H_2O - NaOH having a molar ratio of $Ca(OH)_2/SiO_2$=1

(i)   at $NaOH/SiO_2$ < 1, tobermorite 1.13 nm with a degree of condensation of the siloxo-anions being intermediate between chain-like versus band-like

(ii)   at $NaOH/SiO_2$=1-5, tobermorite with chains of silica anions

(iii)   at $NaOH/SiO_2$>5, sodium substituted tobermorite ($NaCaHSiO_4$) having isolated ortho-anions.

Starting from a given moment, strength development is accompanied by a decrease in the pH value of the liquid phase. This may be attributed to its interaction with the products of destruction of the solid phase with the formation of stable, zeolite-like hydrosilicates of the alkaline or alkali- earth composition, which have the lower solubility compared to hydration products phase of Portland cement stone.

Table I. Influence of Anion of the Alkaline Activator on Strength of Slag Alkali-activated Cement[19]

Cation Na+ / Anion	Strength after steam curing treatment			
	in compression		in flexure	
	MPa	% of reference	MPa	% of reference
OH^-	36.6	100	4.05	100
$OH^- + Cl^-$	38.3	146	4.20	104
$OH^- + SO_2^{2-}$	28.2	77	3.80	94
$OH^- + SiF_6^2$	64.1	175	6.80	168
CO_3^{2-}	45.6	100	5.36	100
$CO_3^{2-} + Cl^-$	48.4	100	5.65	105
$CO_3^{2-} + SO_2^{2-}$	50.2	110	5.80	108
$CO_3^{2-} + SiF_6^2$	55.5	122	5.85	109
SO_2^{2-}	34.8	100	4.20	100
$SO_2^{2-} + Cl^-$	36.7	105	4.38	104
$SO_2^{2-} + SO_2^{2-}$	35.1	101	4.85	115
$SO_2^{2-} + SiF_6^2$	65.9	189	5.20	124
SiO_2^{2-}	81.8	100	8.26	100
$SiO_2^{2-} + Cl^-$	87.8	107	8.40	102
$SiO_2^{2-} + SO_2^{2-}$	95.7	117	8.70	105
$SiO_2^{2-} + SiF_6^2$	100.9	123	9.20	111

At early stages of hydration and hardening (for example, of the slag alkaline cements), the microstructure is caused mainly by the formation and crystallization of the low-basic hydrosilicates and hydrogarnets. The alkaline and alkali-earth hydroalumino•silicates, as a result of their slower crystallization, occur in the later stages. Being formed predominantly in pore spaces, they fill them and promote the formation of strong crystallization contacts with the primary phases, as well as initiate the formation of a more homogeneous and dense microstructure. Furthermore, high pH-values of the medium at which the hydration process takes place block the transformation of the Ca-ions into the solution, thus explaining the absence of $Ca(OH)_2$ and the fact that the resulting calcium hydrosilicate has, as a rule, a basicity of no more than 1.[20] Hence, the phase composition of the slag alkaline cement hydration products is represented predominantly by tobermorite-like, low-basic hydrosilicates of calcium, hydrogarnets of variable composition, alkaline hydroaluminosilicates of the zeolite and mica type, as well as alkaline-alkali-earth compounds which are less soluble, compared to the high-basic compounds and stable as a result of their low basicity (Table 2). This can be expected to provide stability and durability of the cement stone over time, not only in normal service conditions, but under corrosion exposure to various aggressive environments. High performance properties of the alkaline cement stone may be attributed not only to the more perfect phase composition of the hydration products but also to the more perfect features of its structure (Figure 4).

Conditions of Elevated Temperatures

The formation of calcium silicates modified with alkaline ions of Me^+ and alkali-alkaline-earth hydroaluminosilicates (hydronepheline, analcime, natrolite, sodalite) within the reacting anhydrous products was found to predetermine a smooth flow of dehydration processes without disturbing thestructure in the resulting fired cement stone. Final dehydration products are the reaction products of feldspar composition analogs of natural minerals of nepheline, kaliophilite, leucite, albite, orthoclase, etc. This may explain the high thermal or temperature resistance of the alkali-activated cement stone. E.g., a hydration product composition of the system $CaO \cdot Al_2O_3$-$CaO \cdot 2Al_2O_3 \cdot Na_2O \cdot SiO_2 \cdot H_2O$ is represented by high-silica hydrogarnets of the $3CaO \cdot Al_2O_3 \cdot 2SiO_2$-$3CaO \cdot Al_2O_3 \cdot SiO_2 \cdot 4H_2O$ composition, low-basic calcium hydrosilicates of the tobermorite row and alkali-alkaline-earth hydroalumino silicates of gmelinite-analcime

Table II. Comparative Data of Solubility of Hydration Products of Slag Alkali-activated and Portland Cement Stones

Cement Type	Hydration Product		Solubility, kg/cub m
	Mineral	Stoichiometric formula	
Slag alkali-activated cement	CSM(B)	$5CaO \cdot SiO_2 \cdot nH_2O$	0.05
	Xonotlite	$6CaO \cdot 6SiO_2 \cdot H_2O$	0.035
	Riversideite	$5CaO \cdot 6SiO_2 \cdot 3H_2O$	0.050
	Plombierite	$5CaO \cdot 6SiO_2 \cdot 10.5H_2O$	0.05
	Gyrolite	$2CaO \cdot 3SiO_2 \cdot 2.5H_2O$	0.051
	Calcite	$CaCO_3$	0.014
	Hydrogarnet	$3CaO \cdot Al_2O_3 1.5SiO_2 \cdot 3H_2O$	0.02
	Na-Ca hydrosilicate	$(Na,Ca) \cdot SiO_4 \cdot nH_2O$	0.05
	Thomsonite	$(Na,Ca) \cdot Si_2O_3 \cdot Al_2O_3 \cdot 6H_2O$	0.05
	Hydronepheline	$Na_2O \cdot Al_2O_3 \cdot 2SiO_2 \cdot 2H_2O$	0.02
	Natrolite	$Na_2O \cdot Al_2O_3 \cdot 3SiO_2 \cdot 2H_2O$	0.02
	Analcime	$Na_2O \cdot Al_2O_3 \cdot 4SiO_2 \cdot 2H_2O$	0.02
Portland cement	Calcium hydroxide	$Ca(OH)_2$	1.3
	C_2SH_2	$2CaO \cdot SiO_2 \cdot nH_2O$	1.4
	CSH(B)	$5CaO \cdot 6SiO_2 \cdot nH_2O$	0.05
	Tetracalcium hydroaluminate	$4CaO \cdot Al_2O_3 \cdot 13H_2O$	1.08
	Tricalcium hydroaluminate	$3CaO \cdot Al_2O_3 \cdot 6H_2O$	0.56
	Hydrosulfoaluminate	$3CaO \cdot Al_2O_3 \cdot 3CaSO_4 \cdot 31H_2$	high

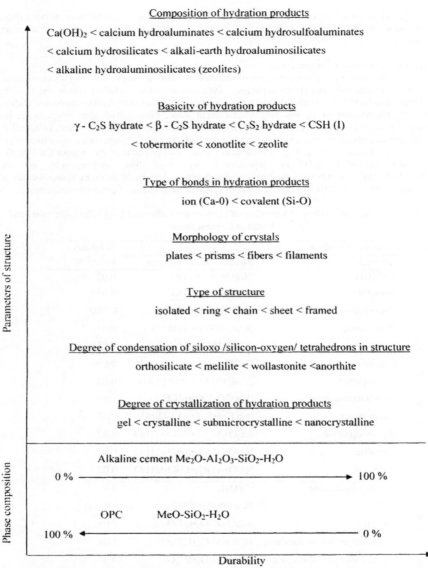

Figure 4. Durability of the cement stone versus hydration product phase and structure.

The formation of strong accretions or aggregates of the anhydrous phases as a result of thermal, high temperature treatment was found to predetermine the high heat resistance of the cement stone, which is expressed as a lack of strength decline (Figure 5).

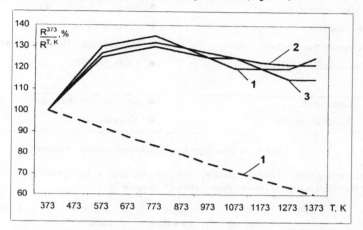

Figure 5. Change in residual strength while heating a cement stone based on calcium aluminates after mixed with water (stroked line) and with sodium silicate solution (solid line):1 - $CaO \cdot 2Al_2O_3$; 2 - slag of aluminum thermic production; 3 - (high) alumina cement.

PRINCIPLES OF COMPOSITIONAL BUILD UP OF ALKALINE CEMENTS

The established possibility to simulate the processes taking place in the Earth's crust, the processes based on an interaction between the decay products of rock-forming minerals, viz., clays and alkalis, followed by synthesis of these minerals, suggested the use of natural soils as starting materials for making binders. This explains why the binders/cementitious materials have been called "soil cements" by V.D.Glukhovsky, and the concretes made with them "soil silicates" to describe their nature. At the beginning these new cementitious materials were called "binders" to differentiate between ordinary Portland cements commonly defined as "cements." Later, when these new materials were brought into the practice of construction they were called "cements" – a general term applied to binding/cementitious/ materials. Below are given different names of these materials under which one can find them (Figure 6).

However, independent of the name applied to these cementitious materials, the following principles of compositional build up of the alkali-activated cements can be emphasized:

(i)   alkalis serve not only as activators but also as structure-forming elements of the resulting products

(ii)   the resulting phase products are characteristic of the presence of new compounds, which are $Me_2O \cdot Al_2O_3 \cdot SiO_2 \cdot H_2O$ and $Me_2O \cdot CaO \cdot Al_2O_3 \cdot SiO_2 \cdot H_2O$, $Me_2O \cdot Al_2O_3 \cdot SiO_2$ and $Me_2O \cdot CaO \cdot Al_2O_3 \cdot SiO_2$

(iii)   the quantities of alkalis to be introduced should meet a stoichiometric composition or stoichiometry requirement for alkaline and alkaline-alkali-earth hydroaluminosilicates and aluminosilicates analogous to natural zeolites.

• Soil cement/alkaline binder	• Glukhovsky 1957
• Geopolymer	• Davidovich, 1973
• Alkali-activated cement	• Narang, Chopra, 1983
• F-cement	• Forss, 1983
• Gypsum-free Portland cement	• Odler, Skalny, Brunauer, 1983
• SKJ-binder	• Changgo, 1991
• Geocement	• Krivenko, Skurchinskaya, 1991
• Alkaline cement	• Krivenko, 1994

Figure 6. Terminology for alkaline cementitious materials

According to the above terminology, at present the most widely spread are the following types of the alkali-activated cementitious materials[20] (Figure 7).

Cement Type	OPC	OPC alkaline cement	Slag alkaline cement	Ash alkaline cement	Geocement
Initial solid phase	OPC clinker	OPC clinker $+Me_2O$	Metallurgical slag $+Me_2O$	Ash of heat power stations (coal combustion) $+Me_2O$	Clay $+Me_2O$
Alkali content, Me_2O, %	< 0,6	1-5	2-8	5-10	10-20

Hydration product

$0\%\ Me_2O\text{-}Al_2O_3\text{-}SiO_2H_2O \longrightarrow 100\ \%$

$100\ \% \longleftarrow MeO\text{-}SiO_2\text{-}H_2O$

Figure 7. Examples of compositional constitution of the alkali-activated cements (reference-ordinary Portland cement (OPC)).

CONCLUSIONS

The presence of alkalis in cementitious materials is found to positively affect a ceramic matrix both under normal conditions of hydration and while heating or firing at conditions of elevated temperatures. The roles played by alkali can be described as the following:

- maintaining high pH value for the flow of hydrolytic destruction of aluminosilicate dispersions
- lowering the basicity of the calcium hydrosilicates which are formed
- lowering the degree of over-saturation of a pore fluid with calcium hydroxide
- participating in the formation of more stable and durable alkali and alkaline-earth

hydroaluminosilicates and aluminosilicates.

REFERENCES

[1] R. Malinowski, Ancient Mortars and Concretes: Aspect of Their Durability, *History of Technology*, **7** S9-101 (1982).

[2] R. Malinowski, A. Slatkine, M. Ben Yair, Durability of Roman Mortars and Concretes for Hydraulic Structures at Caesarea and Tiberias, *Proceed. Int. Symp. on Durability of Concrete*, Prague, 1- 14 (1961).

[3] R. Malinowski, Concrete and Mortars in Ancient Aquaducts, *Concrete International*, 1, **66**-67 (1979).

[4] R. Malinowski, Betontechnische Problemlosung bei antiken Wasserbauten, Leichtweiss Institut, Braunschweig, *Mitteilungeng*, **64**, 7—12 (1979).

[5] C. A. Langdon, D. M. Roy, Longevity of Borehole and Shaft Sealing Materials: Characterization of Ancient Cement-based Building Materials, *Materials Research Society Symposium Proceeding*, Pittsburgh, **26**, 543-549 (1984).

[6] J. Davidovits, Ancient and Modern Concretes: What is the Real Difference? *Concrete International*, **9**, 23- 29 (1988).

[7] V. D. Glukhovsky, Ancient, Modern and Future Concretes, *Proceed. 2nd Int. Seminar*, Gothenburg, Sweden, 53- 62 (1989).

[8] R. Feret, Slags for the Manufacture of Cement, *Rev. Mater. Constr. Trav. Publ.*, 1-145 (1939).

[9] L. Chasserent, *17 me Congr. De Chim* (1937).

[10] P. Budnikov, 1 Znachko-Jiavorsky, Granulated Slags and Slag Cements. Moscow: Promstroizdat Publisher (1953).

[11] H. Kuhl, Zementchemie, Verlag Technik, Berlin, Germany. Band III, (1958), or Zement, 19, (1930).

[12] A.Purdon, The action of Alkalis on Blast Furnace Slag, *Journal of the Society of Chemical Industry*, **59**, 191-202 (1940).

[13] V. D. Glukhovsky, Soil Silicate-Based Articles and Constructions, Budivel'nyk Publish., Kiev, Ukraine (1957)

[14] V. D. Glukhovsky, Soil Silicates (*Gruntosilikaty-in Russian*), Kiev, Budivel'nyk Publisher (1959).

[15] Alkaline and Alkaline–Alkali-earth Hydraulic Binders and Concretes (Edited by V.D.Glukhovsky), Vysschaya Shkola Publisher, Kiev, Ukraine (1979).

[16] I.J.Ginzburg, I.A.Rukavishnikova, Minerals of the Ancient Zone of Weathering of the Ural Mountains, Moscw, USSR (1951)

[17] P. Krivenko, Synthesis of Cementitious Materials of the $R_2O-Al_2O_3-SiO_2-H_2O$ System with Required Properties, DSc(Eng) Degree Thesis, Kiev, Ukraine (1986)

[18] V.Timashev, N. Nikonova, and A. Prostjakov, Some Physico-Mechanical Properties of Calcium Hydrosilicate Crystals, *Proceed. National Seminar on Calcium Hydrosilicates and Their Use*, Kaunas, Lithuania, 128-132 (1980).

[19] P. Krivenko, Special Slag Alkaline Cements, Kiev, Ukraine, Budivelnik Publisher (1995).

[20] F. P. Glasser., T. Hapuarachchi, The Role of Alkalis in Controlling Phase Development in Calcium Aluminosilicate Binders, *Proceed. 1st Int. Conference on Alkaline Cements and Concretes, Kiev, Ukraine*, **1**, 484-492 (1994).

[21] C. Shi, P. Krivenko, D. Roy, Alkali-activated Cements and Concretes, Taylor and Francis, London, 376 (2006).

WILL GEOPOLYMERS STAND THE TEST OF TIME?

John L. Provis[1], Yolandi Muntingh[2], Redmond R. Lloyd[1], Hua Xu[1], Louise M. Keyte[1], Leon Lorenzen[2], Pavel V. Krivenko[3], Jannie S. J. van Deventer[1]*

[1] Department of Chemical & Biomolecular Engineering
University of Melbourne
Victoria 3010, AUSTRALIA.
* Email: jannie@unimelb.edu.au

[2] Department of Process Engineering
University of Stellenbosch
Private Bag X1, Matieland, 7602, SOUTH AFRICA.

[3] V.D. Glukhovskii Scientific Research Institute for Binders and Materials
Kyiv National University of Civil Engineering and Architecture
Kiev, UKRAINE.

ABSTRACT

One of the major unanswered questions in the field of geopolymer technology relates to the long-term performance of geopolymeric materials in cement replacement applications. There are many different mechanisms by which traditional concretes become corroded and eventually fail, and the successful introduction of a new alternative to ordinary Portland cements (OPC) will require that the new product is at least as resistant to each of these forms of attack as the traditional technology. Durability testing is therefore a major research focus at present, and preliminary results are very promising in many areas. Accelerated chloride diffusion testing shows that the resistance of the geopolymer matrix to chloride penetration is orders of magnitude higher than that of OPC matrices, which provides highly significant advantages in prevention of rebar corrosion under aggressive salt-laden environments. By careful design of matrix composition and curing conditions, porosity can be tailored to provide optimal performance. Samples from the former Soviet Union that have been exposed to service conditions for in excess of 30 years are shown to have been remarkably little degraded, and in many cases display mechanical strengths significantly in excess of their original design strengths. The ongoing strength development and high durability are attributed to a number of factors including the ongoing polymerization occurring within the geopolymer gel matrix, as well as the presence of a very disconnected pore network. Geopolymers do therefore appear to stand the test of time in concrete replacement applications and in aggressive environments.

INTRODUCTION

Geopolymers have increasingly been studied over the past two decades due to their potential for application as a high-performance, environmentally friendly construction material [1,2]. The engineering properties of geopolymeric mortars and concretes have also been shown to compare favorably with those of OPC.[3] However, due to the relatively short timeframe in which geopolymer research has been conducted, the availability of data related to the durability of geopolymers under service conditions and for extended periods of time is necessarily limited. Some work has been conducted whereby geopolymers have been exposed to attack by corrosive solutions, however these studies have in general been focused on applications in toxic waste encapsulation or immobilization and so have utilized tests such as the Toxicity Characteristic Leaching Procedure (TCLP) protocol, the Product Consistency Test (PCT), or purpose-designed acid leach protocols.[4,5] The resistance of metakaolin-based geopolymers to attack by aggressive environments including sulfate solutions and simulated seawater has also been investigated.[4] Some zeolite crystallization from the initially amorphous geopolymeric binder was observed, but the mechanical properties of the geopolymer samples remained essentially undiminished. Some samples in fact increased in strength while submerged in aggressive solutions, which was attributed to microstructural developments within the binder phase.

The degradation of cements and concretes under environmental conditions is a major economic burden on modern society. An estimated USD1.6 trillion will be required between 2005-2010 to repair infrastructure to a 'serviceable' level in the USA alone,[5] and a significant proportion of this figure is attributable to the failure of Portland-based cements in service. Concretes degrade by a large variety of mechanisms, with probably the most significant in most parts of the world being chloride attack leading to corrosion of steel reinforcing.[6] Other issues that are significant to a greater or lesser degree in certain parts of the world or under specific service conditions include freeze-thaw cycling and carbonation. The presence of sulfate - at low concentrations in groundwater or in higher concentrations in specialized applications such as sewer pipes – also leads to cement degradation via ettringite and/or thaumasite formation.

The resistance of geopolymeric concretes to damage by several of these mechanisms has long been noted. Resistance to chloride and sulfate attack has always been a primary driver for geopolymer research, with many publications on this topic providing a wide variety of outcomes depending largely on the success (or otherwise) of the various attempts to generate a satisfactorily reacted geopolymer. In general, 'poor' resistance to acid attack is observed for samples with a low extent of geopolymer binder formation,[7,8] with performance significantly exceeding that of ordinary Portland cement (OPC) observed in all except very poorly reacted samples.[9] Similarly, blending 50% geopolymer cement with OPC has been shown to give very significant improvements in resistance to highly concentrated (pH < 1) H_2SO_4 solutions,[10] with the formation of a highly impermeable layer on the geopolymer surface observed in some instances (including in the presence of small amounts of Ca) to provide protection against further corrosion.[11] The lower Ca levels in geopolymeric binders when compared to OPC will generally be beneficial for sulfate resistance, as the formation of the expansive compounds responsible for sulfate degradation of OPC generally requires the presence of significant levels of available Ca.

The permeability of the geopolymeric binder phase has recently been shown to be very much lower than that of Portland cements.[12] This is likely to be of significant value in the development of geopolymer technology for applications ranging from bunkers for nuclear waste storage to coatings for protection of transport infrastructure. However, the main benefit of this low permeability in large-scale applications is the reduction of chloride ingress, which will minimize the rates of chloride attack on steel reinforcing. Geopolymer cements have been shown to provide a sufficiently high pH to maintain steel reinforcing for extended periods under high humidity – whether activated by hydroxide or silicate solutions – with performance in galvanostatic pulse testing comparable to that of OPC.[13] A low permeability is expected not only to reduce chloride penetration, but also to minimize carbonation (and therefore pH reduction) within the binder structure. The very much lower Ca levels of most geopolymers when compared with OPC will mean that any carbonation that occurs will most likely be via formation of sodium (bi)carbonates rather than $CaCO_3$ as is the case in high-Ca cements. This is quite a different mechanism of carbonation, and the exact ramifications of these differences on the rates of rebar corrosion are as yet unknown.

This paper presents a selection of results from investigations of geopolymer samples from a variety of sources, with the aim of providing an overview of geopolymer durability rather than analyzing a single set of samples in detail. An issue of critical importance in the development and international uptake of geopolymer technology is the validation of geopolymer durability for products synthesized from a range of waste material sources under widely differing conditions. The work presented here addresses just a few of the potential cases, but provides at least an indication of the expected performance of waste-based geopolymers in general.

MICROSTRUCTURES AND PROPERTIES OF GEOPOLYMER CONCRETES

Figures 1 and 2 present representative backscattered electron scanning electron microscope (BSE SEM) images of a fly ash-slag geopolymer concrete and a Portland cement (blended with ~20% fly ash) concrete respectively. Each micrograph shows the interface between the binder phase and a large particle of natural siliceous aggregate. The most significant difference between the two microstructures can be observed in the region immediately adjoining the aggregate particle in each case. Figure 1 shows that the geopolymer binder microstructure is essentially unchanged from the bulk up to within a few microns of the aggregate particle – and it is possible that even this difference is attributable to effects from the SEM sample polishing process due to the different hardnesses of the aggregate and binder phases.

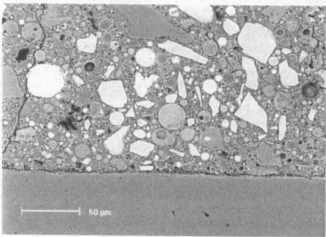

Figure 1. Interface between fly ash-slag geopolymer binder (top) and a large particle of natural siliceous aggregate (bottom)

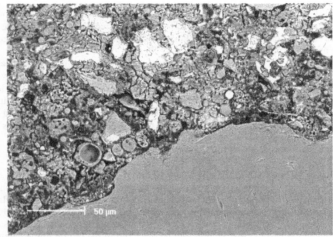

Figure 2. Interface between Portland cement (blended with ~20% fly ash) binder (top) and a large particle of natural siliceous aggregate (bottom)

In contrast, Figure 2 shows a significant change in the microstructure of the Portland cement in the region immediately surrounding the aggregate particle. The darker regions in this BSE SEM micrograph represent regions of higher porosity, and it is seen that the porosity of the binder phase in contact with the aggregate particle – the interfacial transition zone (ITZ) – is much higher than that of the bulk material. This is consistent with the results of previous studies of OPC-based concretes, and has been confirmed by detailed image analysis-based porosity calculations.[14] Previous work on the interface between geopolymers and siliceous aggregates has shown that an ITZ is only formed in fly ash-kaolinite geopolymers if significant (and deleterious) contamination of the mix by added ionic salts occurs prior to setting.[15] Figure 1 confirms that this is also true for the more industrially relevant fly ash-slag systems. The exact microstructure of the geopolymeric binder will of course depend significantly on binder composition and aluminosilicate source mineralogy, with silicate addition to the activating solution resulting in a less open pore structure.[16]

This absence of a significant ITZ region is likely to provide a significant durability advantage for geopolymer concretes when compared with OPC concretes. The porous ITZ of OPC provides an avenue whereby dissolved aggressive species (eg. acids) may more easily attack the binder structure, leading to localized material failure at the binder-aggregate join and creating macroporosity. This will then allow much more rapid penetration of the aggressive dissolved species throughout the material, degrading its mechanical properties more rapidly. In the absence of a porous ITZ, geopolymer concretes will be more resistant to this mode of chemical attack.

PERMEABILITY TESTS ON GEOPOLYMER PASTES AND MORTARS

The chloride permeability measurement of cement by electrically accelerated testing (the accelerated chloride-ion diffusion (ACID) test) is currently covered by ASTM C1202-05. The experiments described in this work were designed in accordance with that Standard, with the exception of the use of slightly different voltages and testing times to those specified by the Standard.[12] The experimental setup is illustrated in Figure 3. Very briefly, a small voltage (~30V) is applied across the sample, and the movement of Cl⁻ through the sample induced by this voltage is monitored. 3M NaOH and 3 wt% NaCl solutions are used, and the passage of Cl⁻ is measured directly via determination of the chloride concentration in the NaOH solution as a function of time.

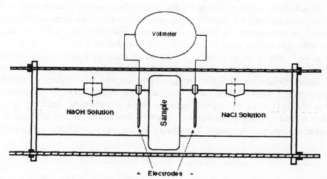

Figure 3. Apparatus used for accelerated chloride-ion diffusion (ACID) and sulfate diffusion tests.

Tests were conducted on a number of fly ash-based geopolymer pastes and mortars, derived from different fly ashes and with variable water and sand content, as well as in the presence of ground granulated blast furnace slags.[12] Two Australian fly ashes (from black coal combustion for electrical power generation) were used as the solid aluminosilicate sources for geopolymer synthesis throughout the ACID tests presented here. Both are Class F fly ashes, containing less than 5% CaO by mass. Specific details of sample formulations and raw materials characterization are given by Muntingh,[12] and will not be reproduced in full here. Instead, sample compositions will be described by reference to the 'base case' outlined in Table I, and the effect of varying each parameter will be presented in terms of deviation from this formulation. The mix design given in Table I is calculated on the basis of an approximate sample density of 2400 kg/m³.

Table I. Reference geopolymer paste formulation

Component	Mix proportion (kg/m^3)
Fly ash	1698
NaOH (solid)	41
KOH (solid)	86
Sodium silicate (PQ Corp., Grade N)	177
Water	398

The chloride diffusion coefficients calculated for different variants on this reference formulation were calculated according to a linearized version of Fick's Law,[12] and the results for one of the ashes are shown in Figures 4-6. Sulfate diffusion testing was also conducted using this cell, with a 3 wt% H_2SO_4 solution replacing the NaCl electrolyte but otherwise maintaining the same testing protocols as used for chloride. An estimate of the experimental uncertainty in these values is obtained by comparing the results of two independent tests conducted on the reference formulation for this ash, which gave D_{Cl} = 2.86 x 10^{-10} and 2.96 x 10^{-10} cm^2/s, and D_{SO4} = 5.31 x 10^{-8} and 3.29 x 10^{-10} cm^2/s respectively. The mean of each of these sets of results is used in Figures 4-6. Note also the differences in units on the vertical scales on each of these plots: the calculated sulfate diffusion coefficients are generally around two orders of magnitude higher than the chloride values. This is most likely attributable to degradation of the geopolymer matrix by the highly acidic (pH < 1) H_2SO_4 test solution used. Conducting similar experiments using Na_2SO_4 solution would provide a potential solution to this issue, as this solution has previously been shown to cause minimal structural degradation of the aluminosilicate geopolymer network.[4]

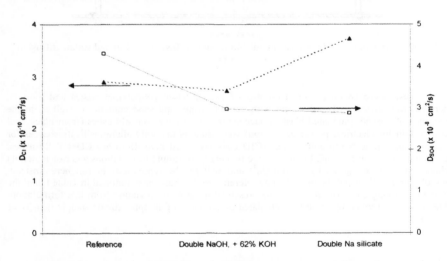

Figure 4. Effect of varying hydroxide and silicate content on measured geopolymer paste chloride (solid triangles) and sulfate (hollow squares) diffusion coefficients. Lines are given as a guide to the eye only.

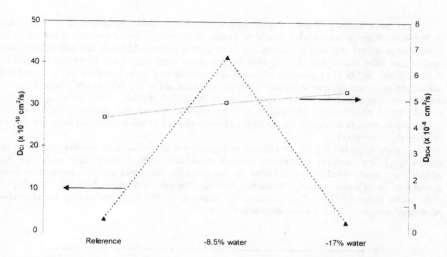

Figure 5. Effect of varying water content on measured geopolymer paste chloride (solid triangles) and sulfate (hollow squares) diffusion coefficients. Lines are given as a guide to the eye only.

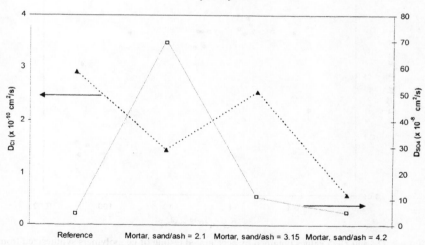

Figure 6. Effect of sand addition on measured geopolymer chloride (solid triangles) and sulfate (hollow squares) diffusion coefficients. The reference sample is a paste, all others are mortars. Lines are given as a guide to the eye only.

It is seen from Figures 4-6 that there appears to be significant variability in the calculated diffusion coefficients across what should be groups of relatively similar samples. It must also be noted that it is very difficult to discern any clear trends between chloride and sulfate diffusion coefficients from these plots. It may also be that there were significant flaws (air bubbles or cracks) in the -8.5% H_2O chloride diffusion test sample in Figure 5 and in the 2:1 sand/ash sulfate diffusion test sample in Figure 6, which may have led to the very much higher diffusion coefficients calculated for these samples. It is clear that further work including detailed refinement of experimental techniques will be required before the trends in diffusivity in geopolymers, even in relatively simple series of samples such as those presented here, are able to be properly understood.

Figure 7, containing the data shown in Figures 4-6 as well as data obtained by the same methodology for samples containing ground granulated blast furnace slag and data for geopolymers synthesized from a different fly ash using the same mix designs, shows that there appears to be at most a weak correlation between the two sets of coefficients. Muntingh[12] additionally presents D_{Cl} values obtained by analysis of PIXE measurements on chloride-penetrated samples, which are also shown in Figure 7.

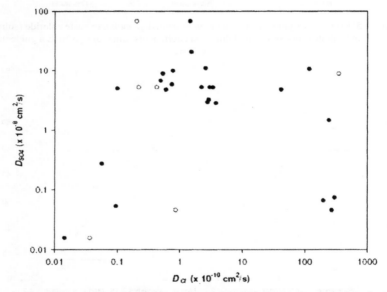

Figure 7. D_{Cl} and D_{SO4} values obtained by accelerated testing of geopolymers synthesized from different ashes, including some containing added slag, as well as variation in other synthesis parameters. Hollow circles are chloride diffusion coefficients calculated from PIXE analysis for selected samples. Data from reference [12].

The extent of correlation that can be seen in Figure 7 provides an estimate of the level of confidence with which these data can be regarded. There appears to be a cluster of three outlying points with a very high D_{Cl} but a low D_{SO4}. It must be noted that all the samples which display very low sulfate diffusion coefficients ($<10^{-8}$ cm^2/s) are those which contain added calcium in the form of slag. Calcium will react with the H_2SO_4 leaching solution to form gypsum ($CaSO_4$), which has previously been observed to precipitate within a leached layer on the surface of a fly ash-slag geopolymer subjected to H_2SO_4 attack.[11] This reaction may be responsible for the very low observed sulfate permeability of these samples, where chemical reaction effects are significantly complicating the overall leaching process.[17] The apparent D_{SO4} values obtained by accelerated testing, particularly for Ca-rich geopolymer systems, will therefore be representing a combination of multiple phenomena (reaction as well as diffusion), rather than being a purely representative diffusional parameter. This may have implications for the transferability of data from accelerated testing to real-world situations, where the timescales available are much longer, so the chemical reaction component of the process will be allowed much more closely to approach equilibrium. Nonetheless, this work provides at the very least a lumped parameter whereby the differences between systems may be analyzed. It is also noteworthy that acid-sulfate leaching of OPC-based materials will show similar complications in analysis of diffusion coefficients due to the even higher levels of Ca present.

Gypsum formation has been observed to be expansive, with gypsum crystals forming within and then filling cracks that develop on the exposed surface of the geopolymer.[11] However, it is seen from Figure 7 that the majority of samples with very low apparent D_{SO4} values also display low D_{Cl} values. Given that a fresh sample is used for every test, it can be confidently stated that the low observed sulfate diffusivities are in large part representative of the material properties (i.e., refined and disconnected porosity) rather than being solely due to the formation of a relatively impermeable gypsum-rich layer. The primary exception to this would appear to be the samples responsible for the points in the lower right-hand corner of Figure 7, which show a very low D_{SO4} but a high D_{Cl}. These are potentially samples in which gypsum formation has significantly reduced the observed ingress of sulfate, reducing D_{SO4} but with no effect on D_{Cl} as the chloride-attacked samples have never come into contact with a sulfate solution to allow gypsum formation. It is not possible from the present data set to correlate straightforwardly this relationship with the mix design of the geopolymers – of the three samples that fall in this cluster, both ash sources are represented, and there is no clear trend with respect to the level of slag addition which would determine whether a slag-containing sample among the current range of test specimens shows a high or a low D_{Cl}.

Another issue of potential significance in the analysis of accelerated diffusion testing data when applied to alkali-activated systems as opposed to OPC concretes is the differing chemistry of the pore solutions in the two materials. The electrical acceleration of the diffusional processes takes place via the generation of potential and charge gradients, which forces the Cl$^-$ ions from the NaCl electrolyte into the pores of the cement. The concentrations of the ions initially present in the pore solutions will therefore be significant in the analysis of the experimental data. Given that the pore solutions present within fresh geopolymeric binders will be of significantly higher ionic strength and pH than those in fresh OPC, this is expected to have an effect on the ionic transport mechanisms controlling the accelerated chloride transport. Direct comparison of OPC and geopolymer results must therefore be viewed with a certain degree of caution, as it may be that the high ionic strengths in geopolymer pore solutions reduce the observed D_{Cl} values compared to OPC. The results presented here must be analyzed in this context, however it is unlikely that these effects will be strong enough to be solely responsible for the large differences in D_{Cl} observed. Muntingh[12] also presents pore volume and BET surface area results for these samples, but there is again no obvious correlation between these data and the calculated diffusivities.

Several significant assumptions were made in the extraction of diffusion coefficients from the cell data, in particular the assumption of Fickian diffusion via the highly simplified equation of Dhir and Sugiyama.[12] It has been shown that the physical reality of the ion migration tests is in fact significantly more complex than this simplified case,[17,18] and it is hoped that future work in this field will lead to the calculation of refined D_{Cl} values by application of more advanced theoretical expressions. However, Figure 7 does show that the values presented here are expected to be, on average, repeatable to within an order of magnitude or thereabouts. This is significant in comparisons between geopolymer cements and OPC, given that a representative D_{Cl} value for OPC is approximately 10^{-8} cm^2/s.[19] Regardless of any degree of experimental uncertainty in the measurements presented here, the fact remains that the chloride and sulfate diffusivities of geopolymer pastes and mortars are lower than those of OPC, with the highest chloride diffusion coefficient of any sample studied here being less than 3×10^{-8} cm^2/s. The exact extent of this advantage, and the relationship between permeability and mix design, remain to be determined in more detail in future work. However, the initial set of results presented here show at least that geopolymer technology displays the potential for the development of very low-permeability, chloride and/or sulfate-resistant cements and concretes.

MICROSTRUCTURE AND DURABILITY OF SLAG-BASED CEMENTS

Due to the relatively recent development of aluminosilicate geopolymer technology, there are no existing samples that have been exposed to environmental conditions for extended periods of time, by which their durability may be analyzed. There have over the past several years been a number of claims that the Pyramids of Egypt were produced by a process resembling modern geopolymerization.[20-22] However, regardless of the validity of this theory, the predominance of calcium compounds in Pyramid stones compared to the aluminosilicate-based geopolymers of primary interest here means that claims of extreme durability of geopolymers based on the continuing existence of the Pyramids of Egypt must be viewed with caution.

However, working under that caveat, it is possible to provide some analysis of the role of slag in geopolymers by investigating high-Ca alkali-activated cement samples. Alkali activated slags have been studied and used on a large scale in the former Soviet Union for around 50 years,[23] and so the durability of these samples – which are known (rather than simply proposed, as in the case of the Pyramids) to have been produced by alkaline activation methods – may be used, up to a point, as a means of understanding and predicting geopolymer durability. Samples of these Soviet slag cements, generated by the reaction of metallurgical slags with alkali carbonate solutions, have been subjected to NMR, SEM, TEM and elemental microanalysis for the first time.[24] These samples were taken from concretes that had been in service under conditions of varying aggressiveness for up to 30 years, including under conditions in which OPC-based concretes had required replacement after less than 3 years' service.[25] In contrast to the rapid failure of OPC under these circumstances, the five slag cement samples studied had actually increased in strength while in service, by factors ranging from 25% to in excess of 100%.[24]

A BSE SEM image of one of these aged slag cement samples – after 20 years' exposure to outdoor conditions in the north-eastern Europe – is given in Figure 8. This micrograph shows unreacted slag particles as bright regions, surrounded by rims of dense Si-rich C-S-H which appear to contain nanocrystalline calcite inclusions.[24] The bulk regions of the sample are comprised of hydrated C-S-H gel, with a Ca/Si ratio very similar to that of the unreacted slag particles.

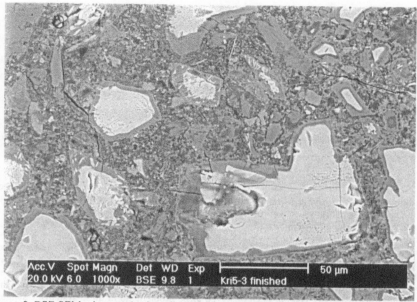

Figure 8. BSE SEM micrograph of an alkali-carbonate activated slag sample following 20 years' exposure to service conditions and freeze-thaw cycling.

The most notable aspect of the performance of the sample shown in Figure 8 is that not only has it not been degraded over 20 years' exposure to aggressive conditions, but that the continuing microstructural development enabled by the use of an alkaline activator has actually led to improved performance. A mechanism has been proposed whereby the nanocrystalline calcite acts as a sink for the alkaline components in a cyclic hydration process, allowing the dense regions surrounding the slag particles to grow and bind more strongly to the 'outer product' regions over time. The extent of polycondensation of the silicates present, as measured by ^{29}Si MAS NMR spectroscopy (Figure 9), is also much higher than is generally observed in OPC systems. Other than the Q^4 silicon peak due to residual quartz sand (large sand and unreacted slag particles were removed where possible from the sample prior to NMR analysis), Figure 9 shows that the primary NMR resonances of this sample are in the region corresponding to Q^2 and Q^3 (chain and branching) silicon centers. This relatively high extent of polycondensation may also be related to the durability of the samples, as the lability of silicate sites decreases significantly with increasing connectivity.

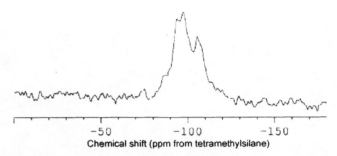

Chemical shift (ppm from tetramethylsilane)

Figure 9. ^{29}Si MAS NMR spectrum of the same aged alkali-carbonate activated slag sample as depicted in Figure 8.

The implications for geopolymerization and the study of geopolymer durability of these slag-alkaline cement investigations may well be significant. It has been shown that the aluminosilicate network in a well-reacted geopolymeric binder has a very high extent of connectivity.[26,27] The geopolymerization reaction sequence is also known to include processes that may continue for a significant period of time following initial setting, which have been identified as the development of a more highly structured gel binder and, under certain conditions, the growth of nanocrystalline inclusions.[1,28] It is not yet known whether this ongoing development contributes to or hinders the very long-term strength and durability of geopolymers. However, the fact that geopolymer gel is known to bind strongly to reactive aggregates,[15] and that alkali-silica reactivity has not been observed to be problematic in these systems,[29] suggests that the ongoing structural development of geopolymeric binders at least provides the potential for tailoring of mix design to minimize strength deterioration over time. Whether or not tailoring for long-term strength increases is possible remains a matter for future investigation, but the indications based on the related slag-alkali systems are positive.

CONCLUSIONS
This paper has discussed a variety of issues to do with the durability of geopolymers. As a relatively young research field, the absence of 30-year-old specimens of geopolymer cement or concrete that have been subjected to environmental weathering means that alternative means of analyzing durability must be used. The results of microstructural analysis and of accelerated corrosion testing of geopolymers have been presented here, and both techniques appear to show significant durability advantages for geopolymer cements and concretes compared to OPC-based materials. Comparison with related alkali-activated slag systems, which have significantly outperformed OPC under highly aggressive environmental conditions for periods of decades or

more, also shows that the durability of geopolymers may realistically be expected to be very good. So, while it is not yet possible to definitively answer the question posed by the title of this paper, 'Will geopolymers stand the test of time?' – the evidence to date appears very positive.

REFERENCES

[1] P. Duxson, A. Fernández-Jiménez, J.L. Provis, G.C. Lukey, A. Palomo, J.S.J. van Deventer, Geopolymer Technology: The Current State of the Art, *J. Mater. Sci.*, In press, DOI 10.1007/s10853-006-0637-z (2007).

[2] P. Duxson, J.L. Provis, G.C. Lukey, J.S.J. van Deventer, The Role of Inorganic Polymer Technology in the Development of 'Green Concrete', *Cem. Concr. Res.*, Submitted (2007).

[3] M. Sofi, J.S.J. van Deventer, P.A. Mendis, G.C. Lukey, Engineering Properties of Inorganic Polymer Concretes (IPCs), *Cem. Concr. Res.*, 37, 251-57 (2007).

[4] A. Palomo, M.T. Blanco-Varela, M.L. Granizo, F. Puertas, T. Vazquez, M.W. Grutzeck, Chemical Stability of Cementitious Materials Based on Metakaolin, *Cem. Concr. Res.*, 29, 997-1004 (1999).

[5] J.W. Phair, Green Chemistry for Sustainable Cement Production and Use, *Green Chem.*, 8, 763-80 (2006).

[6] D.W. Hobbs, Concrete Deterioration: Causes, Diagnosis, and Minimising Risk, *Int. Mater. Rev.*, 46, 117-44 (2001).

[7] T. Bakharev, Resistance of Geopolymer Materials to Acid Attack, *Cem. Concr. Res.*, 35, 658-70 (2005).

[8] T. Bakharev, Durability of Geopolymer Materials in Sodium and Magnesium Sulfate Solutions, *Cem. Concr. Res.*, 35, 1233-46 (2005).

[9] H. Rostami, W. Brendley, Alkali Ash Material: A Novel Fly Ash-Based Cement, *Environ. Sci. Technol.*, 37, 3454-57 (2003).

[10] E. Hewayde, M. Nehdi, E. Allouche, G. Nakhla, Effect of Geopolymer Cement on Microstructure, Compressive Strength and Acid Resistance of Concrete, *Mag. Concr. Res.*, 58, 321-31 (2006).

[11] A. Allahverdi, F. Škvára, Sulfuric Acid Attack on Hardened Paste of Geopolymer Cements. Part 1. Mechanism of Corrosion at Relatively High Concentrations, *Ceram.-Silik.*, 49, 225-29 (2005).

[12] Y. Muntingh, Durability and Diffusive Behaviour Evaluation of Geopolymeric Material, M.Sc. Thesis, University of Stellenbosch, South Africa, 2006.

[13] J.M. Miranda, A. Fernández-Jiménez, J.A. González, A. Palomo, Corrosion Resistance in Activated Fly Ash Mortars, *Cem. Concr. Res*, 35, 1210-17 (2005).

[14] M.K. Head, N.R. Buenfeld, Measurement of Aggregate Interfacial Porosity in Complex, Multi-Phase Aggregate Concrete: Binary Mask Production Using Backscattered Electron, and Energy Dispersive X-Ray Images, *Cem. Concr. Res.*, 36, 337-45 (2006).

[15] W.K.W. Lee, J.S.J. van Deventer, The Interface between Natural Siliceous Aggregates and Geopolymers, *Cem. Concr. Res.*, 34, 195-206 (2004).

[16] P. Duxson, J.L. Provis, G.C. Lukey, S.W. Mallicoat, W.M. Kriven, J.S.J. van Deventer, Understanding the Relationship between Geopolymer Composition, Microstructure and Mechanical Properties, *Colloid. Surf. A*, 269, 47-58 (2005).

[17] E. Samson, J. Marchand, K.A. Snyder, Calculation of Ionic Diffusion Coefficients on the Basis of Migration Test Results, *Mater. Struct.*, 36, 156-65 (2003).

[18] Z. Liu, J.J. Beaudoin, The Permeability of Cement Systems to Chloride Ingress and Related Test Methods, *Cem. Concr. Aggr.*, **22**, 16-23 (2000).

[19] K.Y. Yeau, E.K. Kim, An Experimental Study on Corrosion Resistance of Concrete with Ground Granulate Blast-Furnace Slag, *Cem. Concr. Res.*, **35**, 1391-99 (2005).

[20] J. Davidovits, F. Davidovits, The Pyramids: An Enigma Solved. 2nd Revised Ed., Éditions J. Davidovits, Saint-Quentin, France, 2001.

[21] G. Demortier, PIXE, PIGE and NMR Study of the Masonry of the Pyramid of Cheops at Giza, *Nucl. Instr. Meth. Phys. Res. B*, **226**, 98-109 (2004).

[22] M.W. Barsoum, A. Ganguly, G. Hug, Microstructural Evidence of Reconstituted Limestone Blocks in the Great Pyramids of Egypt, *J. Am. Ceram. Soc.*, **89**, 3788-96 (2006).

[23] C. Shi, P.V. Krivenko, D.M. Roy, Alkali-Activated Cements and Concretes, Taylor & Francis, Abingdon, UK, 2006.

[24] H. Xu, J.L. Provis, J.S.J. van Deventer, P.V. Krivenko, Characterization of Aged Slag Concretes from the Former Soviet Union, *ACI Mater. J.*, Submitted (2007).

[25] V.P. Ilyin, Durability of Materials Based on Slag-Alkaline Binders, in: Proceedings of the First International Conference on Alkaline Cements and Concretes. 1994, P.V. Krivenko, (Ed.), Kiev, Ukraine, VIPOL Stock Company, pp. 789-836.

[26] P. Duxson, G.C. Lukey, F. Separovic, J.S.J. van Deventer, The Effect of Alkali Cations on Aluminum Incorporation in Geopolymeric Gels, *Ind. Eng. Chem. Res.*, **44**, 832-39 (2005).

[27] P. Duxson, J.L. Provis, G.C. Lukey, F. Separovic, J.S.J. van Deventer, ^{29}Si NMR Study of Structural Ordering in Aluminosilicate Geopolymer Gels, *Langmuir*, **21**, 3028-36 (2005).

[28] J.L. Provis, G.C. Lukey, J.S.J. van Deventer, Do Geopolymers Actually Contain Nanocrystalline Zeolites? - A Reexamination of Existing Results, *Chem. Mater.*, **17**, 3075-85 (2005).

[29] K.-L. Li, G.-H. Huang, L.-H. Jiang, Y.-B. Cai, J. Chen, J.-T. Ding, Study on Abilities of Mineral Admixtures and Geopolymer to Restrain ASR, *Key Eng. Mater.*, **302-303**, 248-54 (2006).

NEW TRENDS IN THE CHEMISTRY OF INORGANIC POLYMERS FOR ADVANCED APPLICATIONS

Kenneth J.D. MacKenzie, Suwitcha Komphanchai,
MacDiarmid Institute for Advanced Materials and Nanotechnology,
Victoria University of Wellington,
P.O. Box 600 Wellington,
New Zealand,

Ross A. Fletcher,
Industrial Research Ltd., Lower Hutt,
New Zealand.

ABSTRACT

Conventionally, inorganic polymers (also called geopolymers) composite materials based on these are prepared by reaction of a solid aluminosilicate such as dehydroxylated kaolinite clay (metakaolinite) with an alkali silicate solution under highly alkaline conditions. This paper discusses recent studies aimed at extending the applications of inorganic polymer materials by addressing the following questions:

Can viable conventional aluminosilicate inorganic polymers be formed with improved energy efficiency by eliminating the need for thermal pre-treatment of the solid aluminosilicate source (i.e. by mechanochemical or chemical pre-treatment)?

Can viable inorganic polymers be formed with other tetrahedral network-forming groups such as borate, and what are the likely practical applications for such compounds?

What are the possibilities for forming viable bioactive inorganic geopolymers containing phosphate groups and calcium?

What are the possible synthesis strategies for forming viable inorganic polymers with useful electronic properties, and could they be formed by soft chemical reactions in solution without the need for solid aluminosilicate sources?

INTRODUCTION

The previous history of inorganic polymers (also called geopolymers) is somewhat chequered. Alkali-activated materials of this type have been known for a long time, those most directly related to geopolymers being the building materials developed in the 1960s by Berg et al[1] from kaolinite hardened by reaction with sodium hydroxide. A series of disastrous fires in France in the 1980s led to the quest by a French polymer chemist, Joseph Davidovits, for an inorganic fireproof substitute for the highly flammable organic materials used in buildings. The resulting materials, which he named geopolymers, were aluminosilicates based on reactions of alkali with dehydroxylated kaolinite (metakaolinite). The main difference between these materials and those of Berg was in the use of dehydroxylated kaolinite. Davidovits' background in conventional polymer chemistry led him to view these materials as polymers derived from a geological raw material, for which he coined the name geopolymers[2]. A subsequent suggestion by Davidovits, that he had re-discovered materials which were used by much older civilizations, especially the Egyptians in their monumental buildings, led to his publication with Margie Morris of a controversial book[3] "The pyramids: An Enigma Solved". However, irrespective of

249

any pre-history these materials might have had, they have shown great promise as materials of the 21st century, for the following reasons:

(1) They may provide an ecologically-friendly and energy-efficient substitute for Portland cement-based building materials because they do not require high-temperature processing and do not generate CO_2.

(2) They may provide the basis for lightweight, fireproof body panels for transport vehicles. Already, carbon fibre-reinforced geopolymer panels have been used in specific applications in Formula 1 racing cars and experimental French fighter planes.

(3) They may provide the means of mitigating hazardous radioactive or heavy metal wastes by immobilising them and allowing them to be safely disposed of.

But, as this paper will discuss, there are a number of other exciting possibilities for materials based on inorganic polymerisation reactions.

DEFINING CHARACTERISTICS OF INORGANIC POLYMERS (GEOPOLYMERS)

The widespread and indiscriminate use of these terms in the literature makes it very important to define these materials in terms of their characteristics.

The conventional geopolymers of Davidovits were prepared by condensing a solid aluminosilicate (metakaolinite) with sodium or potassium silicate solutions, or a mixture of the two, under highly alkaline conditions and controlled water content[2]. The latter is particularly important during the solidification (curing) stage. On this basis the geopolymer could be defined as an aluminosilicate which attains strength at ambient temperature. Furthermore, geopolymers prepared in this way are all X-ray amorphous, but are shown by solid-state NMR to contain both their aluminium and silicon in solely tetrahedral coordination (Figure 1).

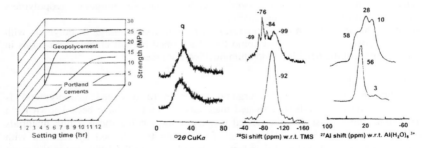

Figure 1. Unique defining characteristics of an aluminosilicate geopolymer in terms of (1) achievement of strength at ambient temperatures, (2) X-ray amorphous diffraction pattern, (3) fully tetrahedral Al coordination, (4) broad tetrahedral Si NMR spectrum centred at the position of $Q_4(3Al)$ units. The upper diffraction pattern and NMR spectra are of the reactant, the lower correspond to the geopolymer product. The sharp XRD peak (q) in the reactant is due to a quartz impurity in the clay.

Whereas the Si:Al ratio was originally thought to be fairly limited, more recently we have shown that inorganic polymers with Si:Al ratios much higher than 1:1 are viable, but above Si:Al of about 24 they are rubbery and become porous on heating due to the evolution of water[4]. At Si:Al ratios <1, the products contain crystalline zeolites and are no longer X-ray amorphous[4].

RECENT WORK TO EXTEND THE RANGE OF POSSIBLE NEW APPLICATIONS OF INORGANIC POLYMERS

Improving the energy efficiency of the synthesis reaction.

A question of some interest is whether it is possible to produce viable inorganic polymers (according to the above definitions) without the need to dehydroxylate the solid aluminosilicate as recommended by Davidovits[2]. Our most recent results indicate that under standard geopolymer synthesis conditions, an undehydroxylated clay will not achieve the full strength and texture of a well-cured geopolymer. The XRD trace of this material indicates some loss of X-ray crystallinity, its ^{27}Al NMR spectrum shows the conversion of some of the octahedral Al to tetrahedral, and its ^{29}Si NMR spectrum shows the formation of new Si-O-Al bonds (-85 ppm). Thus, this material is consistent with a partially-formed geopolymer[5].

Next, as an alternative to thermal dehydroxylation, high-energy grinding of the solid phase (the clay mineral) was tried. Milling in a vibratory mill for 15 min produced a material in which some of the Al(VI) was transformed to Al(V) and Al(IV) and much of the crystallinity was destroyed (apart from the quartz and cristobalite impurities) (Figure 2).

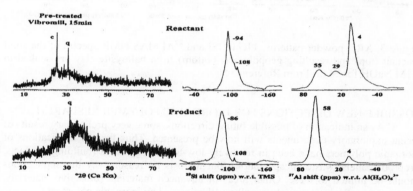

Figure 2. XRD powder patterns, 11.7T ^{29}Si and ^{27}Al MAS NMR spectra of the solid reactant (top) and resulting geopolymer (bottom) from halloysite clay milled in a vibratory mill (Bleuler, Switzerland) for 15 min. Key to the sharp impurity peaks in the X-ray patterns: c = cristobalite, q = quartz. From Reference 5.

The ground product proved to be a much more satisfactory starting material, hardening to forming a product with the X-ray and NMR attributes of a typical geopolymer[5]. However, energy still has to be supplied to the system by grinding, so chemical pretreatment of the solid reactant with acid or alkali was also investigated. Since the solubility of the Al-component appears to be a determining step in the geoplymerisation process, it might be expected that acid treatment of the clay would be effective (acid leaching selectively removes the octahedral component of layer-lattice aluminosilicates[6]). However, treatment of the clay with 0.1M HCl for 24 hr gave the same result as the raw clay, namely, a poorly-formed geopolymer[5].

On the other hand, treatment of the clay with 0.1M NaOH for 24 hr produced a hydrated zeolite (LTA) and converted a significant amount of the Al(VI) to Al(IV)[5] (Figure 3) The geopolymer formed from this material hardens and contains exclusively Al(IV), but is not fully X-ray amorphous, containing a crystalline sodium aluminium carbonate.

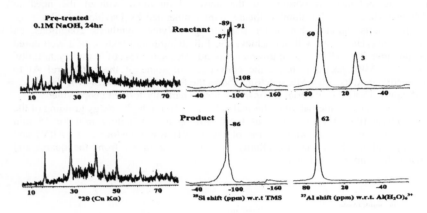

Figure 3. XRD powder patterns, 11.7T ^{29}Si and ^{27}Al MAS NMR spectra of the solid reactant (top) and resulting geopolymer (bottom) from halloysite clay pre-soaked in 0.1M NaOH for 24 hr. From Reference 5.

POSSIBLE NEW DIRECTIONS FOR INORGANIC POLYMER RESEARCH

As an indication of possible future directions, some very preliminary results of recent exploratory experiments will now be presented. Thus far, the applications of inorganic polymers have been in relatively low-technology areas (building products, fireproof panels, waste encapsulation), but there is no reason why the applications should be limited to these. Other areas of advanced materials (bioactive materials, electronic materials, etc) might equally well be served by inorganic polymers.

PRACTICAL APPLICATIONS ARISING FROM THE INCORPORATION OF OTHER STRUCTURAL UNITS

The structures of inorganic polymers can be modified by incorporating tetrahedral borate units[7]. A practical application of this is the use of borate to retard the setting of geopolymer cements made from flyash (an inorganic waste material from the burning of coal in thermal power stations and used in cements for economic reasons). However, New Zealand Class C flyash, which contains much more Ca than Australian Class M flyash, produces geopolymer cements which have initial setting times much too rapid for practical purposes (10 minutes or less). However, we have found that the addition of about 4 wt% of sodium or lithium borates can delay the initial setting time of the cement by up to 50 minutes without a detrimental effect on the compressive strength or high temperature thermal properties[7] (Figure 4). XRD, ^{27}Al and ^{29}Si MAS NMR shows these boron-containing compounds to be conventional geopolymers, while ^{11}B MAS NMR shows that by contrast with the original borate additives, the boron is in solely tetrahedral coordination (Figure 5), indicating that it has entered the geopolymer lattice[7].

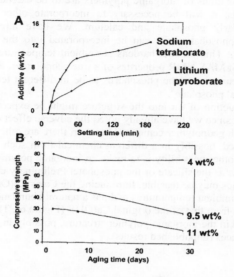

Figure 4. A. Behaviour of a NZ flyash geopolymer containing sodium tetraborate (borax) and lithium pyroborate additives. A. Setting time as a function of additive concentration, B. Compressive strength as a function of aging time for three concentrations of borax additions. From Reference 7.

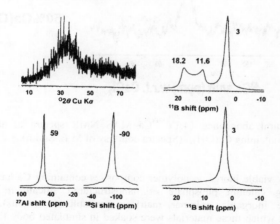

Figure 5. XRD pattern (upper left), ^{11}B, ^{27}Al and ^{29}Si MAS NMR spectra (lower) of an aluminosilicate geopolymer containing borate units. The [11]B spectrum of the reactant (upper spectrum) reveals trigonal and tetrahedral sites whereas the geopolymer spectrum (lower) contains solely tetrahedral units. From reference 7.

DEVELOPMENT OF INORGANIC POLYMERS WITH POSSIBLE BIOACTIVE PROPERTIES

If the applications of inorganic polymers are to be extended into the field of bioactive materials, it will be necessary to incorporate other elements into the structure, particularly phosphate and calcium. We have already been able to demonstrate that phosphate units can be incorporated into the inorganic polymer structure as $AlPO_4$. This product hardens at ambient temperature and shows all the XRD, ^{27}Al and ^{29}Si MAS NMR properties of a true geopolymer. Furthermore, the ^{31}P MAS NMR spectrum shows the phosphorus to be in different tetrahedral sites from those in the original phosphate[8].

The introduction of Ca into the structure might be expected to pose another potential problem, since we have already noted the adverse effect of Ca on the setting properties of flyash geopolymer cements. However, there are other ways in which Ca might be introduced, possibly in a different chemical form such as the carbonate or the hydroxide. Another possibility is to make an inorganic polymer composite with a Ca compound such as the silicate or the phosphate. Preliminary experiments indicate that both approaches may be feasible. Introducing the Ca as $Ca(OH)_2$ gives a material which hardens at ambient temperatures and has a natural abundance ^{43}Ca MAS NMR spectrum different from that of the original $Ca(OH)_2$ (Figure 6). This suggests that the Ca is incorporated in the inorganic polymer structure, possibly in the hydration water filling the pore spaces (unpublished results).

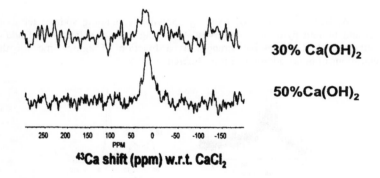

30% Ca(OH)₂

50%Ca(OH)₂

⁴³Ca shift (ppm) w.r.t. CaCl₂

Figure 6. Natural abundance 14.1T ^{43}Ca MAS NMR spectra of aluminosilicate geopolymers containing $Ca(OH)_2$. (Spectra courtesy of M.E. Smith).

Equally viable inorganic polymer composites containing Ca have been made by suspending calcium phosphate or nanostructured silicate in conventional aluminosilicate inorganic polymer matrices (unpublished results). Preliminary experiments in which these materials were soaked in simulated body fluid (SBF) for up to 7 days indicate that these materials possess a degree of bioactivity, since XRD peaks of crystalline phases related to hydroxycarbonate apatite (HCA) appear on the sample surfaces even after 1 day of soaking (Figure 7).

In Vitro exposure to simulated body fluid (SBF) at 37°C

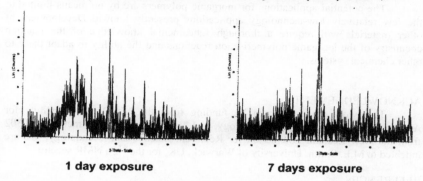

1 day exposure   **7 days exposure**

Figure 7. XRD patterns of potassium aluminosilicate geopolymer containing 50 wt% $Ca_3(PO_4)_2$ soaked in Kokubo simulated body fluid. Note the progressive disappearance of the amorphous geopolymer feature and development of crystalline HCA-related phases with increasing soaking time.

DEVELOPMENT OF INORGANIC POLYMERS WITH POSSIBLE ELECTRONIC PROPERTIES

The most obvious approach which has already proved to have interesting possibilities is to incorporate electrically conducting elements into the inorganic polymer before or after curing. The nanoporosity of a cured inorganic polymer is such as to permit the introduction of pyrrole monomer which can then be polymerised *in situ* into electrically conducting polypyrrole by treatment with an oxidising agent. This oxidised material will act as a reducing agent for solutions of metal ions, allowing the deposition of silver or copper nanoparticles or nanowires throughout the bulk of the inorganic polymer. Silver nanoparticles in particular have biocidal properties which can improve the safety of an implanted bioactive calcium phosphate inorganic polymer (unpublished results).

Another approach to producing electroceramic materials is to adapt the aluminosilicate chemistry of conventional inorganic polymers to produce entirely new classes of compounds containing, for instance, Ga or Ge. As a first step in this direction, a synthesis method has been developed which does not involve solid reactants. Thus, viable aluminosilicate inorganic polymers can now be made from alkaline solutions of sodium aluminate and silicate [9]. These materials harden to give good compressive strengths (about 26 MPa), typical XRD and ^{27}Al NMR spectra, although the ^{29}Si NMR spectra indicate the presence of three better-defined Si environments than the typical broad inorganic polymer resonance centred at about − 90 ppm. One of these Si species (at −109 ppm) arises from unreacted SiO_2, while that at −87 ppm indicates a Si environment associated with a higher number of Al nearest neighbours [9]. Nevertheless, this synthesis strategy opens the way to possible new compounds with a range of potential high-technology applications.

CONCLUSIONS

The potential applications for inorganic polymers are by no means limited to the few relatively low-technology applications presently known. Development of other materials will require a thorough fundamental knowledge of the reaction chemistry of the inorganic polymerisation reactions and the ability to adapt these to other chemical systems.

ACKNOWLEDGEMENTS

This work was supported by funding from the MacDiarmid Institute for Advanced Materials and Nanotechnology, and by funding from contract CO8X0302 from the New Zealand Foundation for Research, Science and Technology. We are indebted to M.E. Smith, University of Warwick, UK, for the ^{43}Ca NMR spectra.

REFERENCES

[1] L.G. Berg, B.A. Demidenko, V.A. Remiznikova and M.S. Nizamov, "Unfired finishing building materials based on kaolinite", *Stroit. Mater.*, **10**, 22, (1970).

[2] J. Davidovits, "Geopolymers: Inorganic polymeric new materials". J. *Thermal Anal.*, **37**, 1633-56, (1991).

[3] J. Davidovits and M. Morris, "The Pyramids. An Enigma Solved" Dorset, NY, (1990).

[4] R.A. Fletcher, K.J.D. MacKenzie, C.L. Nicholson and S. Shimada, "The composition range of aluminosilicate geopolymers", *J. Eur. Ceram. Soc.*, **25**, 1471-7, (2005).

[5] K.J.D. MacKenzie, D.R.M. Brew, R.A. Fletcher and R. Vagana, "Formation of aluminosilicate geopolymers from 1:1 layer-lattice minerals pre-treated by various methods: a comparative study", *J. Mater. Sci.*, (in press).

[6] K. Okada and K.J.D. MacKenzie. "Nanoporous Materials from Mineral and Organic Templates", In *Nano Materials: From Research To Applications*, Ed. H. Hosono, Y. Mishima, H. Takezoe and K.J.D. MacKenzie, Elsevier, Oxford. (2006), pp. 349-82.

[7] C.L. Nicholson, K.J.D. MacKenzie, B.J. Murray, R.A. Fletcher, D.R.M. Brew and M. Schmücker. "Novel geopolymer materials containing borate structural units", *Proc. World Geopolymer Conf., St. Quentin, Paris*, 31-33 (2005).

[8] K.J.D. MacKenzie, D.R.M. Brew, R.A. Fletcher, C.L. Nicholson, R. Vagana and M. Schmücker. "Towards an understanding of the synthesis mechanisms of geopolymeric materials", *Proc. World Geopolymer Conf., St. Quentin, Paris*, 41-44 (2005).

[9] D.R.M. Brew and K.J.D. MacKenzie. "Geopolymer synthesis using fume silica and sodium aluminate", *J. Mater. Sci.*, (in press).

INFLUENCE OF GEOPOLYMER BINDER COMPOSITION ON CONVERSION REACTIONS AT THERMAL TREATMENT

A. Buchwald[1], K. Dombrowski[2], M. Weil[3]

[1] Bauhaus-University Weimar, Chair of building chemistry, Weimar, D-99421, Germany

[2] Freiberg University of Mining and Technology; Institute of Ceramic, Glass and Construction Materials; Freiberg, D-09596, Germany

[3] Forschungszentrum Karlsruhe; Department of Technology-Induced Material Flow (ITC-ZTS), Germany

ABSTRACT

Geopolymeric binders and cements are made by means of an alkaline activation of materials reactive in this respect. Concerning the "reactive material" there is a wide range of possible raw materials. Beside primary resources like clay (after thermal activation) and natural pozzolan for instance volcanic ashes, a considerable quantity of secondary resources like ashes and slags from different processes can be used. The alkaline activator may also vary in a wide range for instance in the concentration, kind of used alkali ion or added silicate solution. Concerning the huge number of the possible geopolymer binder compositions, the performance of each single geopolymer may differ as well because it is strongly correlated to its composition. Beside the main components, side components like reactive fillers or a calcium containing additive will play an enormous role on the conversion reaction at high temperatures.

The presentation will sum up results of different model investigations about the influence of different binder compositions on the strength of the binder as well as the composition of reaction products after hardening and conversion reaction. Model mixtures base on metakaolin or fly ash were prepared under variation of the activator concentration and the addition of calcium. The reaction products of the binders were investigated by X-Ray diffraction and NMR Spectroscopy, the reaction products after thermal treatment of 1000°C were determined by X-Ray diffraction and Rietveld refinement. Additionally the strength tests on hardened binder samples were performed.

INTRODUCTION
Geopolymer
Geopolymer binders are inorganic 2-component systems, consisting of

1. a *reactive solid* component that contains SiO_2 and Al_2O_3 in sufficient amount and in reactive form (e. g. ashes, active clays, pozzolana, slags etc.) and
2. an *alkaline activation solution* that contains (apart from water) individual alkali hydroxides, silicates, aluminates, carbonates and sulfates or combinations thereof.

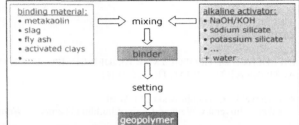

Figure 1: Initial constituents and process path for the manufacture of a geopolymer

When the solid and the activator components come into contact with each other, hardening results due to the formation of an aluminosilicate network ranging from amorphous to partial crystalline aluminosilicate, which is water resistant. The geopolymer can contain crystalline zeolites [1], the amorphous reaction product consist the same (sub-) structures than zeolites, but without long range order. Depending on composition, homogeneity and curing conditions several crystalline and amorphous phases may coexist in the hardened phase. Figure 1 illustrates the definition on the process path for the manufacture of a geopolymer.

Thermal resistance of geopolymers
Geopolymers have excellent thermal stability as shown by several authors[2, 3, 4]. Depending on the chemical composition of the geopolymeric binder itself, especially the kind of alkali ion, different high temperature phases are built. The phase transformation (conversion reaction) at high temperatures may also be influenced by the kind of other reactive additives as well as of filler materials. Especially aggregates with small grain size and a certain amount of reactive phase can react with the binder material during the alkaline activation or the thermal treatment as shown in investigations on geopolymer stabilized clays[5]. The thermal stability of geopolymer binders is therefore strongly correlated to its real composition.

The necessity of a certain thermal resistance depends on the special application of the geopolymer material. Three main groups of application can be divided according to the frequency and duration of thermal load:

1. Building materials which resist fire disaster have to be designed to the long life property at normal climate temperatures. In the cases of fire disaster they should
   - protect the construction behind from high temperatures (low thermal conductivity, foamed materials as plates or cover).
   - be fire resistent in the case of a massive construction material (wall, column etc.).
2. The geopolymer material has to pass a high temperature production step
   - as an additive like in stabilized foam clays [5].
   - in the production of special ceramics and glass-ceramics via "geopolymer route" [2, 6, 7].
3. High temperature application materials like refractory materials[2] and adhesives[8] as well as filter materials[9] have to withstand permanent temperature load under special conditions.

The material profile may be different in all three groups of application, but – according to the thermal properties – both, the change of the physical properties (shrinkage, strength under and after thermal load, etc.) and the change of the phase composition of the geopolymer are of interest.

The main focus of the investigation presented here are the interdependence of the geopolymer composition and the phase composition before and after heating at 1000°C. Conversion reaction at thermal treatment

First information about the thermal stability of geopolymers can be derived from the thermal stability of zeolites due to the near structural relation of geopolymers and zeolites. Generally zeolites undergo different structural changes during heat treatment including a volume contraction due to water removal and the structural breakdown seen in a complete amorphization or recrystallization of new phases. The temperature of these changes is related to the zeolite composition as the Si/Al-ratio and the kind of alkali[10]. Sodium based zeolites recrystallize to nepheline at about 800-1000°C, either with or without the formation of the intermediate phase carnegieite low or via an amorphous state[11, 12, 13]. At temperatures above 1400°C nepheline converts to carnegieite high. The nepheline structure is based on a tridymite-like structure of the aluminosilicate framework and carnegieite structure on a cristobalite-like structure respectively.

Beside the simple molar composition of Na-nepheline with Na:Si:Al=1:1:1, nepheline may contain a certain amount of potassium as well as higher amounts of silica up to a Si/Al-ratio of 1.5 [13, 14]. Higher Si/Al ratios did not lead to a nepheline recrystallization[11].

Basically the reactions of geopolymers at thermal treatment follow the same scheme. Own thermal analysis investigation showed as well two exotherms in the temperature range between 800-1000°C indicating the intermediate crystallization of carnegieite low followed by nepheline in low the Si/Al range[15]. But there is not necessarily intermediate carnegieite crystallization in low Si/Al systems before starting of nepheline crystallization as shown by DUXSON[16].

In contrast to zeolites the Si/Al ratio does not mean a certain homogeneity in one single phase, as it is usually reached by adding of silicate solution to the alkaline activator. Therefore geopolymers that contain different amounts of silicate solution vary significantly in their microstructure, which influences extremely the thermal behavior[17]. Additionally the initial dissolution of the solid raw material may be incomplete and unreacted phases as well as side phases may exist in geopolymer binders. What happens with the unreacted Metakaolin inside the binder during temperature treatment? How influences a calcium addition the conversion reaction of geopolymers? These questions will be discussed basing on the results presented here.

INVESTIGATIONS

Geopolymer binders were prepared by activating either Metakaolin (Metastar 402, Imerys GB) or class F-fly ash with sodium hydroxide (laboratory grade). The influence of calcium addition was investigated either by mixing the metakaolin with granulated blast furnace slag[18] or by addition of calcium hydroxide[19].

The chemical and physical properties of the solid materials are documented in Table 1. Samples of $1 \times 1 \times 6$ cm³ geometry were prepared and kept inside the moulds at 40°C for one day. Afterwards the samples were stored in closed boxes over water at room temperature.

The compressive strength was measured on six broken half's of the prism with a compression area of 1×2 cm². The real density was measured on the dried, pulverized samples by helium pycnometer.

The amount of reaction product was investigated by dissolving the dried and ground binders in hydrochloric acid following the instructions given in[20]. For the test, 1.00 g of the sample was suspended in 250 ml HCl (1:20) and homogenized for three hours at room temperature. Afterwards the solution was filtered. The residue was washed and dried at 100°C.

To determine the X-ray diffraction pattern the dried material was mixed with a standard material (ZnO about 10 % per weight) and ground under addition of isopropanol in a bar mill (McCrone-Micronizer, McCrone Ltd., UK) for about 1 minute. This mill was chosen to avoid preferred orientation (texture) and the introduction of stress. For the X-Ray measurement CuKα radiation was used. Quantification was carried out using Rietveld refinement (Autoquant®).

Table 1: Properties of the used solid materials, surface area measured by Blaine test or nitrogen absorption* (BET)

w.-%	Metakaolin MK	Fly ash SFA	Slag S
Loi	0.80	1.9	4.91
SiO₂	52.10	50.7	31.34
Al₂O₃	43.00	27.4	10.39
Fe₂O₃	0.70	6.2	1.08
CaO	0.00	3.9	42.61
MgO	0.30	2.3	7.52
MnO	0.01	0.1	-
TiO₂	0.02	1.1	-
K₂O	2.50	2.7	0.66
Na₂O	0.12	1.1	0.25
SO₃	0	0.4	1.08
density [g/cm³]	2.55	2.61	2.9
surface area [m²/g]	11.6*	2.9	8.1

Solid state NMR experiments were performed with a Bruker Avance™ 400 MHz WB, the resonance frequency of ²⁹Si and ²⁷Al was 79.52 and 104.29 MHz, respectively. For the ²⁹Si MAS NMR spectra the dried and ground samples were packed in 4 mm rotors and spun at 55 kHz under the magic angle. The chemical shifts were recorded relative to external tetramethylsilane (TMS). The single pulse technique was used; a typical number of scans were 12000. For the ²⁷Al MAS NMR spectra the samples were packed in 4 mm rotors and spun at 15 kHz under the magic angle. The chemical shifts were recorded relative to external $Al(H_2O)_6^{3+}$. The single pulse technique was used; the number of scans was 1000. The signal patterns of the spectra were deconvoluted with the Seasolve PeakFit® software using the Lorentz curve shape.

RESULTS AND DISCUSSION
Influence of activator concentration

The influence of the activator concentration on the strength performance was determined by activating metakaolin with NaOH. The mixture compositions are given in Table 2. The water content was kept constant, whereas the NaOH content in the solution was varied between 9 and 36 w.-percent, which corresponds with a Na/Al molar ratio between 0.3 and 1.3.

Figure 2 gives an overview about some properties of the samples measured at 28 days age. The content of reaction product increases with increasing NaOH concentration up to 83 w.-% of the dried binder. This performance is reached already with a Na/Al ratio of 0.9.

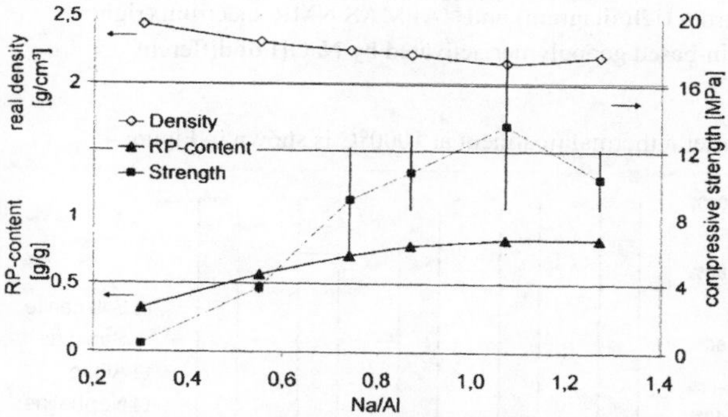

Table 2: Mixture composition [g]

Na/AL	metakaolin	NaOH	water
1.3	100	40.46	70.6
1.1	100	31.62	70.6
0.9	100	25.5	70.6
0.7	100	21.4	70.6
0.6	100	15.32	70.6
0.3	100	7.1	70.6

Figure 2: Metakaolin-based geopolymer with pure NaOH activator - Influence of activator concentration on strength performance after 14 days; content of reaction product (RP) and density

The reacted binders, shown in figure 2, were investigated by X-Ray diffractometry and NMR spectroscopy to determine the phase composition.

It was found that the crystallinity was highest in the case of high NaOH concentration. Different crystalline zeolites (zeolite A, zeolite X and herschelite) could be grown out of the amorphous aluminosilicate phase. Binders with NaOH content lower than 15 w.-percent in the solution show no crystalline reaction products and a big amorphous hump between 15 and 40° 2Θ. The center of that amorphous hump changes from lower to higher concentration to larger angels indicating a larger amount of built reaction product. The amorphous hump of the binders is mostly dominated by the unreacted metakaolin that could not be dissolved by the low NaOH concentration. This fact could be verified by the results of the NMR spectroscopy shown in figure 3. It can be seen in the ^{29}Si MAS NMR spectra the broad peak of the metakaolin at about −105 ppm. Depending on the NaOH concentration the metakaolin has been dissolved almost completely (Na/Al = 1.1; silicon content in unreacted metakaolin = 4%) or only partly (Na/Al = 0.3; silicon content in unreacted metakaolin = 51%). The ^{29}Si MAS NMR spectrum of the binder with Na/Al = 1.1 appears of mainly one signal at about −85 ppm according to an aluminosilicate network consisting of mostly Q_4(4Al)-units. The ^{29}Si MAS NMR spectrum of binders made with less NaOH concentration feature clearly more than one Q_4-signals. The ^{27}Al MAS NMR spectra of these binders differ only slightly showing one main peak centered about 60 ppm as typical for an aluminosilicate network.

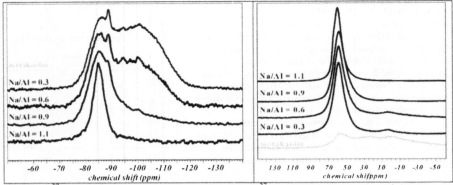

Figure 3: ^{29}Si MAS NMR spectrum (left diagram) and ^{27}Al MAS NMR spectrum (right diagram) of metakaolin-based geopolymer activated by NaOH of different concentration

The phase composition after a thermal treatment at 1000°C is shown in Figure 4. Surprisingly 100 % of the binder with Na/Al-ratio of 1.1 was transferred into nepheline; no other phase could be detected. With decreasing activator concentration (and therefore Na/Al-ratio) feldspars (sanidine and albite) and amorphous phase appear. Samples with very low activator concentration and therefore with a very low reaction degree contain no crystalline nepheline but an enormous amount of amorphous phase and additionally sillimanite was built. It was expected from the knowledge about the firing of kaolinite the formation of mullite. Being aware the proximate reflexes of mullite and sillimanite due to little difference in cell volume, the refinement with sillimanite resulted in a better quality factor.

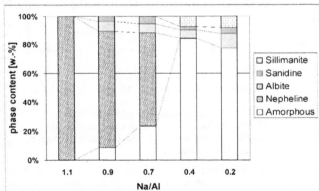

Figure 4: Phase composition of the metakaolin binders after thermal treatment at 1000°C, measured by XRD and Rietveld refinement

Influence of calcium addition

Some of the raw materials especially slags contain a remarkable amount of calcium in the glassy phase. Other raw materials for instance clays and natural pozzolana may contain about 5-

20 % CaO in reactive manner. The reaction products from the alkaline activation differ significantly. Slags which contain a lot of calcium oxide in the reactive glass form CSH- and CAH-phases similar to such formed by the hydration of Portland cement, apart from incorporated tetrahedral aluminum into the dreierketten structure of the calcium silicate hydrates [21]. Pure aluminosilicate materials like metakaolin or aluminosilicate fly ashes form aluminosilicate polymer networks. The question occurs: What happens in between these extreme different calcium contents? It could be shown on model mixtures that both reaction products (aluminosilicate and CSH/CAH-phases) coexist in different proportions depending on concentration of the alkaline activator [22]. What kind of crystalline reaction product will be built after thermal treatment?

Therefore several investigations were carried out about the influence of calcium in the raw material on the structure and phase composition of the geopolymer binder. These investigations were separated in three parts. In each part the calcium was added in a different way, in the first part as calcium containing raw material (blast furnance slag [1]), in the second part as calcium hydroxide [19] and in the third part incorporated into the glassy phase [23].

Calcium addition as blended binders

Metakaolin (M) and blast furnace slag (S) were blended together (50%/50% and 25%/75% blends) and the combination activated by sodium hydroxide solution. The boundary mixtures were set by the activation of the pure slag and pure metakaolin powders. The mixture compositions are given in Table 3. The Na/Al ratio of all mixes is 0.4. Details about casting, test regimes and results are also described in [1].

The ^{29}Si MAS NMR spectrum and the ^{27}Al MAS NMR spectrum of the raw materials and alkali-activated binders are shown in Figure 5. The alkaline activated

Table 3: Composition of alkali-activated blends

Name/ w.-%	metakaolin	slag	NaOH	water
AA-M	54.5	0.0	7.1	38.4
AA-M50/S50	31.2	31.2	5.1	32.4
AA-M25/S75	16.8	50.5	3.9	28.7
AA-S	0.0	73.2	2.5	24.4

metakaolin (AA-M) reacted to form an aluminosilicate network (ASN) showing, as expected, only Q_4-Signals in the ^{29}Si MAS NMR spectrum and four-coordinated aluminum in the ^{27}Al MAS NMR spectrum. The appearance of signals from structures other than the Q_4(4Al) is generally not observed. This can be explained by the low activator concentration as was shown in [24]. The alkaline activated slag (AA-S) reacted to form C-S-H-phases (with incorporated aluminum) revealed as Q_1 and Q_2 in the ^{29}Si MAS NMR spectrum and four-coordinated aluminum in the ^{27}Al MAS NMR spectrum, as well as to hydrotalcite which was seen as six-coordinated aluminum in the ^{27}Al MAS NMR spectrum and confirmed by X-ray diffraction.

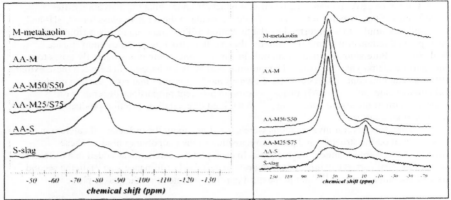

Figure 5: ^{29}Si MAS NMR spectra (left diagram) and ^{27}Al MAS NMR spectra (right diagram) of the raw materials and alkali-activated binders

The deconvolution results of the ^{29}Si MAS NMR spectrum measured from the 50%/50%-blend are shown in Figure 7. At the first, the reaction product of the alkali-activated blends (AA-M50/S50 and AA-M25/S75) seems to be only the sum of each single activated powder. This means no dominance of one partner could be found. Considering the low activator concentration chosen, the dominance of formation of C-S-H phases was expected according to the results of PALOMO [22]. However, in this case both partners contributed independently. Even for the alkali-activated blend AA-M25/S75, with the lowest activator concentration of metakaolin containing mix, the reaction degree of the metakaolin is much higher than that of the pure activated metakaolin binder (AA-M). Contrary to the investigations of PALOMO [22], both materials of the blend possess all ingredients necessary for reaction. The reason for the essentially parallel reaction may be found in the different time and place of reaction. Metakaolin and the components of the slag may have different kinetics for their dissolution/ precipitation. These reactions take place at a nanoscale around the grains.

A closer inspection reveals that there is in fact an interaction between the single powders. Slag incorporates aluminate from the metakaolin as can be seen from the change in the Si/Al molar ratio which decreases from 6 (pure slag binder) to 2.8 (50/50 blend). This causes a longer average chain length which increases from 6 to 13. The simultaneous increase of the Si/Al ratio of the aluminosilicate network from 1.07 to about 1.2 underlines the consistency of the measurement and interpretation. The effect of elongation the silicate chains in C-S-H by increasing aluminum content was described by Andersen et al. [25] as well. Schneider et al. [26] mentioned the decrease of the Si/Al molar ratio affected by the increasing sodium

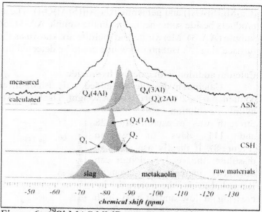

Figure 6: ^{29}Si MAS NMR spectrum and interpretation of the alkali-activated blend 50% metakaolin/50% slag (AA-M50/S50)

hydroxide concentration. As proved in parallel investigation [27], both average chain length and Si/Al ratio remain unaffected by the change of alkaline concentration in a pure slag system. Only the reaction degree could be increased by a higher concentration.

After a thermal treatment at 1000°C the different reaction products as well as the unreacted raw materials will transformed in new phases as shown in Figure 7. As expected from the results shown before only amorphous phase, feldspars and sillimanite/mullite could be built in the pure metakaolin sample (AA-M) due to the very low activator concentration. The blend with 50 % metakaolin and 50 % slag per mass allowed the metakaolin a much better performance since a remarkable amount of nepheline could crystallize. This is remarkable because the activator concentration was lower than for the pure metakaolin sample. Nevertheless the result is consistent with the measured

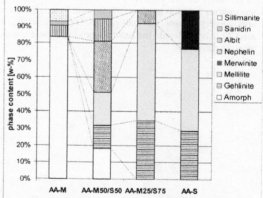

Figure 7. Phase composition of the binder blends after thermal treatment at 1000°C, measured by XRD and Rietveld refinement

reaction degree by NMR spectroscopy [18]. The amount of feldspars and the amorphous content is lower in the AA-M50/S50-blend. Additionally gehlenite and melilite were found. Both disilicates are structurally related in the way that melilite is a solid solution between akermanite

($Ca_2Mg[Si_2O_7]$) and gehlenite ($Ca_2Al[(SiAl)O_7]$). These two phases are the main reaction products beside some nepheline in the sample AA-M25/S75 and some merwinite in the pure slag sample (AA-S). Merwinite and melilite are known as the mineral phases of slowly cooled blast furnace slag [28]. But no merwinite could be detected in the blends.

Calcium addition with calcium hydroxide

Calcium hydroxide was added to fly ash in different amounts so that the powder mixture contained 0 %-20 % $Ca(OH)_2$ per weight. The blend was then activated with 8 mol/l sodium hydroxide solution.

Strength was measured after 28 and 111 days of reaction (Figure 8). If the fly ash starts to dissolute, the ready calcium can obviously increase the strength. Remarkable is the strength increasing at low and at very high calcium contents (8 % and 30 % $Ca(OH)_2$ in the powder mix).

The reaction products were determined or characterized by X-Ray diffraction, thermal analysis and NMR spectroscopy. The details are published before [19, 29].

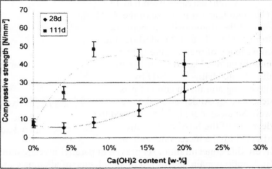

Figure 8: Influence of $Ca(OH)_2$-content on strength

In summary almost exclusive amorphous reaction products were built, beside a little amount of crystalline sodalite at all ages and in every sample. Crystalline C-S-H (hillebrandite) was found in the 20 % and 30 % sample after 111 days. Beside the reaction product, some unreacted calcium hydroxide could be detected after 28 days.

The results of the ^{29}Si MAS NMR spectroscopy of the fly ash binders at the age of 28 days are given in Figure 9. The broad fly ash hump centered at about -100 ppm indicates a very low reaction degree. The main peak of the reaction products are found between -70 and -90 ppm. As the calcium content increases, the center of this peak seems to shift from -87 to -85 ppm, connected with a spreading/broadening. If the peak at about -87 ppm of the Ca(OH)$_2$-free sample is labeled to be the Q$_4$(4Al)-signal of the built aluminosilicate network, then the shifting might be caused by the overlapping of this peak with the Q$_2$-signals (at about -84 ppm) from the C-S-H-phase. A deconvolution was determined based on these considerations. The spectra of the unreacted fly ash was fitted as two separate peaks, one broad peak of the glass centered at about -102 ppm and a second of the crystalline mullite at -88 ppm [30]. The parameters of these peaks (center and width) was used to fit the Ca(OH)$_2$-free sample. The results are shown in Figure 10, left diagram. Four peaks of the aluminosilicate network were then fitted with centers at -86.6 ppm/Q$_4$(4Al), -93 ppm/ Q$_4$(3Al), -96 ppm/ Q$_4$(2Al) and -100 ppm/ Q$_4$(1Al). This deconvolution used peaks with a Lorentz shape and identical peak widths.

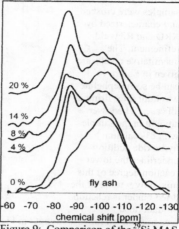

Figure 9: Comparison of the ^{29}Si MAS NMR spectra of fly ash binders (28 d)

The parameters determined by this procedure were used to deconvolute the 20 % Ca(OH)$_2$-sample using known peak positions of C-S-H from [31]. The results are shown in Figure 10. The Si/Al molar ratio of the aluminosilicate network should be slightly larger than 1.0 because of the detected signals up to −100 ppm, indicating not only aluminum tetrahedra in the next nearest neighborhood. It can be concluded that both – aluminosilicate phases and C-S-H-phases – can be proved in the matrix of the calcium containing samples, especially at high calcium hydroxide contents of about 20 %.

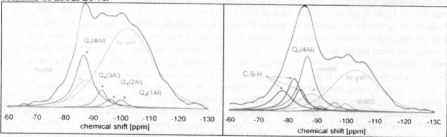

Figure 10: Peak separation of the ^{29}Si MAS NMR spectra of fly ash binders (28d) with 0 % Ca(OH)$_2$ content (left diagram) and with 20 % Ca(OH)$_2$ content (right diagram)

After a thermal treatment of 1000°C the samples were crushed and characterized by XRD and Rietveld refinement. The quantitative results are given in Figure 11. It can be seen nepheline was built after firing in all cases. The lower amount in the sample with 8% calcium hydroxide addition underlines the lower reaction degree of this mix, only 9 w.-% of the dried binder is reaction product (also

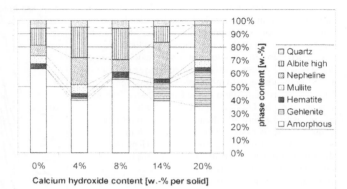

Calcium hydroxide content [w.-% per solid]

Figure 11:   Phase composition of the calcium hydroxide added binders after thermal treatment at 1000°C, measured by XRD and Rietveld refinement

recognizable on the low strength in Figure 8). Beside the nepheline, a remarkable amount of gehlenite could be crystallizing in the samples with calcium hydroxide addition of 14 % and 20 % possibly by consumption of portlandite and lime respectively. But the samples with lower calcium content contain a little, but detectable amount of gehlenite as well. In all samples out of the sample with 20 % calcium hydroxide, a high amount of albite could be detected. The absent of mullite in the samples with 8 % and 14 % calcium hydroxide is somehow unclear, because mullite, quartz and hematite are present in the origin fly ash.

CONCLUSION

The results presented showed clearly the stringent influence of geopolymer composition on the reaction products built by hardening and during thermal treatment. In this pure sodium based systems without any addition of silicate solution the reaction product of the conversion reaction at 1000°C are nepheline via a possibly intermediate phase of carnegieite low. The building of nepheline is not related to the amount of crystalline zeolites. It underlines the thesis that geopolymers contain the same aluminosilicate (sub-)structures than zeolites. Unreacted metakaolin were partly transformed into mullite. Higher temperatures or even a longer heating may complete this conversion. The additionally produced silica obviously reacts either with nepheline or preforms of it to albite in accordance to natural socialization.

If calcium in a reactive manner and a reasonable amount enters the aluminosilicate system C-S-H phases with incorporated aluminate tetrahedra are the additional reaction product of the alkaline activation. These C-S-(A)-H phases transform to gehlenite after heat treatment of 1000°C. The higher SiO_2/Al_2O_3-ratio of gehlenite compared to common C-S-(A)-H phases might be compensated by incorporation of more aluminate from the neighbor aluminosilicates. If this is not possible, the building of silicate rich melilite is conceivable. On the other hand melilite and merwinite can be crystallizing out of the unreacted slag glass. These considerations are shown graphically in Figure 12.

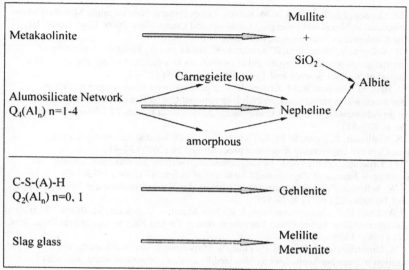

Figure 12:   Conversion reaction in geopolymer systems with and without calcium addition

The results show a clear influence of the structural performs on the conversion reaction in the way that tecto-silicates transform to tecto-silicates and (defect) chain silicates transform to soro-silicates. Multiple phase materials will still consist several phases after thermal treatment.

REFERENCES

[1] J.L. Provis, G.C. Lukey, J.S.J. van Deventer: Do Geopolymers Actually Contain Nanocrystalline Zeolites? A Reexamination of Existing Results. Chemistry of Materials. 17 (2005) 3075-3085

[2] D.C. Comrie, W.M. Kriven: Composite cold ceramic geopolymer in a refractory application. American Ceramic Society (ed.) Advances in ceramic matrix composites IX : proceedings of the Ceramic matrix Ceramic transactions. 153. (2003) 211 - 225

[3] P. Duxson, G.C. Lukey, J.S.J. van Deventer: The effect of alkali metal type and silicate concent ration on the thermal stability of geopolymers. Proceedings of the 4th International conference on Geopolymers, St. Quentin, France, 2005. (2006) 189-193

[4] K. Dombrowski, A. Buchwald, M. Weil: The influence of calcium content on the structure formation and thermal performance of fly ash based geopolymers. Journal of Material Science. SPECIAL EDITION: "ADVANCES IN GEOPOLYMER SCIENCE & TECHNOLOGY". in press, online available at DOI 10.1007/s10853-006-0532-7

[5] A. Buchwald, M. Hohmann, Ch. Kaps: Stabilisierung von Schaumtonmassen mit Geopolymerbindern für die Herstellung hochwärmedämmender Ziegel. Ziegeljahrbuch, (2004) 104-114

[6] M. Gordon, J.L. Bell, and W.M. Kriven. Geopolymers: New Ceramic Materials Made by Gel Formation. in Novel and Emerging Ceramics and Composites. 2006. Kona Coast, HI: Engineering Conferences International, Brooklyn, NY

[7] R. Vallepu, Y. Nakamura, R. Komatsu, K. Ikeda, and A. Mikuni: Preparation of forsterite by the geopolymer technique - gel compositions as a function of pH and crystalline phases. Journal of Sol-Gel Science and Technology, 35 (2005) 107-114

[8] J. Bell, M. Gordon, W.M. Kriven: Use of geopolymeric cements as a refractory adhesive for metal and ceramic joins. in 29th International Cocoa Beach Conference and Exposition on Advanced Ceramics & Composites. 2005. Cocoa Beach, FL: American Ceramic Society. 26: p. 407-413

[9] S. Mallicoat, P. Sarin, W.M. Kriven: Novel, alkali-bonded, ceramic filtration membranes. Ceramic engineering & science proceedings. 26 (2005) 37-44

[10] G. Cruciani: Zeolites upon heating: Factors governing their thermal stability and structural changes. Journal of Physics and Chemistry of Solids. 67 (2006) 1973-1994

[11] W. Schmitz: Zum Hochtemperaturverhalten der Natriumzeolithe der Faujasitgruppe. Kristall und Technik. 12 (1977) 467 - 471

[12] W. Lutz, H. Fichtner-Schmittler, J. Richter-Mendau, G. Becker, M. Bülow: Formation of a Special Intermediate during Phase Transformation of Zeolite NaA to Nepheline. Cryst. Res. Technol. 21 (1986) 1339 - 1344

[13] R. Dimitrijevic, V. Dondur, P. Vulic, S. Markovic, S. Macura: Structural characterization of pure Na-nephelines synthesized by zeolite conversion route. Journal of Physics and Chemistry of Solids. 65 (2004) 1623 - 1633

[14] W.A. Dollase, W. M. Thomas: The Crystal Chemistry of Silica-Rich, Alkali-Deficient Nepheline. Contributions to Mineralogy and Petrology. 66 (1978) 311 - 318

[15] A. Buchwald, M. Vicent, R. Kriegel, Ch. Kaps, A. Barbera: Geopolymeric binders with different fine filler aggregates - Influence on the phase transformation at high temperatures. in preparation

[16] P. Duxson, G.C. Lukey, J.S.J. van Deventer: Evolution of Gel Structure during Thermal Processing of Na-Geopolymer Gels. Langmuir. 2006, 22 (21), 8750-8757

[17] P. Duxson, G.C. Lukey, J.S.J. van Deventer: The thermal evolution of metakaolin geopolymers: Part 1 – Physical evolution, Journal of Non-Crystalline Solids. 352 (2006), 52-54, Pages 5541-5555

[18] A. Buchwald, H. Hilbig, Ch. Kaps: Alkali-activated metakaolin-slag blends – performance and structure in dependence on their composition. Journal of Material Science. Advances in Geopolymer Science & Technology. 42 (2007) 9, 3024-3032

[19] K. Dombrowski, A. Buchwald, M. Weil: The influence of calcium content on the structure formation and thermal performance of fly ash based geopolymers. Journal of Material Science. Advances in Geopolymer Science & Technology. 42 (2007) 9, 3033-3043

[20] A. Fernandez-Jimenez, A.G. de la Torre, A. Palomo, G. Lopez-Olmo, M.M. Alonso, M.A.G. Aranda: Quantitative determination of phases in the alkaline activation of fly ash. Part II: Degree of reaction. Fuel 85 (2006), 14-15, Pages 1960-1969

[21] P.J. Schilling, L.G. Butler, A. Roy, H.C. Eaton: 29Si and 27Al MAS-NMR of NaOH-Activated Blast-Furnance Slag. Journal of American Ceramic Society. 77 (1994) 9, 2363-2368

[22] A. Palomo, T. Blanco-Varela, S. Alonso, L. Granizo: NMR study of alkaline activated "Calcium Hydroxide - Metakaolin" solid mixtures. Proceedings of 11th International Congress on the Chemistry of Cement (ICCC), 2003 Durban, South Africa. (2003) 425 - 434

[23] A. Buchwald, A. Kehr, K. Dombrowski, M. Weil: Einfluss des Calciumgehaltes alumosilikatischer Gläser auf Erhärtungsverlauf und Phasenbestand nach alkalischer Aktivierung. Proceedings of the 16th Ibausil 2006, Weimar

[24] P.S. Singh, M. Trigg, I. Burgar, T. Bastow: ^{27}Al and ^{29}Si MAS NMR Study of Aluminosilicate Polymer Formation at Room Temperature. in Proceedings of the 3rd Int. Conf. on Geopolymer, Melbourne, October 2002, edited by G.C. Lukey (CD)

[25] M.D. Andersen, H.J. Jakobsen, J. Skibsted: Incorporation of Aluminium in the Calcium Silicate Hydrate (C-S-H) of hydrated Portland Cements: A High-Field ^{27}Al and ^{29}Si MAS NMR Investigation. Inorg. Chem. 42 (2003) p. 2280-2287

[26] J. Schneider, M.A. Cincotto, H. Panepucci: ^{29}Si and ^{27}Al high-resolution NMR characterization of calcium silicate hydrate phases in activated blast-furnace slag pastes. Cem. Con. Res. 31 (2001) p. 993-1001

[27] H. Hilbig, A. Buchwald: The effect of activator concentration on reaction degree and structure formation of alkali-activated ground granulated blast furnace slag. Journal of Material Science, Volume 41 (2006) 6488-6491

[28] C. Schneider, B. Meng: Bedeutung der Glasstruktur von Hüttensanden für ihre Reaktivität. proceedings of the 14th Ibausil 2000, Weimar, 1-0455 - 1-0463

[29] A. Buchwald, K. Dombrowski, M. Weil: The influence of calcium content on the performance of geopolymeric binder especially the resistance against acids. 4th International conference on Geopolymers, 29.6.-1.07.05, St. Quentin, France; 35-40; ISBN: 2-9514820-0-0

[30] K.J.D. Mackenzie, M.E. Smith: Multinuclear solid-state NMR of inorganic materials. R.W. Cahn (ed.) Pergamon materials series. Amsterdam : Pergamon (2002)

[31] G.K. Sun, J. Francis Young, R. James Kirkpatrick: The role of Al in C–S–H: NMR, XRD, and compositional results for precipitated samples. Cement and Concrete Research 36 (2006) pp. 18 – 29

LASER SCANNING CONFOCAL MICROSCOPIC ANALYSIS OF METAKAOLIN-BASED GEOPOLYMERS

Jonathan L. Bell, Waltraud M. Kriven
Department of Materials Science and Engineering
University of Illinois at Urbana Champaign
1304 W. Green St.
Urbana, IL 61801, USA

Angus P. R. Johnson and Frank Caruso
Department of Chemical and Biomolecular Engineering,
University of Melbourne
Victoria 3010, AUSTRALIA

ABSTRACT

Laser scanning confocal microscopy (LSCM) is a microscopy technique that has the capability of transmitting light through thin sections of material, to visualize a 2-dimensional plane of the sample interior. Computer software is then used to form 3-dimensional representations of the sample from multiple 2-dimensional scans. In this study, LSCM was employed to resolve the distribution of the formed geopolymer phase and metakaolin precursors in geopolymers with varying molar ratios ($M_2O \cdot Al_2O_3 \cdot xSiO_2$, where x = 0, 1, 2 and M = Na$^+$, K$^+$, or Na/K = 1) and make correlations between the observed distributions and the degree of reaction between the two precursors. LSCM analysis agreed with results in NMR and microstructural studies presented in literature- increasing concentrations of silica and higher sodium to potassium ratios resulted in less reaction between metakaolin particles and alkali-silicate solutions. When potassium is used, the resultant microstructure consisted of finer precipitate sizes and porosity regardless of the amount of silica added. The use of LSCM allowed for a 3-D microstructural perspective of pores, unreacted metakaolin, and formed geopolymer precipitates.

INTRODUCTION

A "geopolymer" is an aluminosilicate binder formed by mixing aluminosilicate materials (e.g. metakaolin, fly ash) with alkali or alkali-silicate solutions.[1] The set time and resultant properties are dependant on the processing conditions, materials used, and the $M_2O:SiO_2:Al_2O_3:H_2O$ molar ratios (M = Li$^+$, Na$^+$, K$^+$, Rb$^+$ or Cs$^+$). Typically Na$^+$ or K$^+$ is used as the alkali source due to cost and performance concerns. After mixing the metakaolin and the alkali-silicate solution, the resultant gel is typically sealed to prevent dehydration and allow proper chemical setting while being cured between 40 – 80°C. Setting times are rapid, 1 – 3 hours, however maturation time may take days or weeks, depending on conditions. The aluminosilicate source material must be sufficiently reactive in the alkali-silicate solution to promote leaching of dissolution and release of aluminum. For this reason, crystalline kaolin is typically dehydroxylated via calcination into amorphous metakaolin to ensure proper reactivity.[2] Although dependent on composition and processing conditions, metakaolin-based geopolymers are heterogeneous in nature. In addition to the formed geopolymer phase, metakaolin-based geopolymers contain porosity, unreacted metakaolin, chemical impurities, and weakly and strongly bound water.[3] Geopolymers consist of four coordinated (IV) interlinked Si and Al, although metakaolin contains Al(IV), Al(V), and Al(VI).[1, 4] The relative degree of reaction between alkali-silicate solutions and metakaolin precursors was explored by Duxson *et. al.* using MAS NMR. Duxson et al. compared the amount of Al(VI) (attributed to residual metakaolin) present in geopolymers synthesized with varying silica contents and alkali chemistries.[4] Duxson et al. concluded that the degree of reaction between metakaolin and alkali-silicate solutions decreased as the ratio of Si/Al and the ratio of Na/K in the alkali-silicate solutions increased.[4] SEM analysis of geopolymers made with metakaolin and sodium silicate solutions indicated that the geopolymer microstructure became more homogenous with increasing silica content.[5]

The majority of the work done to visually study the geopolymer structure has been based on electron and conventionally optical-light microscopy. Although useful, these techniques observe only the surface of a material; the former operates in low vacuum, which can damage the microstructure of hydrated systems (e.g. cements and geopolymers). Confocal microscopy has been used to generate three-dimensional topographic maps of cement fracture surfaces to generate fractal dimensions and roughness parameters which were then related to porosity, compressive strength, and fracture properties.[6-9] Laser scanning confocal microscopy (LSCM) has several advantages over traditional confocal microscopy. Laser light can be transmitted through thin sections of a sample to capture 2-dimensional images of the sample interior. The focal planes are adjusted by computer controlled stepping motors, allowing for high resolution images higher (~ 0.2 μm horizontal and ~ 0.5 μm vertical).[10] The individual slices can then be processed using imaging software to generate a three-dimensional representation of the sample. Starting materials can be dyed with appropriate fluorescing agents to enhance visualization of the sample interior. LSCM has been used to perform wet-chemistry studies and surface characterization of cements and *in-situ* monitoring of the alkali/silicate reaction near the cement/aggregate interface.[11, 12] LSCM can be used to collect a variety of information: reaction products at the interface, three-dimensional representations of the aggregate, the development of cracks, and quantitative measurements of the gel ring thickness around aggregate particles. This ability to collect three-dimensional information *in-situ* without disruption of the hydration process separates LSCM from conventional microscopy methods.

In this study, metakaolin-based geopolymers were studied LSCM. Attempts were made to correlate the distribution of the fluorescing dyes to the final geopolymer microstructure and the nature of the alkali-silicate/metakaolin reaction as function of silica content and alkali chemistry.

EXPERIMENTAL PROCEDURE

Nine alkali-silicate solutions (Table 1) were prepared by dissolving Cab-O-Sil fumed silica (Cabot Corp.) into solutions of potassium hydroxide (KOH), sodium hydroxide (NaOH) and mixed alkali (NaOH/KOH) solutions. When fumed silica was dissolved into the alkali solutions, the resultant alkali-silicate solutions were allowed to mature for at least 24 hours to ensure complete silica dissolution.

Table 1: Composition of alkali-silicate solutions.

Composition (Molar Oxide Ratio)			
Na_2O	K_2O	SiO_2	H_2O
1			11
1		1	11
1		2	11
	1		11
	1	1	11
	1	2	11
0.5	0.5		11
0.5	0.5	1	11
0.5	0.5	2	11

Fluorescein isocyanate (FI) solutions were prepared by adding 1 g/L of FI powder to pre-made 1 N alkaline solutions. Alkali-silicate solutions (Table 1) were made fluorescent by mixing with pre-made solutions of FI dye in volume ratios of 95:5 (alkali-silicate, FI solution). It was found that FI was degraded by the alkali solutions when no silica was present; the initial yellow color of the solutions became light blue after two days. Metakaolin powder for the experiment was also dyed by mixing the metakaolin with a dyeing solution of rhodamine-B (RB) in a ratio of 10 mL of RB solution per 1 g of metakaolin. The RB solution was prepared by mixing 10 mg of RB powder per liter of de-ionized water. After mixing, the dyed metakaolin slurry was centrifuged at 1300 RPM. The supernatant was discarded. The dyed metakaolin was collected and dried at 40°C for 3 days. Geopolymers were made by hand mixing the dyed metakaolin and alkali-silicate solutions. The molar ratios of all the geopolymers were $1M_2O:1\ Al_2O_3$. (It was assumed that the metakaolin had a stoichiometric composition of $Al_2O_3:2\ SiO_2$.) Geopolymer slurry was placed between two glass slide cover slips and cured at 40°C for 36 hours. The geopolymers samples were 2 – 8 µm thick, however no attempt was made to control the thickness of the samples.

The geopolymer samples were analyzed using a laser scanning confocal microscope (LSCM) equipped with an argon laser (Leica Microsystems Pty Ltd., Gladesville, NSW Australia). Both reflected and transmitted pictures were taken simultaneously. At the beginning of each run, the photomultiplier tube gain value was adjusted in scanning mode to minimize image saturation. The step size in the z (thickness) direction was 0.5 µm. 3D representations of the geopolymer samples were constructed using image analysis software. Attempts were made to observe the evolution of the geopolymer microstructure during curing, however useful data could not extracted due to rearrangement of metakaolin particles and flow of the alkali solutions.

Samples were removed from the glass cover slips sputter coated with a gold-palladium alloy and examined using a scanning electron microscope (Hitachi S-4700 SEM).

RESULTS

Three-dimensional reconstructions of the analyzed geopolymers are given in Figs. 1 to 3. Regions rich in rhodamine-B (RB) are shown in the grey areas in the top row of figures. The presence of fluorscein isocyanate (FI) is indicated by the grey regions in the bottom row of figures in Fig. 2. The intensity of the color (in the color photographs) is proportional to the concentration of dye in the geopolymers. No color indicates a region completely devoid of dye, and hence the nature and distribution of porosity can be discerned from the lighter regions (in the black and white photographs). SEM analyses showed that geopolymers made from alkali solutions without dissolved silica (Fig. 4) were significantly more porous than geopolymers formed from alkali silicate solutions (Figs 5 and 6), but did not have a pronounced effect on the low silica geopolymers (Fig. 4). Low silica sodium-based geolymers were comprised of equiaxed precipitates, approximately 0.5 μm in diameter (Fig. 4). The morphology of the precipitates became plate-like as sodium was replaced by potassium (Fig. 4).

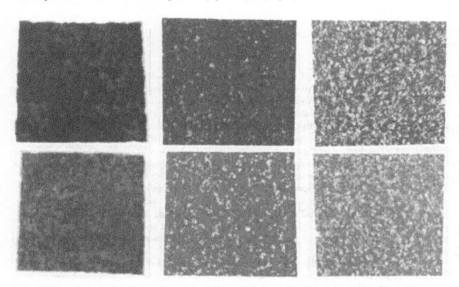

Fig. 1. 3D representations of geopolymers made with alkali solutions ($M_2O : 0\ SiO_2$), showing the fluorescing signal from rhodamine-B (top) and fluorescein isocyanate (bottom). Na^+ based geopolymers (left), Na^+/K^+ geopolymers (center), K^+ geopolymers (right). All of the above pictures are 60 μm x 60 μm with varying thickness.

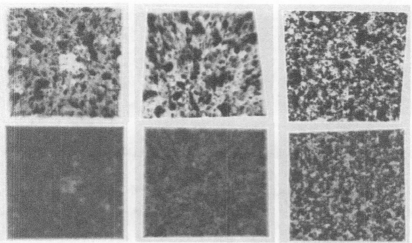

Fig. 2. 3D representations of geopolymers made with alkali solutions (M_2O : 1 SiO_2), showing the fluorescing signal from rhodamine-B (top) and fluorescein isocyanate (bottom). Na^+ based geopolymers (left), Na^+/K^+ geopolymers (center), K^+ geopolymers (right). All of the above pictures are 60 μm x 60 μm with varying thickness.

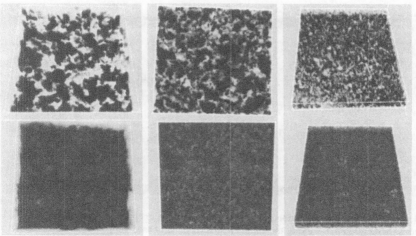

Fig. 3. Fig. 3: 3D representations of geopolymers made with alkali solutions (M_2O : 2 SiO_2), showing the fluorescing signal from rhodamine-B (top) and fluorescein isocyanate (bottom). Na^+ based geopolymers (left), Na^+/K^+ geopolymers (center), K^+ geopolymers (right). All of the above pictures are 60 μm x 60 μm with varying thickness.

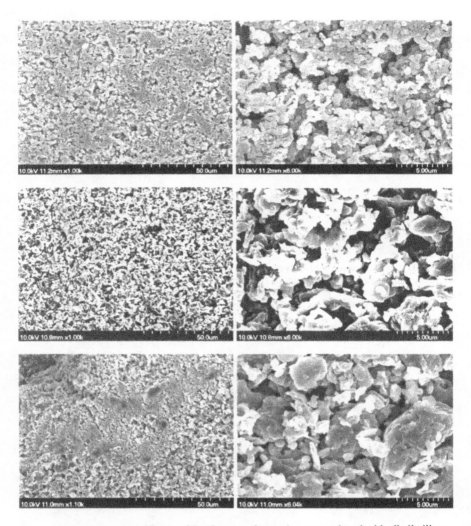

Fig. 4. High and low magnification SEM images of geopolymers produced with alkali-silicate solutions (M_2O : 0 SiO_2), varying the alkali chemistries. Na^+ based geopolymers (top), Na^+/K^+ geopolymers (center line), K^+ geopolymers (bottom line).

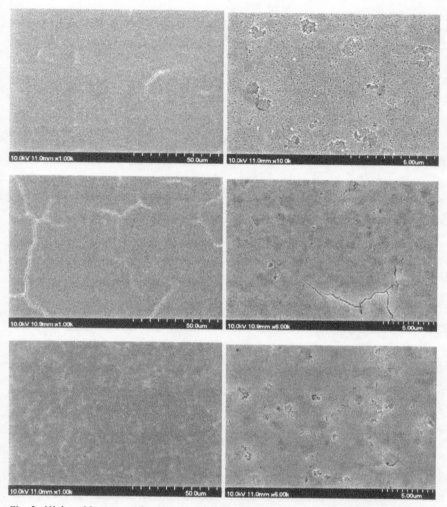

Fig. 5. High and low magnification SEM images of geopolymers produced with alkali-silicate solutions (M_2O : 1 SiO_2), varying the alkali chemistries. Na^+ based geopolymers (top), Na^+/K^+ geopolymers (center line), K^+ geopolymers (bottom line).

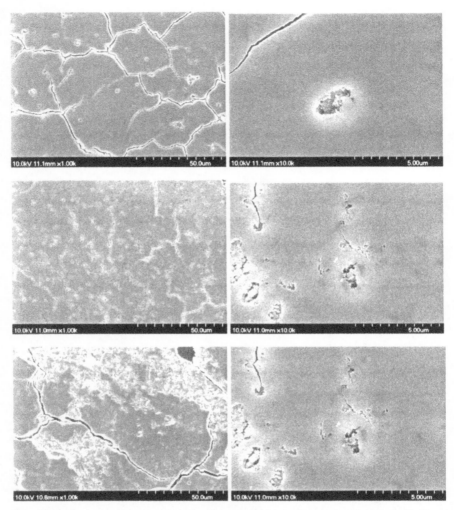

Fig. 6. High and low magnification SEM images of geopolymers produced with alkali-silicate solutions (M$_2$O • 2 SiO$_2$), varying the alkali chemistries. Na$^+$ based geopolymers (top), Na$^+$/K$^+$ geopolymers (center line), K$^+$ geopolymers (bottom line).

DISCUSSION

Large clusters of RB dye in the LSCM data were representative of unreacted metakaolin particles. Conversely, increased levels of reaction between metakaolin and alkali-silicate solutions were characterized by RB dye being distributed throughout the geopolymers. Results from LSCM images indicated that as the silica content was increased, the amount of unreacted metakaolin increased in geopolymers made from sodium and mixed alkali-silicate solutions. LSCM images also indicated that potassium-silicate solutions reacted more readily than sodium or mixed alkali-silicate solutions. Both trends, the affect of silica content and alkali chemistry, are consistent with MAS NMR results in literature.4

Overlapping of the LSCM scans showed that void regions in low silica geopolymers, where the alkali to silica ratio ($M2O:SiO2$) within the activating solution was ≤ 1 had neither dye present, indicating pore space. In the highest silica ($M2O : 2 SiO2$) geopolymer sample, however, it was apparent that FI dye was distributed throughout the microstructure, even in regions that were poor in metakaolin (no RB dye was present). These regions were presumably pores as well, however in the case of higher silica systems, the pores are filled with fluid (or silica gel) carrying the FI dye. SEM analysis showed the microstructures of the geopolymers became homogenous when silica was added to the alkali solution, agreeing with the findings of Duxson et al.5

CONCLUSIONS

LSCM has the potential for being a powerful tool to measure the degree of reaction between the constituent aluminosilicate powders and the activating alkali-silicate solutions. LSCM results agree with findings in literature; reactivity between alkali-silicate solutions and metakaolin particles was decreased when higher amounts of dissolved silica were present in the alkali-silicate solutions and when the ratio of potassium to sodium in solution was increased.

ACKNOWLEDGEMENTS
This work was supported by the US Air Force office of Scientific Research under an STTR Stage II grant, number FA 95550-04-0063. This work was carried during a student exchange visit by Jonathan Bell to the Department of Chemical and Biomolecular Engineering at the University of Melbourne, Victoria, Australia. We gratefully acknowledge Prof. Jannie van Deventer, of the Department of Chemical and Biomolecular Engineering and Dean of Engineering at the University of Melbourne, Australia, for hosting the student exchange visit. Use is also acknowledged of some of the facilities at the Center for Microanalysis of Materials in the Frederick Seitz Materials Research Laboratory at the University of Illinois at Urbana-Champaign, which is partially supported by the U.S. Department of Energy under grant DEFG02-91-ER45439.

REFERENCES
[1]J. Davidovits, "Geopolymers and Geopolymeric Materials," *Journal of Thermal Analysis*, **35**[2] 429-441 (1989).

[2]H. Rahier, B. Wullaert and B. Van Mele, "Influence of the Degree of Dehydroxylation of Kaolinite on the Properties of Aluminosilicate Glasses," *Journal of Thermal Analysis and Calorimetry*, **62**[2] 417-427 (2000).

[3]I. Lecomte, M. Liegeois, A. Rulmont, R. Cloots and F. Maseri, "Synthesis and Characterization of New Inorganic Polymeric Composites Based on Kaolin or White Clay and on Ground-granulated Blast Furnace Slag," *Journal of Materials Research*, **18**[11] 2571-2579 (2003).

[4]P. Duxson, G. C. Lukey, F. Separovic and J. S. J. van Deventer, "Effect of Alkali Cations on Aluminum Incorporation in Geopolymeric Gels," *Industrial and Engineering Chemistry Research*, **44**[4] 832-839 (2005).

[5]P. Duxson, J. L. Provis, G. C. Lukey, S. W. Mallicoat, W. M. Kriven and J. S. J. van Deventer, "Understanding the Relationship Between Geopolymer Composition, Microstructure and Mechanical Properties," *Colloids and Surfaces A-Physicochemical and Engineering Aspects*, **269**[1-3] 47-58 (2005).

[6]A. B. Abell and D. A. Lange, "Fracture Mechanics Modeling using Images of Fracture Surfaces," *International Journal of Solids and Structures*, **35**[31-32] 4025-4033 (1998).

[7]D. A. Lange, H. M. Jennings and S. P. Shah, "Analysis of Surface-Roughness using Confocal Microscopy," *Journal of Materials Science*, **28**[14] 3879-3884 (1993).

[8]D. A. Lange, H. M. Jennings and S. P. Shah, "Relationship between Fracture Surface-Roughness and Fracture-Behavior of Cement Paste and Mortar," *Journal of the American Ceramic Society*, **76**[3] 589-597 (1993).

[9]D. A. Lange, C. Ouyang and S. P. Shah, "Behavior of Cement Based Matrices Reinforced by Randomly Dispersed Microfibers," *Advanced Cement Based Materials*, **3**[1] 20-30 (1996).

[10]S. Paddock, *Methods in Molecular Biology*. Confocal Microscopy: Methods and Protocols, Edited by J. Walker. Vol. 122, Humana Press Inc., Totowa, New Jersey, 1999.

[11]K. E. Kurtis, N. H. El-Ashkar, C. L. Collins and N. N. Nalk, "Examining Cement-based Materials by Laser Scanning Confocal Microscopy," *Cement & Concrete Composites*, **25**[7] 695-701 (2003).

[12]C. L. Collins, J. H. Ideker and K. E. Kurtis, "Laser Scanning Confocal Microscopy for In Situ Monitoring of Alkali-silica Reaction," *Journal of Microscopy-Oxford*, **213** 149-157 (2004).

SYNTHESIS AND CHARACTERIZATION OF INORGANIC POLYMER CEMENT FROM METAKAOLIN AND SLAG

L. Cristina Soare, A. Garcia-Luna,
GCC Technology and Processes,
1A, Avenue des Sciences
Yverdon les Bains, Switzerland, CH-1400

ABSTRACT

Two types of inorganic polymer systems were prepared by alkali activation of metakaolin and a mix of metakaolin and amorphous slag GGBS. Inorganic polymers were synthesized at room temperature (25°C) in a highly alkaline medium (NaOH and Na_2SiO_3), followed by curing starting from room temperature up to 60°C. During this study the ratio liquid to solid was kept constant during all syntheses. The strength of the resultant products was examined by comparing the compressive strengths, Fourier - transform infrared spectroscopy (FTIR) spectra, X-ray diffraction (XRD) diffractograms and scanning electron microscopy (SEM).

The temperature, the times of curing together with the metakaolin/slag ratio are some of the variables that were studied. These variables have been shown to considerably influence the development of the mechanical strength for the final product.

The mechanical strength of inorganic polymers conducted in the presence of slag, followed by curing at 60°C for 3h up to 24h, exhibited an increase in strength by 10MPa. However, for a such low amount of slag of 1%wt and cured at 60°C for 24h the inorganic polymers reached the strength of 84MPa.

INTRODUCTION

The term "inorganic polymer" was first introduced by Davidovits[1] (1979) to designate a new class of three dimensional alumino-silicate materials. Inorganic polymers or "inorganic polymers" emerge as a new class of engineering materials that offer the potential to fulfill important requirements that are not met by the organic polymers. Inorganic polymers can be synthesized from aluminosilicate reactive materials, such as metakaolin, fly ash and slag in strong alkaline solutions, such as NaOH or KOH, then cured at room temperature or high temperature. Numerous studies have indicated inorganic polymeric gel formation[2,3], by dissolution of aluminosilicate, polycondensation and structural re-organization. A schematic diagram of typical geochemical reactions associated with mineral polymer (polysialate) and inorganic polymer formation is provided in the following reaction:

$$n(Si_2O_5, Al_2O_3) + 2nSiO_2 + 5nH_2O + 2nMOH \rightarrow M^+_{2n}- (O-Si-O-Al-O-Si-)_{2n} + 6nH_2O$$

(Si-Al materials)

amorphous geopolymer

In addition, most inorganic polymers synthesized from different raw materials (mixture of kaolinite and stilbite) are mixtures of crystalline, semicrystalline and amorphous particles of alumino-silicates, this was possible to understand using the technique of electrons transmission. Few researches have been made in this field due to the complexity of the structural composition[4]. Furthermore, the work carried out by Xu et al[5] also explained the effect of structural composition on the mechanical strength of inorganic polymers, such as a function of ratio of kaolinite/stilbite which, when decreasing makes the strength increase. The latter is correlated to the dissolution of crystalline kaolinite particles during inorganic polymerisation. However, the chemical and physical nature of these phases has not much been subjected to a detailed analysis, and it has been very difficult to determine due to the complex and intergrown nature of the binder phases and the presence of significant quantities of unreacted raw materials.

The work carried out by D. P. Dias et. al[6] showed the evolution of strength in function of the samples' crystallinity. The inorganic polymers cements were synthesized from the metakaolin amorphous structure and fly ash high crystallinity structure. The results indicated that the larger the amount of fly ash, substituting the metakaolin on the inorganic polymeric cement, the smaller the compressive strength of the mortar because of the high crytallinity degree and of the low fineness of the fly ash. The compressive strength is also strongly dependent also on Si:Al and Na:Al ratios[14].

Some research was also conducted on the comprehension of the relationship between the chemical composition of the inorganic polymers' synthesis and their properties such as the compressive strength and fire resistance[7]. The inorganic polymers were synthesized from a mixture of metakaolin and granulated blast furnace slag in the presence of potassium hydroxide and sodium silicate. In this work two behaviors of strength evolution were observed: first, increasing the amount of metakaolin leads to strength increase and the reason could be that the more metakaolin added, the more Al gels were formed in the system, therefore, giving a higher degree of inorganic polymer reaction; secondly, the strength evolution could be the KOH concentration which decreases the mechanical properties due to too much K^+ ions in the framework cavities.

For the inorganic polymer characterization most of the techniques used in the literature are: scanning electron microscopy (SEM)[3,8] for the internal structure, Fourier transform infrared spectroscopy (FT-IR)[10] for the analysis of its sensitivity to structures of short range structural order, X-ray diffraction (XRD)[9], nuclear magnetic resonance (NMR)[3,10] for the analysis of a structure with a high heterogeneity and lack of structural order. The thermal gravimetric analysis (TGA)[7] is also used for water adsorption, mercury intrusion porosity for measure of densities[7].

A possible mechanism of inorganic polymer formation is well described by nucleation; the aluminosilicate dissolves in the alkaline solution. This first step is significantly affected by the thermodynamic and kinetics parameters[15].

The purpose of this work is therefore to investigate the effect of metakaolin and slag mixes on mechanical properties and compared them with the composition of an inorganic polymer prepared only with metakaolin. The effect of curing regime on final properties of the inorganic polymer has also been investigated. The objective of these mixes and curing regime is to develop an approach for semiquantitative characterization of the inorganic polymeric materials possessing both crystalline and amorphous phases. XRD, SEM, FT-IR techniques have been applied to analyze the structure of the inorganic polymers.

SAMPLE CHARACTERIZATION

The morphology and internal structure of samples have been investigated using HRSEM (High resolution scanning electron microscopy) XL 30 FEG Sirion (FEI, Eindhoven, Netherlands) operating at 3kV, with a working distance of 5 mm and a spot size of 3. The samples were analyzed without any further special treatment, such as gold and/or carbon coating, in order to avoid visual artifacts at a nanodomain scale.

Particle size distribution (PSD): The measurements of the particle size of raw materials have been carried out by laser diffraction (CILAS 1064L). A given amount of powder has been dispersed in isopropanol and then applied to an ultrasonic bath for 5 minutes in order to disperse the agglomerates and get more realistic values of the particle size. For particle analysis the Fraunhofer method[11] has been used.

X-ray diffraction (XRD): The diffraction patterns have been obtained using an X-pert Pan Analytical Philips with copper radiation. wavelength $\lambda(Cuk_{\alpha 1}) = 0.1506$ nm, generated as U = 40 kV and I = 40 mA. The inorganic polymers have been rotated in order to determine the phase under investigation and then to estimate the crystalline and amorphous parts using the Rietveld simulations.

X-ray fluorescence specttroscopy (XRF): The chemical compositions of raw materials have been investigated using the X-ray fluorescence spectroscopy, AXIOS - Sequential XRF spectrometer, brand Pan Analyical, Netherlands.

Fourier transformed infrared (FT-IR): Infrared spectrums have been carried out on samples suspended in KBr using Bruker, TENSOR 27 spectrometer in adsorption mode. All spectrums have been obtained with a sensitivity of 4 cm^{-1} and 16 scans per spectrum taken, and the atmospheric conditions have thus been compensated.

Compressive strength: the testing of strength was performed using a Control device moving at a constant cross-head displacement of 500kPa per second. Testing has been conducted immediately after inorganic polymers have been removed from the tank curing. Inorganic polymer masses have been measured to check any possible fabrication error. At least three inorganic polymers have been tested at each chemical composition to compute the average strength.

Raw materials characterization: Metakaolin was purchased under the brand name MetaMax provided by Engelhard, it was used for all synthesis of inorganic polymers. The chemical phase composition, obtained from XRF measurements is listed in table I. The specific surface area was measured by nitrogen absorption to be 12.9 m^2/g, the relative density was measured to be 2.56 g/cm^3.

Table I: chemical composition of the precursor (Metakaolin) (% wt) determined by XRF analysis

Sample	SiO$_2$	Al$_2$O$_3$	TiO$_2$	Fe$_2$O$_3$	CaO	Na$_2$O	K$_2$O	P$_2$O$_5$	MgO	LOI
MK	51.11	45.45	1.54	0.45	0.09	0.18	0.12	0.05	-	0.7
Slag	31.36	13.50	0.97	0.39	36.59	0.23	0.35	-	7.92	0.07

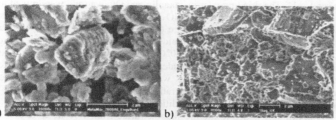

a)   b)

Figure 1: Micrographs: (a) metakaolin X 16000; (b) slag GGBS X 8000

Figure 1 shows the layer by layer metakaolin substructure. The particles' shape is more like plates with a size of around 3 μm or less. In addition, some small particles on the top with more sphere-like morphology can be observed. The size of the spherical particles is less than 500 nm, which could be formed during the ultrasonic bath treatment. Furthermore, it has been observed (it is not showed in the present paper) that there are agglomerates with a size larger than 10 μm. These different types of size correlated well with the particle size distribution measured by laser diffraction (see table II).

Table II: Particle size distribution of metakaolin

Sample name	$d_{v10}(\mu m)$	$d_{v50}(\mu m)$	$d_{v90}(\mu m)$
Metakaolin	0.91	3.81	12.64
Slag GGBS	1.20	9.32	26.52

The diffraction patterns of metakaolin is mostly amorphous 98,2 % wt. However, it also contained approximately 1.8 % wt crystalline phase like anatase (TiO_2, JCPDF no. 084-1285). The X-ray analysis made by Rietveld simulation correlated well with the XRF analysis such as for the single crystalline phase, such as anatase (see table no. 1). The Rietveld simulations were carried out with an average error of ±0.4 % wt for the crystalline phase and respectively ±2% wt for the amorphous phase.

The slag GBBS (Ground Granulated blast Furnace Slag) was supplied from Civil and Marine Slag Cement Limited U.K. This cement shows the following crystallinity: 3.30% wt calcite (JSPDF no. 086-2334), melelite 10% wt (JCPDF no.04-0682) and 86.70% wt amorphous phase. Most of this amorphous part is made of alumiosilicates. The slag shows large particles like a block and small particles on the top at around 500 nm. The XRF analysis shows 36.59 % wt of CaO. The specific surface area measured by nitrogen adsorption is 1.8m²/g, which corresponds to an agglomerate powder.

Sodium silicate solution was supplied by PQ Europe, Netherlands, under the brand name Na5/Si2 with the following chemical composition: 29 wt % SiO_2 and weight ratio SiO_2/Na_2O = 3.22. Sodium hydroxide solution was prepared by the dissolution of 540g NaOH pellets (Merck, 99.99 %) in 1000 ml de-ionized water, with all containers kept sealed wherever possible so as to minimize the contamination by atmospheric carbonation.

Inorganic polymer synthesis: Inorganic polymers were synthesized by mechanical mixing (HOBART) of stoichiometric amounts of metakaolin and activator solution (sodium silicate and sodium hydroxide). The ratio liquid to solid was kept constant at a value of 0.27 by adding de-ionized water. The Ottawa Sand was added in the same weight ratio as metakaolin (metakaolin/sand = 1). The dried powder metakaolin and Ottawa sand were mixed together for 1 minute and then the activator solution (Na_2SiO_3, NaOH and distillated water) was added. The synthesis was carried out at room temperature (25°C). The resulted slurry was cast into steel molds measuring 50 x 50 x 50 mm and the molds were then transferred into a curing equipment instantly to keep the temperature constant to 60°C. The time curing varied from 3 hours up to 24 hours. After curing, the samples were demolded and analyzed by the classical techniques of materials' characterization. The micrographs were performed on fracture surfaces. Table III provides the chemical composition: the molar ratio of SiO_2/Al_2O_3, Na_2O/Al_2O_3, the weight ratio of liquid and solid and the samples curing conditions.

The amount of slag added to the metakaolin varied from 1 % wt up to 5 % wt. The slag was used due to an important amount of calcium oxide and previous results showed an increase in strength as a result of the increase in slag amount[7].

Table III: Conditions of inorganic polymer synthesis: molar ratio of oxides: SiO_2/Al_2O_3 and Na_2O/Al_2O_3, weight ratio of liquid and solid, temperature of curing and curing time

Sample name	SiO_2/Al_2O_3	Na_2O/Al_2O_3	L/S	Temperature	Time of curing
MKRT	2.67	0.57	0.27	RT	24h
MK60	2.67	0.57	0.27	60°C	24h
MKS5	2.77	0.81	0.27	RT	24h
MKS5_3h	2.77	0.81	0.27	60°C	3h
MKS5_24h	2.77	0.81	0.27	60°C	24h
MKS1	2.63	0.58	0.32	RT	24h
MKS1_3h	2.63	0.58	0.27	60°C	3h
MKS1_24h	2.63	0.58	0.27	60°c	24h

RESULTS AND DISCUSSION

The compressive strength tests were carried out on a number of cured samples. The values of strength are given in table no. IV. The inorganic polymers prepared with slag are compared with products prepared in the same condition without slag (MKRT, MK60).

Table IV: Compressive strength (MPa) of inorganic polymer at a different ratio of slag, temperature and curing time

MKRT	MK60	MKS5	MKS5_3h	MKS5_24h	MKS1	MKS1_3h	MKS1_24h
65	73	44	68	78	25	74	84

The values of compressive strength of products prepared only with metakaolin shows an increase in strength from 65MPa to 73MPa depending on temperature curing. Moreover, the 73MPa value is higher than that well described in literature using metakaolin raw material[12,13].

The samples cured at 60°C during 24 hours shows an increase in strength from 73 MPa without slag, to 84 MPa with only 1 % wt of slag. Using 5 % wt of slag it was obtained an intermediate value of 78 MPa. One must notice an increase in strength with increase of the curing time independently of the slag amount. However, such a difference in strength is strongly dependent on both Si:Al and Na:Al moral ratio of material and the ratio of amorphous and crystalline phases[14]. Some research explains that the lower the sodium content is, the lower the compressive strength becomes[14]. This study shows the inverse effect and it also reports a high compressive strength at low sodium content[19]. However, the total amount of the crystalline phase diminish the strength value, as already remarked elsewhere[6].

The scanning electron microscopy (SEM) micrographs show only the samples prepared with 1% wt up to 5%wt slag. The micrographs in this paper depict the microstructure developed by inorganic polymer formation at different curing times.

Low Magnification   High Magnification

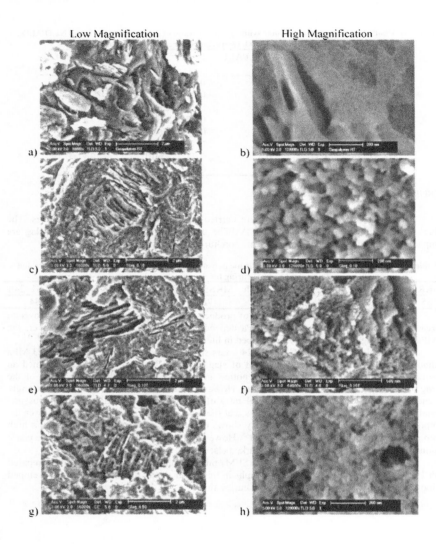

 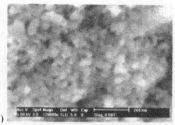

Figure 2: HRSEM micrographs of inorganic polymers: (a,b) synthesis at room temperature MKRT; synthesized at 60°C in the presence of slag: (c, d) inorganic polymer with 1% wt slag, cured for 3 hours; (e, f) inorganic polymer with 1% wt slag, cured for 24 hours; (g, h) inorganic polymer with 5% wt slag, cured for 3 hours; (i, j) inorganic polymer with 5% wt slag, cured for 24 hours

Figure 2 (a. b) shows the microstructure of the inorganic polymer prepared with metakaolin and cured at room temperature. At room temperature the inorganic polymers are formed by particles in a range a few of nanometers. Figure no. 2 (c, d, e, f, g, h, i, j) shows the internal structure of inorganic polymers prepared with slag GBBS as added material. The conditions are given in table I. The micrographs at low magnification show particles with a layer by layer structure. These types of particle morphology correspond to the unreacted metakaolin raw materials[14]. Going further to the internal substructure of inorganic polymer formation (figure 2 d, f, h, j), the inorganic polymers are built up of 50 nm nanoparticles. The structure comes from the effect that the inorganic polymers are formed of the dissolution of metakaolin in the presence of a high alkaline solution, followed by the formation of nucleus and then the growth occurs before polycondensation starts and forms a chaotic three-dimensional network polysodium aluminosilicate. This behavior of inorganic polymer formation has already been seen by A. Frenadez et. al[15] in a system carried out in the presence of fly-ash. It was remarked the co-existence of raw material (fly ash, metakaolin) particles which have reacted with the alkali dissolution with some unreacted spheres and even some other particles partially covered with reaction products. Here the micrographs of samples prepared at room temperature with slag GGBS are not presented. Due to the increase in curing, the size of the primary particles increases from a few of nanometers (figure 2 a) up to 50 – 100 nm. Furthermore, at high magnification. metakaolin-based gel inorganic polymer resembles aluminosilicate particles that are connected and form nanochanels and nanopores[16].

It is noted that the particle size of the starting materials is within the micron range (see figure no. 1), whilst the scale of the inorganic polymer matrix microstructure is below 100 nm. However, the microstructures of the specimens with different ratio of slag and different curing time are similar.

The FTIR spectra of samples activated for 3 up to 24 hours at 60°C and at room temperature (25°C) in the presence of 5% wt of slag, its show important differences when compared with the spectrum of starting metakaolin (see figure no. 4). The metakaolin spectrum (figure 4 a, black curve) is a typical one, with broad bands at about 3450 cm^{-1} and 1650 cm^{-1} from the adsorbed water, contains adsorption at 1087 cm^{-1} due to the stretching vibration of Si-O, Si-O-Al vibration at 810 cm^{-1} and Si-O bending vibration at about 464 cm^{-1}[17,18].

The broadness of the adsorption peaks reveals the amorphous character of the material. The shift of 1087 cm^{-1} band, to lower wave number of the coupled asymmetric Si(Al)-O stretching vibration at about 1000 cm^{-1} is due to the increasing Al content. This movement of the band is higher when the sample is cured during 24 hours at 60°C, which gives a high value of strength. The same behavior was observed for the sample prepared with only 1% wt of slag. The typical vibration of 810 cm[-1] due to the Si-O-Al completely disappears and it is replaced by several weaker bands within the range of 800 – 600 cm[-1]. The broad adsorption, again typical amorphous materials, around 3400 cm[-1] is ascribed to both bound and free water, and is thus not suitable for quantifying the amount of bond water[19]. The data obtained from FTIR confirmed that when the time and curing temperature of reaction increases the strength is increased[20].

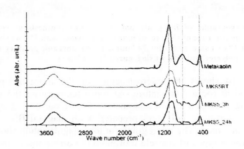

Figure 3: FT-IR spectra of metakaolin and inorganic polymers obtained in different conditions of formation: 60°C

A selection of typical powder X-ray diffraction patterns of the starting material and inorganic polymer is shown in figure no. 4. The XRD patterns of inorganic polymer shows an amorphous behavior. Their formation is characterized by the shift in the scattering peak of the inorganic polymers which is different from that of the original metakaolin and slag GGBS. When the starting material is activated with an alkaline silicate solution, the scattering peaks move from ≈ 22 to ≈ 28° in 2 theta. The XRD patterns of the cured inorganic polymers showed no change from the pattern of the precursor materials. The diffraction peaks present are due to quartz, anatase, carbonate of the raw materials (metakaolin and slag GGBS). There are no significant differences in the XRD patterns of samples cured at 60°C from 3 and respectively 24 hours. These X-ray data suggest that irrespective of the development of good mechanical properties, the structure of all the polymers is typically glass-like, consisting of randomly developed Al-Si polyhedral with a lack of periodically repeating atomic order[10].

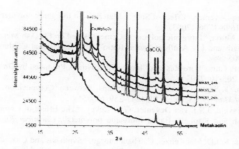

Figure 4: XRD patterns of inorganic polymers synthesized in the presence of slag and their comparison with metakaolin raw material

CONCLUSIONS

The conclusions are directly related to the mechanical strength. A significant factor affecting the mechanical strength is always the temperature. The longer the time of curing, the higher the strength is. Mechanical strength of inorganic polymers prepared with slag GGBS cured at 60°C for 24h are much higher 84MPa than those prepared only with metakaolin 73MPa due to the ratio Si:Al, Na:Al and the high amount of amorphous amount.

The microstructural observations using the HRSEM reveal that metakaolin-based inorganic polymers show particles with a size raging from a few nanometers to 100 nm and nanosize pore characteristics. Furthermore, the temperature bears an influence on the inorganic polymers formation.

X-ray diffraction patterns and FTIR show amorphous inorganic polymers formed with no change in the ratio between the crystalline and amorphous phase; however the FTIR data show a reorganization of the structure with new bond formation in function of the time and temperature. Researches are conducted with a view to understand the physico-chemical properties of the inorganic polymers on the microstructure influence on mechanical properties.

ACKNOWLEDGEMENTS: The authors thank Dr. Helene Minard for the Rietveld simulations.

REFERENCES

[1] J. Davidovits. "Inorganic polymers and Inorganic polymeric Materials", *J. of Thermal Analysis*, **35**, 429-41 (1989)

[2] J.G.S. Van Jaarsveld, J.S.J. Van Deventer, "The Potential Use of Inorganic polymeric Materials to Immobilize Toxic Materials: Part I Theory and Applications ", *Min. Eng.* **10**, 659-71 (1997)

[3] H. Xu. J.S.J. Van Deventer, "The Inorganic polymerization of Alumino-Silicate Minerals", *Int. J. Miner. Process*, **59**, 247-53 (2000)

[4] H. Xu, J.S:J. Van Deventer, "Microstructural Characterization of Inorganic polymers Synthesized from Kaolinite/Stilbite Mixtures using XRD, MAS-NMR, SEM/EDX; TEM/EDX, and HREM, *Cement and Concrete Research*, **32**, 1705-16 (2002)

[5] H. Xu, J.S.J. van Deventer, "Effect of Source Materials on Inorganic polymerization", *Industrial Engineer Chemistry Research*, **42**, 1698-1706, (2003)

[6] D.P. Dias, F.L. Murta, "Inorganic polymeric Cement Based on Fly Ash/Metakaolin: Correlation Between Axial Compressive Strength and Fly Ash/Metakaolin Ratio", 4th International Conference of Inorganic polymers, Saint-Quentin, France, 2005

[7] T.W. Cheng, J.P. Chiu, "Fire-Resistant Inorganic polymers Produced by Granulated Blast Furnace Slag", *Mineral Engineering*, **16**, 205-10, 2003

[8] M. Stevenson, K.Sagoe-Crentsil, "Relationships Between Composition, Structure and Strength of Inorganic Polymers", *J. of Materials Science*, **40**, 4247-259, (2005)

[9] J.G.S. van Jaarsveld, J.S.J. van Deventer, G.C. Lukey, "The Effect of Composition and Temperature on the Properties of Fly-Ash and Kaolinite-Based Inorganic polymers", *Chemical Engineering Journal*, **89**, 63-3, (2002)

[10] V.F.F. Barbosa, K.J.D. MacKenzei, C. Thaumaturgo, "Synthesis and Characterization of Materials Based on Inorganic Polymers of Alumina and Silica: Sodium Polysialate Polymers", *Int. J. of Inorganic Materials*, **2**, 309-17, (2000)

[11] P. Bowen, "Particle Size Distribution Measurement from Millimeters to Nanometers and from Rods to Platelets" *J. of Dispersion Science and Technology*, **23**, 631-62 (2002)

[12] M. Schmucker, K.J.D. MacKenzei, "Microstructure of sodium polysialate siloxo geopolymer", *Ceram. Internat.*, **31**, 433-37, (2005)

[13] W. M. Kriven, J.L.Bell, M. Gordon, "Microstructure and microchemistry of fully-reacted geopolymers and geopolymers matrix composites", *Ceram. Trans.*, **153**, 227-50, (2003)

[14] M. Rowles, B. O'Conner, "Chemical optimization of the compressive strength of aluminosilicate geopolymers synthesized by sodium silicate activation of metakaolinite", *J. of Materials Chemistry*, **13**, 1161-65, (2003)

[15] A. Fernadez-Jimenez, A. Palamo, M Criado, "Microstructure development of alkali-activated fly ash cement: a described model", *Cement and Concrete Research*, **35**, 1204-9, (2005)

[16] J.P. Hos, P. G. MacCormick, "Investigation of a synthetic aluminosilicate inorganic polymer", *J. of Materials Science*, **37**, 2311-16, (2002)

[17] R.W. Parker, R.L.Frost "The application of DRIFT spectroscopy to the multicomponent analysis of organic chemicals adsorbed on montmorillonite" *Clays Clay Mins*, **44**, 32-0, (1996)

[18] R.L. Frost, P.M.Frederick, H.F. Shurvell, "Raman microscopy of some kaolinite clay minerals" *Canadian J. Applied Spectroscopy*, **41**, 10-4, (1996)

[19] H. Rahier, W. Simons, B.Van Mele, "Low-temperature synthesized aluminosilicate glasses", *J. of Materials Science*, **32**, 2237-47 (1997)

[20] A. Palomo, M.W. Gruzeck, M.T.Blanco, "Alkali-activated fly ashes, A cement for the future", *Cement and Concrete Research*, **29**, 1323-29, (1999)

RECENT DEVELOPMENT OF GEOPOLYMER TECHNIQUE IN RELEVANCE TO CARBON DIOXIDE AND WASTE MANAGEMENT ISSUES

Ko Ikeda
Department of Advanced Materials Science and Engineering
Yamaguchi University
Ube 755-8611, Japan

ABSTRACT

Recent trends of geopolymers have been reviewed from point of view of suppressing carbon dioxide emissions and recycling use of wastes. First of all, the principle of geopolymers was briefly explicated in view of dissolution properties of filler materials, metakaolin and fly ash. Then, sleeping kaolin resources, specifically in China and Tunisia were introduced. Then, application of wastes as filler mixed with fly ash was described in view of waste management issues. Finally, spreading strategies of geopolymer process world over were mentioned to make use of sulfate resistance and fire proof properties of geopolymers.

INTRODUCTION

There are several gases whose emissions are a serious warning to so called "green house effect" of the earth's atmosphere. It is well-known that CO_2, CH_4, N_2O and flons are the big four warning gases other than H_2O. Usually, H_2O vapor is not taken into account in the category of warning gases, since H_2O vapor is constantly emitted due to the daily activities of mankind and the suppression of vapor is very much difficult. Talking about the green house effect, the factor of each gas contributing to the green house effect is more important than the share of each gas. As shown in Fig. 1 drawn on the basis of literature[1] in which this factor is taken into account, methane having 21 factor against carbon dioxide shares 19.8% and is steadily increasing step by step probably due to bio-origin relating to the increasing world populations of mankind. Flons including so many varieties called HFCs and PFCs share 13.5% with a tremendously big factor in average. Therefore, the emission of flons should urgently be suppressed by replacing with non-harmful alternatives, since flons are purely artificial-origin applied to cooling and cleaning media of air conditionings and hi-tech industries, both supporting comfortable modern society. Dinitrogen oxide gas shares 6.2% and has a large factor of 310 that is very big. Suppressing the emission of dinitrogen gas is also urgently requested, since this gas might be automobile-origin. Although modern cars are equipped with catalyzer devices to reduce NOXs emission, urgent introduction of full-electric cars are desirable to avoid the high temperature combustion of engines generating NOXs. Nevertheless, it is believed that the green house effect is caused by increasing emissions of carbon dioxide sharing 60.1%. This means tremendously large amount of carbon dioxide emissions, although this gas has only unity factor.

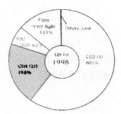

Fig. 1. Cumulative effects of warning gases from the period of industrial revolution to 1998, in which the factor in parentheses are taken into account.

Source and sink of carbon dioxide is represented in Fig. 2 based on literature[2]. Annual emissions of the world due to the combustion of fossil fuels are estimated to reach 6.3 billion tons-C as carbon base in early 1990s, of which 1.7 billion tons-C are absorbed by the ocean and 1.4 billion tons-C by the forest. The rest 3.2 billion tons-C are automatically obliged to go to the atmosphere, since the absorption capacities of the ocean and the forest are almost constant. Therefore, increasing emission of carbon dioxide straightly means increasing accumulation of carbon dioxide in the atmosphere. Thus, abrupt increasing rate has been observed since 1960 due to a rapidly growing industrialization occurring world over. At the epoch of so called "Industry Revolution" date back to the year of 1800 regarding as a pivot, the carbon dioxide level in the air is believed to be 282 ppm. The next turning point is the year of 1860 close to so called "Meiji Restoration" in Japan, on which modern Japan is inaugurated, and at that time the level was 286 ppm. Another important turning points are 1960 of 318 ppm and 2000 of 370 ppm. Most recently, 377ppm was reported in 2004 and 458 ppm will be reached in 2050 due to rapidly growing emission rate as seen in Fig. 3 drawn on the basis of literature[3]. That is, increasing rate of carbon dioxide in the period of 1800-1860 is 0.07 ppm per annum and very low. That in 1860-1960 is 0.32 ppm. That in 1960-2000 is 1.30 ppm, which is 4 times larger than that of the previous period including World War II. Therefore, it is no doubt that this extremely high emission rate is closely related to the devouring energy consumptions to support the heavy industries and subsequent luxury ways of living of mankind occurring world over in late 20th century. This rate is now advancing year after year and the rate turns up to 1.75 ppm per annum only encompassing last 4 years, 2000-2004. Reducing of 6% emission is a target of "COP3" so called Kyoto Protocol issued in 1996. Nevertheless, this target is hardly possible at the moment.

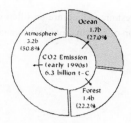

Fig. 2. Source and sink of carbon dioxide

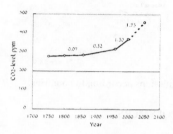

Fig. 3. Averaged variations of carbon dioxide levels in air and increasing rate in ppm per annum.

As for cement industries, energy saving is almost saturated after the introduction of suspension preheater kilns installed world over in the latter half period of last century. It is well-known that 1 ton production of Portland cement clinkers emits 1 ton CO_2. The source materials comprise a large amount of limestone, mainly $CaCO_3$, which can be regarded as a fossil carbon dioxide consisting of the atmosphere of the earth date back to geologic time. Global share of CO_2 emissions from cement industries was 5% in 1988 nearly corresponding to the overall emission share of Japan in 1994 that was 4.9% as seen in Fig. 4[4]. However, recently it is estimated to reach 7 % corresponding to the overall emission share of Russia due to increasing Portland cement productions in developing countries including China. Portland cements have been developed by supporting with chemical and technical improvements for more than 180 years and nowadays have spread to every inch of modern world. Suppose that Portland cements are thoroughly replaced with geopolymers, the COP3 target can easily be cleared, since geopolymers are not depending on limestone, which are so often called "calcium free cements" and have a capability of 80% reduction of carbon dioxide, when compared with Portland cements.

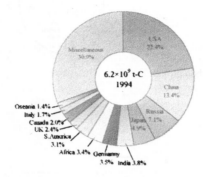

Fig. 4. Carbon dioxide emissions according to countries.

CHEMISTRY OF GEOPOLYMERS

Polycondensation of Si-Complexes

As shown in Fig. 5, the principle of geopolymers is based on polycondensation of Si-complexes. In alkali silicate solutions, so often called water glass, Si^{4+} ions are present in forms of tetrahedral complexes coordinated by OH-ligands. This state is very much stable and no polycondensation takes place to yield polymers under general conditions. However, if foreign metallic ions are present other than alkali ions, these Si-complexes become silicate polymers so easily bridged with these foreign ions. This phenomenon can typically be seen in the course of hardening of terrestrial sedimentary rocks caused by rain falls onto the earth.

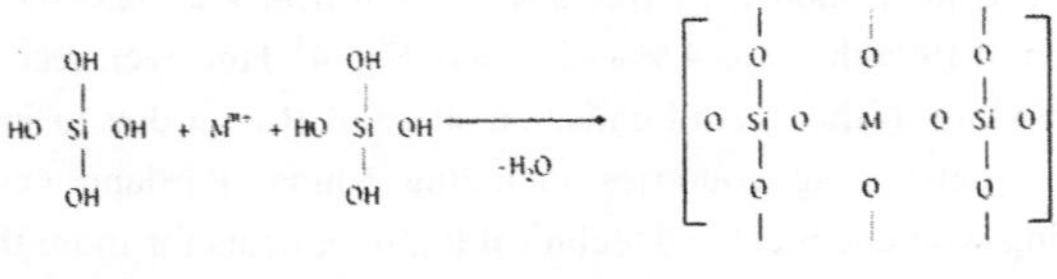

Fig. 5. Polycondensation of Si-complexes by bridging of foreign metallic ions.

There has been another traditional term in advance to the term of "geopolymers" in the discipline of cements, which is called "alkali activated cements". They are very similar each other as long as CaO-components are concerned. As mentioned in preceding section, however, pure geoplymers should be free from lime. Traditional cementitious binders are based on hydrations of lime compounds, whereas geopolymer binders are based on dehydrations of silicate polymers that are polycondensations. now being studied so extensively as appeared in SCOPUS.

Currently, two big streams exist to prepare monolithic bodies of geopolymers. One is "metakaolin based" and the other is "coal fly ash based". The metakaolin based bodies use metakaolin as filler, which is prepared by calcining raw kaolin at temperatures around 750°C. When the metakaolin filler immersed in water glass solutions, some metallic ions such as Al^{3+} and Si^{4+} dissolved out of the filler. Thus, these ions act as bridging foreign metallic ions and contribute to the polycondensation of the Si-complexes and finally the aqueous solution in a mold solidify together with the filler into a monolithic body as seen in Fig. 6. The dissolution properties of calcined kaolin were studied in caustic soda solutions by Mikuni et al.[5]. They found dissolution maxima of kaolin calcined at temperatures around 600°C and 900°C for Al^{3+}, and extraordinary high dissolution around 1100°C for Si^{4+} due to the amorphous silica formation caused by phase transition and disintegration of the original kaolin as seen in Fig. 7. They also studied dissolution properties of pyrophyllite. a kaolin like clay mineral prevailed in western Japan.[6]

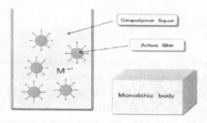

Fig. 6. A schematic picture of dissolutions of filler and a solidified body obtained after demolding.

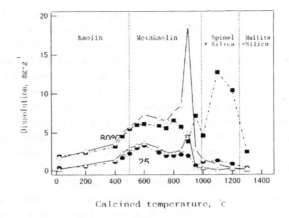

Fig. 7. Dissolution properties of heat-treated kaolin in alkali condition, Al^{3+}(open signs), Si^{4+}(solid signs).

Talking about coal fly ash, there are two types of boilers currently applied world over. One is called "pulverized coal combustion boilers" and the other "fluidized bed combustion boilers".[7] The former is equipped with three stage traps, disintegration of NOXs, fly ash recovery and capture of SOXs as gypsum, and is applied to low sulfur coals pulverized and burned at very high temperature around 1500°C so that the ash called fly ash becomes glassy globules comprising some mullite and quarts in the glassy matrices. The limit of sulfur content of coal is 0.3 %, if not, this type of boiler cannot capture whole sulfur to pollute air. To cope with this problem, the latter type of boilers is applied to high sulfur coal. First of all, limestone pieces are premixed with coal pieces as fuel. Then, the mixture is burned at temperature around 900°C so that the ash, also called fly ash, is not melted and no NOXs generation takes place due to low temperature combustion. The fuel mix is circulating in the boiler and the ash is finally collected. Consequently, sulfur is recovered mainly as anhydrite so that this type of ash is rich in CaO and SO_3 components. The problem of this kind of boiler is low efficiency of coal burning and generally 20% carbon remains intact in the ash. To solve this problem, "pressurized fluidized bed combustion type boilers (PFBC)" appeared recently to improve power generation efficiency as well equipped with two turbines, so that remaining carbon is reduced down to 5 %, comparable to pulverized type boilers.[8]

Dissolution properties of fly ash originated from the both types of boilers, pulverized and pressurized were studied also by Mikuni et al.[9] in strong caustic soda solutions as seen in Figs. 8 and 9. The results show that dissolution amounts of Al^{3+} and Si^{4+} increase with concentrations of caustic soda at 80°C. Over 5M conditions, however, dissolution rate becomes mild. It is generally said that at least 8M concentration is necessary for stable solidification of fly ash. At 25°C, however, no so high dissolution is noted. Therefore, fly ash solidification at room temperature is considered mainly due to the Ca^{2+} dissolution, which is generally included in very small amount for fly ashes of bituminous coal origin collected from pulverized coal boilers in Japan.

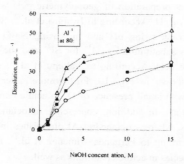

Fig.8. Dissolutions of Al^{3+} for some fly ashes in comparison with water quenched granulated blast furnace slag in solid squares. Open circles, PFBC fly ash.

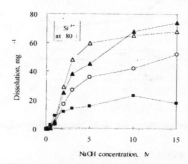

Fig. 9. Dissolutions of Si^{4+} for some fly ashes in comparison with water quenched granulated blast furnace slag in solid squares. Open circles, PFBC fly ash.

Chemical Composition of Geopolymer Gels

Analysis of geopolymer matrices act as binders is very difficult primarily due to being amorphous, secondly due to very small in size and irregular in shape and thirdly due to high contents of water that all prohibit detailed quantitative analysis. Therefore, spectroscopic techniques such as MAS-NMR and FTIR are used other than EDX technique. According to Palomo et al.[10], the matrices consist of herschelite-like composition that is a kind of zeolites. According to Zhang et al.[11], however, sodalite structures such as nepheline and leucite are referred to. According to Feng et al.[12], different compositions from zeolites and sodalites are reported and enrichment of CaO and Fe_2O_3 components are detected in matrices, when mixed with fly ash filler. Therefore, chemical compositions of geopolymer binders are still controversial.

On the other hand, direct preparation of geopolymeric gels was attempted in simplified system in vitro as a function of pH. According to Iwahiro et al.[13], SiO_2 and Al_2O_3 components behaves reversely each other on rising pH and increasing Na_2O component is noted with increasing pH. This indicates substitution of Al^{3+} for Si^{4+} and the charge is compensated by incorporation of Na^+ into the gels. According to Vallepu et al.[14], same behavior is noted on potassium bearing gels as shown in Fig. 10. Furthermore, after their MAS-NMR results, Q(2Al+1Al) is predominant with small presence of Q^1 in acidic conditions, whereas Si(4Al) is predominant in basic conditions. In addition, copresence of octahedral and tetrahedral Al is observed in acidic conditions, whereas octahedral Al diminishes in basic conditions and only tetrahedral Al is found. According to Fernandez-Jimenez et al.[15], nearly the same results are reached, but they found Q^0 species in acidic conditions as well.

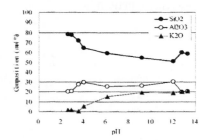

Fig. 10. Variations of chemical compositions for K-Al bearing silicate gels as a function of pH.

KAOLIN RESOURCES IN THE WORLD

Kaolin resources are distributed unevenly in the world. Consequently, this is a big problem to the countries having little kaolin resources. To solve this problem, fly ash-based geopolymers[12,16,17] are extensively studied as alternative filler in addition to kaolin-based geopolymers[4,18,19]. The Georgia kaolin in USA is well-known and UK, Portugal and New Zaland are also major production countries. Such kaolin minerals are traditionally very important to paper, brick and porcelain industries and little afford to geopolymers generally. In these circumstances, therefore, metakaolin-based geopolymers are suitable for the production of high grade materials such as decoration parts of architectures as well as artifacts of souvenirs to make use of whiteness of kaolin. According to Pinto[20], this type of geopolymers is also suitable for repairing stony architectures so as to avoid so called patch works by regulating colors.

Meanwhile, good news is coming from overseas. China is one of the big coal production countries as well as USA. It is of great benefit that coal gangues are very rich in kaolin and discarded as huge amount of pyramid-like piles over so many years. It is said that the piles are nearly 1500 in number. The total amount is estimated 3.4 billion tons and is increasing at a rate of 1.7 billion tons per annum. Moreover, kaolin beds containing some coal are embedded alternately with coal beds and develop around the coal mine area. Therefore, the kaolin resources are almost infinitive in China. Recently, the author visited one of such districts located in north eastern part of China and collected some specimens. The XRD charts are in Fig. 11. It is obvious that these gangues generally comprise much amount of kaolin associated with quartz and sericite in small amount. Some specimens do not show any evidence of kaolin because of fire that spontaneously took place in a big pile to yield natural metakaolin (No. 6). However, mullite and cristobalite is found in some specimens due to too high temperature of fire (Nos. 3 and 5). In addition, one of the specimens coming from a power plant where they use gangues still containing a combustible level of coal as fuel is worthy to note. Since the plant is a fluidized bed type, the combustion temperature is as low as 900°C. This temperature is almost in agreement with the Mikuni's results[5] having high dissolution as mentioned. In this plant limestone is not premixed due to comprising some calcite in the gangue (No. 2) so that the discharged fly ash consists of some anhydrite other than much amount of metakaolin (No. 7), strictly saying, that may be not metakaolin but a mixture of silicate spinel and silica. Application of coal gangues to geopolymers in China, where Portland cement production is increasing at an enormous rate now, will reduce tremendous amount of carbon dioxide emissions, leading to a remarkable contribution to save the earth from the green house effect.

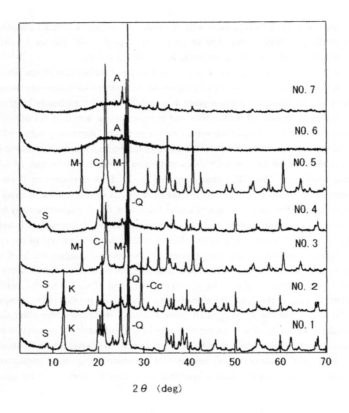

Fig. 11. XRD charts of coal gangues, kaolin(K), sericite(S), quartz(Q), mullite(M), cristobalite(C), anhydrite(A) and calcite(Cc).

Another good news is from Tunisia in northern Africa that is a country of desserts. Recently, it was found that rocks in dessert areas comprise so much amount of kaolin. Traditionally, they build houses using bricks made by traditional firing process and paint them with plasters to make the houses white so called "maison blanche" in French. If they use geopolymer process to make such bricks without burning, tremendous reduction of carbon dioxide emissions is also expected. This project is now under way by the overseas development aid program of Japanese government in Tunisia. In this context, it is of interest to exploit desert areas in the world for much more kaolin.

APPLICATION TO WASTE MANAGEMENTS

Mineral powders such as metakaolin and fly ash are called "active filler"[21], since this group of powders has relatively high dissolution into alkali solutions. To the contrary, mineral powders of waste origin generally discharged in muddy forms called slime or sludge holding so much amount of water in it are called "inactive filler", since this group of powders generally has little dissolution into alkali solutions. Accordingly, the active filler has a capability to solidify itself into monolithic bodies, whereas the inactive filler has not such capability under general conditions. However, when mixed together with the active filler in a certain amount, the mixture can have a capability to solidify itself into monolithic bodies. Applying this technique, Ikeda et al.[22,23] succeeded to obtain monolithic bodies from red mud, a waste from bauxite refineries comprising much hematite as well as soda. Red mud usually containing 40% water and the troublesome drying process of red mud can be skipped by applying more condensed geopolymer liquors. It is a routine measure that caustic soda is premixed to the water glass solution to prevent shrinkage of obtained monolithic bodies and the role of the caustic soda is considered to act as a kind of reactant to dissolve filler particles much more to prevent shrinkage[18,24], probably due to expanding volume. Such a solution is often called "activator" and the mixture of sodium silicate and caustic soda solution is often called "geopolymer liquor" as seen in Fig. 6.

The filler mix technique was also applied to solidifying stony dust generated from crushing plants of concrete aggregates as shown in Fig. 12.[25] Although some depression of strength is observed, the dust could substitute fly ash up to 25% or so. In addition, nowadays in Japan, dumping slag wastes generating from fusion process of urban incineration ash as well as sewage slime are becoming urgent issues. Although the urban incineration ash is heavily contaminated with hazardous metals such as Pb, Zn, Cu and so on, these metals can be diminished into low levels in the fusion process to sink onto the bottom of smelters. It would be of interest to use geopolymers to solidify this kind of glassy wastes also called "slags". Obtained monolithic bodies might be used for constructions, specifically as fish reefs to accumulate and to grow fish in off shore seas where boundless areas can accommodate much amount of this kind of wastes in safe, since heavy metals remaining in the slags in small amount could be immobilized in the geopolymer binder matrices of the monolithic bodies, so far studied extensively in laboratories not only to ordinary wastes but also to radioactive wastes.[26-32]

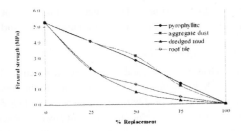

Fig.12. Results of mechanical strength for some monolithic bodies of geopolymers prepared from mineral wastes mixed with fly ash, steam-cured at 80 °C.

CONCLUSIONS

It is said that replacement of traditional Portland cements with geopolymer binders reduce carbon dioxide emission as much as 80%. However, entire shift from the traditional Portland cement process to the geopolymer process is not so easy, since the former has superior techniques improved and established in a long period of time and is now so widely accepted in the world. Therefore, to spread the geopolymer process world over, it should be strongly recommended to make use of the advantages of geopolymer binders first. That is acid resistance and fire proof as well as light weight (~1.5 g/cm^3). Sewage pipes always subjected to sulfuric acid conditions have already been attempted and exhibited in a conference venue of Melbourne in 2000 and they say that the diameter has been enlarged up to 1.5 m now. Marine constructions are also suitable to prevent sulfate attack. Interior panels to decorate ships have been already manufactured in Europe and installed in a certain ship where fire proof is highly requested. It is our recent memory in Japan that a fire broke out in a luxury cruise ship in the course of a welding work of her final decoration stage in a shipyard, since the luxury housing of deck was made of flammable organic polymers. Application to subway trains is also recommended to protect passengers from fire. It should be bear in mind that we can prepare geopolymer materials, but we do not know how to sell them. This fact is our circumstances of geopolymer world.

Finally, the cost of geopolymers should be mentioned. Generally, in Japan, water glass is sold in a form of solutions. Commercial water glass, $Na_2O \square nSiO_2$ is classified into 4 categories according to SiO_2/Na_2O molar ratio designated as JIS No.1(n=2), No. 2(n=2.5), No. 3(n=3) and No. 4(n=4). Generally, No. 3 cullet having an advantage of lower melting point is preferentially manufactured and then, Nos. 1 and 2 solutions are prepared by dissolving caustic soda to No. 3 solutions. Traditionally, 40 % solutions are shipping as for No. 1, but recently this concentration has been changed to 45%. The price of No. 1 solution depends on the units of purchasing, gallon cans, dram cans and tank lorries. If we purchase in the largest unit, the cost of geopolymer liquor including extra caustic soda as activator becomes 175% to that of ordinary Portland cement and is very costly. Therefore, introduction of superplasticizers is strongly recommended to reduce the amount of geopolymer liquor applied, for instance, say W/S from 0.40 to 0.20, the cost becomes half, being comparable to ordinary Portland cement. This technique has recently been published from Curtin University of Technology, Perth, using naphthalene sulphonate for the preparation of geopolymer concrete.[33]

ACKNOWLEDGEMENT

Thanks are due to Dr. A. Mikuni for his help, specifically for X-ray identifications of coal gangues from China.

REFERENCES
[1]Kankyo Hakusho, of Japanese Government 2005, available from
http://www.iae.or.jp/energyinfo/energydata/data5005.html
[2]The Asahi, Feb. 20 (2005).
[3]CDAC ORNL-HP, available from *http://www.iae.or.jp/energyinfo/energydata/data5005.html*
[4]J. Davidovits, "Geopolymers: Inorganic Polymeric New Materials," *J. Therm. Anal.*,**37**, 1633-56 (1991).
[5]A. Mikuni, C. Wei, R. Komatsu, and K. Ikeda, "Thermal Alteration of Kaolins and Elution Properties of Calcined Products in Alkali Solution," *J. Soc. Inorg. Mat. Japan*, **12**, 115-121 (2005).
[6]A. Mikuni, C. Wei, R. Komatsu, and K. Ikeda, "Thermal Alteration of Pyrophyllites and Elution Properties of the Calcined Pyrophyllite in Alkali Solution," *J. Soc. Inorg. Mat. Japan*, **12**, 191-199 (2005).
[7]K. Ikeda, C. Tashiro, and T. Tomisaka, "Advanced Utilization of Clayey Solid Wastes – Development of Novel Porous Materials as Heat Insulators," *Kobutsugaku Zasshi*, **20**, 63-70 (1991). In Japanese.
[8]H. Maenami, "Study on Hydrothermal and Room Temperature Solidification of Inorganic Wastes," *PhD Dissertation of Toa University*, pp. 175, (2001). In Japanese.
[9]A. Mikuni, R. Komatsu, and K. Ikeda,"Dissolution Properties of Some Fly Ash Fillers Applying to Geopolymeric Materials in Alkali Solution," *J. Mat. Sci..* In press.
[10]A. Palomo, S. Alonso, A. Fernandez-Jimenez, I. Sobrados, and J. Sanz, "Alkaline Activation of fly ashes: NMR Study of the Reaction Products," *J. Am. Ceram. Soc.*, **87**, 1141-45 (2004).
[11]Y. Zhang, W. Sun, and Z. Li, "Infrared Evidence on the Structural Nature of the Geopolymeric Cement Products," *Proc. 6th Inter. Symp. Cem. Conc., Xi'an*, **1**, 690-697 (2006).
[12]D. Feng, A. Mikuni, Y. Hirano, R. Komatsu, and K. Ikeda, "Preparation of Geopolymeric

Materials from Fly Ash Filler by Steam Curing with Special Reference to Binder Products," *J. Ceram Soc. Japan*, **113**, 82-86 (2005).

[13]T. Iwahiro, R. Komatsu, and K. Ikeda, "Chemical Compositions of Gels Prepared from Sodium Metasilicate and Aluminum Nitrate Solutions," *Proc. Geopolymers 2002, Melbourne*, 176-183 (2002).

[14]R. Vallepu, A. Fernandez-Jimenez, T. Terai, A. Mikuni, A. Palomo, K. J. D. MacKenzie, and K. Ikeda, "Effect of Synthesis pH on the Preparation and Properties of K-Al Bearing Silicate Gels from Solution," *J. Ceram. Soc. Japan*, **114**, 624-629 (2006).

[15]A. Fernandez-Jimenez, R. Vallepu, T. Terai, A. Palomo, and K. Ikeda, "Synthesis and Thermal Behavior of Differnet Aluminosilicate Gels," *J. Non-Cryst. Solid*, **352**, 2061-66 (2006).

[16]K. Ikeda, and Y. Nakamura, "Consolidation of Fly Ash Powder by the Geopolymer Technique," *Conference Papers, PacRim2, Cairns*, Symp. 18, 227-231 (1999).

[17]T. Kihara, A. Mikuni, Y. Nakamura, R. Komatsu, and K. Ikeda, "Consolidation of Pressurized Fluidized Bed Combustion Ash (PF-Ash) by the Geopolymer Technique at Ambient Temperature," *CIMTEC 2002, Florence, Science for New Technology of Silicate Ceramics*, 163-168 (2002).

[18]A. Palomo, A. Macias, M. T. Blanco, and F. Puertas, "Physical, Chemical and Mechanical Characterization of Geopolymers," *Int. Cong. Chem. Cem. 9th, New Delhi*, 5, 505-511 (1992).

[19]A. Teixeira-Pinto, P. Fernandes, and S. Jalali, "Main Problems When Using Concrete Technology," *Int. Conf. Geopolymers 2002, Melbourne*, 167-174 (2002).

[20]A. Teixeira-Pinto, Oral Communications.

[21]K. Ikeda, "Consolidation of Mineral Powders by the Geopolymer Binder Technique for Materials Use," *Shigen-to-Sozai*, **114**, 497-500 (1998). In Japanese.

[22]W. Cundi, Y. Hirano, T. Terai, R. Vallepu, A. Mikuni, and K. Ikeda, "Preparation of Geopolymeric Monoliths from Red Mud-PFBC Ash Fillers at Ambient Temperature," *Proc. World. Cong. Geopolymer 2005, St-Quentin*, 85-89 (2005).

[23]K. Ikeda, D. Feng, Y. Hirano, T. Terai, R. Vallepu, and A. Mikuni, "Preparation of Geopolymeric Materials from Red Mud Mixed with Fly Ash," *Proc. Inter. Workshop on Geopolymers and Geopolymer Concrete, Perth*, fGGC2005 CD-ROM (2005).

[24]K. Ikeda, T. Terai, R. Vallepu, and A. Mikuni, "Preparation of Geopolymer Monoliths in Various Proportions of Sodium Disilicate and Caustic Soda," *Proc. 6th Inter. Symp. Cem. Conc., Xi'an*, 1, 674-678 (2006).

[25]U. Rattanasak, R. Vallepu, A. Mikuni, and K. Ikeda, "Preparation of Monolithic Materials from Some Fillers by the Geopolymer Technique," *Proc. Inter. Conf. Pozzolan, Concrete and Geopolymer, Khon Kaen*, 40-44 (2006).

[26] A. Palomo, and J. I. Lopez de la Fuente, "Alkali-Activated Cementitious Materials: Alternative Matrices for the Immobilization of Hazardous Wastes, Part I, Stabilization of Boron," *Cem. Concr. Res.*, **33**, 281-288 (2003).

[27] A. Palomo, and M. Palacios, "Alkali-Activated Cementitious Materials: Alternative Matrices for the Immobilization of Hazardous Wastes, Part II, Stabilization of Chromium and Lead," *Cem. Concr. Res.*, **33**, 289-295 (2003).

[28] J. G. S. Van Jaarsveld, J. S. J. Van Deventer, and L.Lorenzen, "The Potential Use of Geopolymeric Materials to Immobilise Toxic Metals: Part I, Theory and Applications," *Miner. Eng.*, **10**, 659-669 (1997).

[29] A. Fernandez-Jimenez, E. E. Lachowski, A. Palomo, and D. E. Macphee, "Microsturactural Characterization of Alkali-Activated PFA Materials for Waste Immobilization," *Cem. Concr. Composites*, **26**, 1001-06 (2004).

[30] A. Fernandez-Jimenez, and A. Palomo, "Fixing Arsenic in Alkali-Activated Cementitious Materials," *J. Am. Ceram. Soc.*, **88**, 1122-26 (2005).

[31] A. Fernabdez-Jimenez, D. E. Macphee, E. E. Lachowski, and A. Palomo, "Immobilization of cesium in alkaline activated fly ash matrix," *J. Nuclear Mat.*, **346**, 185-193 (2005)..

[32] D. S. Perera, E. R. Vance, Z. Aly, J. Davis, and C. L. Nicholson, "Immobilization of Cs and Sr in geopolymers with Si/Al molar ratio of ~2," *Ceram. Trans.*, **176**, 91-96 (2006).

[33] D. Hardjito, and B. V. Rangan, "Development and Properties of Low-Calcium Fly Ash-Based Geopolymer Concrete," Research Report GC1, Faculty of Engineering, *Curtin University of Technology, Perth Australia*, pp.94 (2005). CD-ROM available.

AQUEOUS LEACHABILITY OF GEOPOLYMERS CONTAINING HAZARDOUS SPECIES

D. S. Perera, S. Kiyama, J. Davis, P. Yee and E. R. Vance
Australian Nuclear Science and Technology Organisation
Private Mail Bag 1, Menai, NSW 2234, Australia

ABSTRACT

Geopolymers made from metakaolin by activating using a commercial sodium silicate solution and having Na/Al ~1 and Si/Al ~2 molar ratios were used to immobilize Ag, Cd, Cr and Ba. The geopolymers were cured at 60^0C for 24 hours and subjected to the toxicity characterisation leaching procedure (TCLP) after 7 days. The preliminary investigation showed 4.6 mass% Ba added as the hydroxide or nitrate and 1 mass% Cr^{3+} added as the nitrate passed the TCLP limit for landfills in the USA. The samples with 1.0 mass% of Cd and Ag (both added as nitrates) and Cr (added as the dichromate) failed the test.

INTRODUCTION

Inorganic polymers formed from naturally occurring aluminosilicates have been termed geopolymers by Davidovits[1]. Various sources of Si and Al, generally in reactive glassy or fine grained phases, are added to concentrated alkaline solutions for dissolution and polymerisation to take place. Typical aluminosilicate precursors used are fly ash, ground blast furnace slags, and metakaolin made by heating kaolinite at ~ 750^0C for 6-24 h to render it X-ray amorphous and more reactive. The alkaline solutions are typically a mixture of hydroxide (e.g. NaOH, KOH) and silicate (Na_2SiO_3, K_2SiO_3). The solution dissolves Si and Al ions from the precursor to form $Si(OH)_4$ and $Al(OH)_4^-$ oligomers in solution[2]. The OH^- ions of neighbouring molecules condense to form an oxygen bond between metal atoms and release a molecule of water and then polycondense at near ambient temperatures ($20-90^0C$) to form a rigid polymer with interstitial unbound water[2]. The geopolymers thus consist of amorphous to semi-crystalline three dimensional aluminosilicate networks[1].

The physical behaviour of geopolymers is similar to that of Portland cement and they have been considered as a possible improvement on cement in respect of compressive strength, resistance to fire, heat and acidity, and as a medium for the encapsulation of radioactive and hazardous waste[3-10]. As for Portland cement usage, fine particles (e.g. sand) can be added to the paste to make a mortar, or aggregates can be added to make a concrete[11].

For industrial applications, the use of fly ash as a precursor gives a cost advantage over metakaolin. However to understand the science of the polymerisation process, the use of the latter is preferred, because of the complexity of fly ash in terms of the various distinct crystalline and non-crystalline phases present . In the literature the use of different fly ashes has been published extensively; see for example, Van Jaarsveld et al.[12]. The use of several different aluminosilicate minerals has also been reported[13]. Although the use of metakaolin is limited apart from the early work by Davidovits[1] because most workers are interested in industrial applications, recently for more scientific study metakaolin has been used on account of its near-constant composition and high reactivity with alkaline media; see for example Rahier et al.[14], Kriven et al.[15] and Barbosa et al.[16]. Several workers have studied the variation of physical and mechanical properties on composition of fly ash-based geopolymers[12,17,18] and similarly on metakaolin-based geopolymers[19,20].

The widespread use of geopolymers is currently restricted due to lack of long term durability studies, detailed scientific understanding and lack of reproducibility of raw materials[21]. One aspect of understanding the long term durability of geopolymers is the integrity of geopolymers under the influence of aqueous leaching. This is of particular interest for the use of geopolymers exposed to wet environments in applications as sewer pipes, railway sleepers, and outdoor paths for instance. Another application in which the dissolution behaviour has regulatory impact is when the geopolymer is a candidate for immobilisation of radioactive or hazardous waste. Here there are specific leach test methods which relate to disposal of solids containing toxic (eg. TCLP[22]- see also below) or radioactive waste[23,24], and from a practical point of view the regulatory test methodologies will give insight on how to look at dissolution behaviour in the other applications. Also, the intrinsic scientific approach to mineral dissolution processes[25,26] in aqueous media gives further clues so a valuable approach should be to study the kinetics of the transfer into solution of the principal cations as a function of solid surface area/solution volume, temperature, pH, solution composition etc.

In recent work we have leached a range of metakaolin- and fly ash-based geopolymers by immersion in deionised water (DIW) at ~20°C as a function of time[27]. Since alkalis are the most leachable entities in geopolymers[28,29] we followed the dissolution kinetics of the alkalis using ion-selective electrodes (ISE). The most leachable species in solutions derived from geopolymers exposed to deionised water are alkalis. Leach rates at room temperature derived from Na or K ion-specific electrode studies were in broad agreement with values obtained using PCT-B tests. The equivalent leach rates derived from these tests are ~ 1 $g/m^2/day$. Thus at first sight, typical aqueous durabilities are akin to those of borosilicate glasses, rather than cement. However it has also been found that MCC-1[30] type leach rates at 25°C are in the order of tens of $g/m^2/day$ in the first 10 days of leaching[27]. These tests are characterised by much lower geometrical surface area to volume ratios (0.1 cm^{-1} vs ~ 20-30 cm^{-1} in PCT-B tests). PCT-B releases on a geopolymer were substantially time-independent over a 1-90 day time period and no crystalline products were formed; these results suggest that equilibration between the solid and liquid phase is rapid[27].

The toxicity characterisation leaching procedure (TCLP)[22] is one of the four characteristics used to identify whether a particular hazardous waste is immobilized in solid for landfill disposal, other characteristics being ignitability, corrosivity and reactivity. This test is designed to simulate the climatic leaching expected to occur in landfills. It identifies 8 metal species and 25 organic compounds (pesticides and herbicides etc.) likely to leach into ground water. It is always possible to pass the TCLP test by sufficiently diluting the toxic species. Of the metal species immobilization of Pb in geopolymers has been reported previously by several workers[9,10,31].

As part of a systematic study of immobilizing hazardous metal species, we selected Cd, Ag, Cr and Ba for the present study. In the literature we found only the immobilization of Cr and Cd in a geopolymer has been achieved previously[32,33]. Deja[34] stabilized Cd^{2+}, Cr^{6+} in alkali activated slags containing sand but the batch composition was not given. The bars of size 40x40x160 mm were placed in DIW of volume 2 dm^3. The degree of immobilization was estimated as better than 99 % for 1 wt % addition of the cations to the matrix up to 12 months. Cd, Cr^{3+} and Cr^{6+} have been stabilized in a cement matrix of water to cement ratio of 0.4[35]. However, TCLP tests were not performed so it is difficult to compare with geopolymers made in our work. Singhal et al.[36] stabilized Cr^{6+} containing 652.3 mg/L in an acid sludge in an OPC matrix and found that the TCLP release was 4.6 mg/L thus passing the test. It was not possible to calculate the mass % of the cation in the cement matrix due to insufficient data. However, Ba has been encapsulated with other metals in a geopolymer associated with Canadian mine tailings using a test very similar to TCLP. In the work reported here we added the metals as[37] hydroxides or nitrates, but in the case of Cr in addition to nitrate it was added as a dichromate.

EXPERIMENTAL

The basic batch composition of the geopolymer consisted of: MK (34.9 mass%); sodium silicate D (PQ Corporation, Australia; - SiO_2/Na_2O: 2.0 and Al_2O_3/Na_2O: 3.8 – 65.1 mass%) having Si/Al and Na/Al molar ratios of 2.0 and 1.0 respectively. To this geopolymer composition Cd, Ag, Cr and Ba were added separately as listed in Table I. MK was produced by heating kaolin (Kingwhite 80, Unimin Ltd., Australia) at 750°C for 15 h in air. After mixing the required additives with MK, the mix was added to the sodium silicate solution, mixed for 5 min, poured into a 40 mm diameter polycarbonate cylindrical jar and covered with a screw cap. This was shaken for 5 min on a vibrating table to de-air it. Each batch weighed ~ 40 g and had dimensions of ~ 40 mm diameter x 20 mm thick. They were held at ambient for 24 h followed by curing at 60°C for 24 h and removed from the jar after 3 days. All the samples had been prepared for more than 7 days when subjected to further experimentation, to maximise the chances of complete polymerisation.

Table I. Compositions of geopolymers with additions of hazardous elements.

Sample	Composition, mass %	Elements, mass*%
Cd-GP	MK: 34.4; D: 64.3; Cd(OH)₂: 1.3	1.0 Cd
Ag-GP	MK: 34.3; D:64.0; AgNO₃: 1.7	1.0 Ag
CrN-GP	MK: 32.3; D: 60.3; Cr(NO₃)₃ 9H₂O: 7.4	1.0 Cr
CrD-GP	MK: 33.6; D: 63.7; K₂Cr₂O₇: 2.7	1.0 Cr
BaH-GP	MK: 31.1; D: 58.3; Ba(OH)₂ 8H₂O: 10.6	4.6 Ba
BaN-GP	MK: 31.5; D: 59.8; Ba(NO₃)₂: 8.7	4.6 Ba

*Equivalent element additions.

About 10 volume % of each sample was crushed to a fine powder and analysed by X-ray diffraction (XRD; Model D500, Siemens, Karlsruhe, Germany) using CoKα radiation. Scanning electron microscopy (SEM) was carried out with a JEOL (Model 6400, Tokyo, Japan), operated at 15 keV and fitted with an X-ray microanalysis system (Voyager IV, Tracor Northern, Middleton, WI, USA). Solid samples were embedded in resin and polished to a nominal 1μm diamond finish, although a very fine polish was not actually achieved, due to various amounts of pullout in the matrix.

As per the designated protocol, the TCLP test was carried out on a < 9 mm and > 0.1 mm size fraction of crushed geopolymer. According to the test, the extraction fluid of pH 4.9 ± 0.05 should be used, but geopolymers release sodium and alkaline pore water and thus the extraction fluid quickly achieves a high pH (~ pH 10.5). Therefore following an alternative test protoco lanother TCLP protocol was used in which an extraction fluid was prepared as 5.7 mL glacial acetic made up to 1 L of distilled water (pH 2.88). It was added to the Teflon extraction vessel in the ratio of 20 times the sample mass and the pH was measured after the test. The vessel was rotated at 30 ± 2 rpm for 18 h. Extraction fluid was taken out and analysed by ICP-MS. All the samples listed in Table 1 were subjected to the TCLP test.

RESULTS AND DISCUSSION

The TCLP results are listed in Table II and the degree of leaching as a % of the inventory is listed in Table III. It is seen in Table II that only 3 samples passed and others failed by fractions of 0.04 for Cd, 0.22 for Ag and 0.11 for Cr^{6+}. Assuming fractional extraction is independent of concentration samples that failed would pass if diluted to mass % 0.04, 0.2 and 0.1 for Cd, Ag and Cr^{6+} respectively. It is interesting to note that both barium samples passed. However with Cr, the nitrate added sample passed while the dichromate sample failed.

Table II. TCLP results on geopolymers listed in Table 1.

Sample	Elements, mass*%	Leached amount mg/L	Regulatory level mg/L	Result
Cd-GP	1.0 Cd	27.8	1	failed
Ag-GP	1.0 Ag	22.3	5	failed
CrN-GP	1.0 Cr	0.5	5	passed
CrD-GP	1.0 Cr	44.0	5	failed
BaH-GP	4.6 Ba	31.9	100	passed
BaN-GP	4.6 Ba	59.5	100	passed

*Equivalent element additions.

Table III. Degree of leaching of elements in geopolymers and solution pH.

Sample	Elements, mass*%	Degree of leaching %	pH
Cd-GP	1.0 Cd	5.6	5.37
Ag-GP	1.0 Ag	4.4	5.38
CrN-GP	1.0 Cr	0.1	8.62
CrD-GP	1.0 Cr	9.0	5.12
BaH-GP	4.6 Ba	1.4	7.92
BaN-GP	4.6 Ba	2.6	9.52

*Equivalent element additions.

The XRD analyses results are listed in Table IV. The Cd sample did not show any crystalline phases. Ag_2O was seen in the Ag-GP sample. The $AgNO_3$ forms Ag_2O in the presence of stoichiometric or sub-stoichiometric amounts of NaOH. The other nitrate containing samples showed the presence of $NaNO_3$. The barium hydroxide containing sample showed some $BaCO_3$ formed by the reaction of barium hydroxide with atmospheric CO_2. The quartz and anatase present were from the original clay. The potassium dichromate added sample showed the presence of the $K_2Cr_2O_7$ with a few unidentified weak lines.

The SEM images are shown for the unleached geopolymers listed in Table IV in Figs. 1-6. The sample Cd-GP shows an area of probably Cd hydroxide and a low Cd-containing area (Fig. 1). The quartz is from the original clay used to prepare MK. The amount of Cd present is either insufficient to be visible as a crystalline species in the XRD trace or is in the amorphous state. The EDS analyses of the matrix shows very little Cd (Table V), hence the major fraction of

Cd is not part of the GP matrix. Therefore leaching of a higher amount of Cd may be expected, depending of course on its speciation. The EDS analysis ignores the water present.

Table IV. XRD analyses of geopolymers before TCLP.

Sample	Crystalline phases
Cd-GP	-
Ag-GP	Ag_2O (weak lines)
CrN-GP	$NaNO_3$, quartz
CrD-GP	K_2CrO_7, anatase, weak lines
BaH-GP	Anatase, $BaCO_3$
BaN-GP	$NaNO_3$, Anatase

Note: All the samples showed the presence of an amorphous phase as deduced by the broad diffuse peak at ~ 0.32 nm.

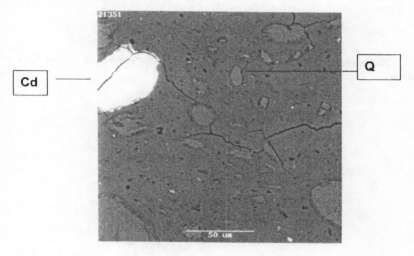

Figure. 1. Backscattered SEM image of Cd-GP sample (white specks in the matrix are anatase from the original clay).

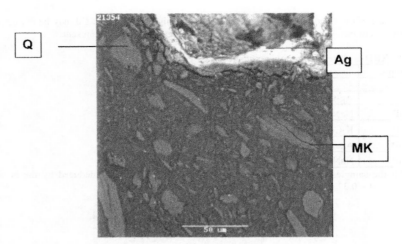

Figure. 2. Backscattered SEM image of Ag-GP sample (white specks are anatase from the original clay.

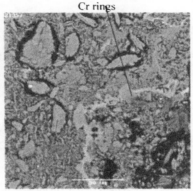

Figure. 3. Backscattered SEM image of CrN-GPsample ((white specks are anatase from the original clay.

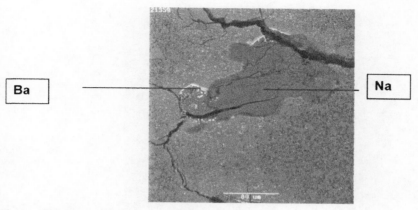

Figure. 4. Backscattered SEM image of BaH-GP sample.

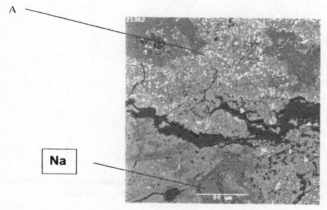

Figure. 5. Backscattered SEM image of BaN-GP sample. A = Ba-rich phase (BaCO₃) growing on the surface; Na = Na-rich phase (NaNO₃).

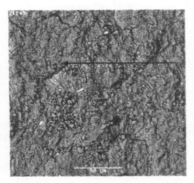

Figure. 6. Backscattered SEM images of CrD-GP sample, (top) polished surface and (bottom) fracture surface. Arrow indicates an area rich in K and Cr; white specks are anatase.

The added Ag is clearly visible over a large area (Fig. 2) and its presence in the matrix is very low (Table V). It is likely the Ag is present as Ag_2O and also $NaNO_3$ would be expected. However, $NaNO_3$ was not detected by XRD or EDS. Here again, Ag_2O would be expected to leach out in acidic or alkali water[40] as listed in Tables II and III, because it is not immobilised in the matrix. There is also quartz and unreacted MK relics in the microstructure.

The Cr added as the nitrate (CrN-GP) forms a ring pattern (Fig. 3) in the back scattered SEM image and hence the fraction of the Cr inventory present in the matrix is low (Table 5). The leach rate is much lower (Table II) compared to those of all the other elements. In the CrD-GP, K and Cr were found on EDS analysis (Fig. 6) agreeing with the XRD analysis (Table IV) where $K_2Cr_2O_7$ was observed.

Table V. EDS analyses of elements in the matrix of geopolymers* (at%).

Element	Cd-GP	Ag-GP	CrN-GP	CrD-GP	BaH-GP	BaN-GP
Na	6.2	6.9	4.9	2.9	4.5	1,4
Al	8.6	8.4	9.4	11.1	11.6	13.1
Si	23.0	22.8	23.0	21.9	21.1	21.4
Add.**	0.2-Cd	0.2-Ag	0.2-Cr	0.9-Cr	0.8-Ba	0.5-Ba
O	62.0	61.7	62.5	63.2	62.0	63.6

* Neglecting the water (~ 30%)
**Added elements.

In the sample BaH-GP the presence of a region of Na was observed (Fig. 4), which is likely to be Na_2CO_3 (formed from atmospheric CO_2) because no nitrate was present to form $NaNO_3$. However, it was not observed by XRD, thus it may have formed on the surface on storage. Ba was present as white specks of $BaCO_3$ presumably and also in the matrix (Table 5). In the sample where Ba was added as the nitrate there were areas where $BaCO_3$ was growing on the polished surfaces and $NaNO_3$ region (Fig. 5). The XRD analysis showed the presence of $NaNO_3$ but not any $BaCO_3$. Here again Ba was present in the matrix (Table V).

Palomo and Palacias[32] found that the addition of 2.6 mass% Cr added as a CrO_3^{3-} (Cr^{3+}) prevented the setting of a fly ash-based geopolymer and formed $Na_2CrO_4.4H_2O$ (Cr^{6+}) within the geopolymer. They also found hydroxysodalite ($Na_6Al_6Si_6O_{24}.8H_2O$) in the geopolymer. However, they did not carry out any leaching on their material because it readily crumbled. None of the samples produced here crumbled.

Xu et al.[33] added 0.1 mass% Cd and Cr (added as nitrates) to geopolymers and under the TCLP test their leach rates were $< \sim 50$ µg/L after 7 days. This value on a pro rata basis for 1 mass% could be estimated as 0.5 mg/L and it is much less than the values we obtained for Cd and equal to Cr (Table II). Looking at the microstructure for our Cd-containing sample, Cd evidently remains as lumps of hydroxide (Fig. 1) but as they were exposed to the acidic leaching solution (Table III) some Cd solubility would be expected if Cd acetate formed.

Cr is present in a hexavalent state in the dichromate and in a trivalent state in the nitrate. The former is soluble in acidic aqueous solutions but not in alkali solutions[38] and is highly toxic, but the latter is insoluble in alkali and is less toxic[32]. Both would be expected to be present as hydroxides in the presence of alkali. However, Cr^{6+} is acidic (see Table III). It should be noted that the dichromate-containing material showed the presence of $K_2Cr_2O_7$ (Table IV) and this is soluble in water. In OPC 1 mass% Cd added as $CdCl_3.H_2O$ was found to give the least metal release compared to As, Pb and Cr^{6+} in a non standard test using DIW at 50°C over different time intervals[35].

In the Canadian mine tailings[37], one of the tailings had a maximum of 0.46 mg/L of Ba when leached according to the Regulation 309 of the Environmental Protection Act of Ontario, 1980, which is very similar to the TCLP test. When this tailing was stabilized and solidified in a geopolymer the resulting solid leached a maximum of 0.29 mg/L of Ba (60% of the inventory). The Regulation 309 specifies a value less than 100 mg/L, thus easily passing. We used a much higher loading of Ba (4.6 mass%) in our work. It should also be noted that more mass% Ba was added than the other metals in our work. $BaCO_3$ is present in the Ba(OH)$_2$ added sample and the former is insoluble in alkaline solutions thus restricting the Ba^{2+} release (Table II).

It is clear as seen in Table III, the species that passed had an alkali solution after the test whereas the ones that passed were acidic. Cd, Ag and Cr^6+ acetates are soluble in water. To immobilize these either they have to be diluted or make the solutions more alkali to pass the TCLP test.

CONCLUSIONS

The preliminary investigation showed 4.6 mass% Ba added as the hydroxide or nitrate and 1 mass% Cr^{3+} added as the nitrate to geopolymers made from metakaolin by activating using a commercial sodium silicate solution and having Na/Al ~1 and Si/Al ~2 molar ratios, passed the TCLP limit for landfills in the USA. The samples with 1.0 mass% of Cd and Ag (both added as nitrates) and Cr (added as the dichromate) failed the test by fractions of 0.04, 0.22 and 0.11 respectively. Assuming fractional release independent of concentration, to pass the test mass % Cd, Ag and Cr^{6+} should be 0.04, 0.2 and 0.1 respectively. In the dichromate Cr is present as Cr^{6+} which is soluble in water, whereas Cr^{3+} is less soluble.

ACKNOWLEDGMENT

We thank Zaynab Aly and Ted Roach of ANSTO for technical assistance.

REFERENCES

[1]J. Davidovits, Geopolymers: Inorganic Polymeric New Materials, *J. Therm. Anal.*, **37**, 1633-56 (1991).
[2]P. G. McCormick and J. T. Gourley, "norganic Polymers – A new Material for the New Millenium, *Materials Australia*, 23 16-18 (2000).
[3]J. Davidovits, Geopolymers: Man Made Rock Geosynthesis and the Resulting Development of Very Early High Strength Cement, *J. Mat. Edu.*, **16**, 91-139 (1994).
[4]J. Davidovits,"Chemistry of Geopolymeric Systems, Terminology," Geopolymere '99, Geopolymer International Conference, Proceedings, 30 June – 2 July, 1999, pp. 9-39, Saint-Quentin, France. Edited by J. Davidovits, R. Davidovits and C. James, Institute Geopolymere, Saint Quentin, France (1999).
[5]A. Allahverdi and F. Skvara, Nitric Acid Attack on Hardened Paste of Geopolymeric Cements, *Ceramics-Silikatay*, **45**, 81-8, (2001).
[6]M. Y. Khalil and E. Merz, Immobilization of Intermediate-level Wastes in Geopolymers, *J. Nucl. Mater.*, **211**, 141-148, (1994).
[7]A. P. Zosin, T. I. Priimak and Kh. B. Avsaragov, Geopolymer Materials Based on Magnesia-Iron Slags for Normalization and Storage of Radioactive Wastes, *Atomic Energy*, **85**, 510- 514, (1998).
[8]D. S. Perera, E. R. Vance; Z. Aly, K. S. Finnie, J. V. Hanna, C. L. Nicholson, R. L. Trautman, and M. W. A. Stewart, "Characterisation of Geopolymers for the Immobilisation of Intermediate Level Waste," Proceedings of ICEM'03, Sepetember 21-25, 2003, Oxford, England, Laser Options Inc., Tucson, USA, 2004, CD, paper no. 4589.
[9]M. Palacios and A. Palomo, Alkali Activated Flyash Matrices for Lead Immobilisation: A comparison of different leaching tests, *Adv. Cement Res.* 16 , 137-44, (2004).
[10]D S Perera, Z. Aly , E R Vance and M. Mizumo, Immobilisation of Pb in a Geopolymer Matrix, *J. Am Ceram Soc.* **88**, 2586-8, (2005).
[11]D. Hardjito, S. E. Wallah and B. V. Rangan, Study on Engineering Properties of Fly ash Based Geopolymer Concrete, *J. Aust. Ceram. Soc.*, **38**, 44-47, (2002).
[12]J. G. S. van Jaarsveld, J. S. J. van Deventer and G. C. Lukey, The Characterisation of Source Materials in Fly ash-Based Geopolymers, *Mater. Lett.*, **57**, 1272-1280, (2003).

[13]Hua Xu and J. S. J. Van Deventer, The Geopolymerisation of Alumino-silicate Minerals, *Int. J. Miner. Proc.*, **59**, 247-266, (2000).

[14]H. Rahier, B. Van Mele, J. Wastiels. Low Temperature Synthesized, Aluminosilicate glasses. Part I. Low-Temperature Reaction Stoichiometry and Structure of a Model Compound, *J. Mater. Sci.*, **37**, 80-85, (1996).

[15]W. M. Kriven, J. L. Bell and M. Gordon, Microstructure and Microchemistry of Fully-Reacted Geopolymer Matrix Composites, *Ceram. Trans.* **153**, 227-250, (2003).

[16]V. F. F. Barbosa, K. J. D. MacKenzie and C. Thaumaturgo, Synthesis and Characterisation of Materials Based on Inorganic Polymers of Alumina and Silica: Sodium Polysialate Polymers, *Int. J. Inorg. Mat.*, **2**, 309-317, (2000).

[17]A. Palomo, P. F. G. Banfill, A. Fernandez-Jimenez and D. S. Swift, Properties of Alkali-Activated Fly Ashes Determined from Rheological Measurements, *Adv. Cem.Res.*, **17**, 143-151, (2005).

[18]M. Stevson and K. Sagoe-Crentsil, Relationship Between Composition, Structure and Strength of Inorganic Polymers, Part 2: Fly Ash-Derived Inorganic Polymers, *J. Mater. Sci.*, **40**, 4247-59, (2005).

[19]R. A. Fletcher, K. J. D. MacKenzie, C. L. Nicholson and S. Shimada, The Composition Range of Aluminosilicate Geopolymers, *J. Euro. Ceram. Soc.*, **25**, 1471-7, (2005).

[20]M. Rowles and B. O'Connor, Chemical Optimisation of the Compressive Strength of Aluminosilicate Geopolymers Synthesised by Sodium Silicate activation of Metakaolinite, *J. Mater. Chem.*, **13**, 1161-1165, (2003).

[21]G. C. Lukey and J. S. J. van Deventer, "Geopolymer Technology: the Formation of Cost-Competitive Construction Materials from Coal Ash and Blast Furnace Slag," Proceedings of the GGC 2005 (International Workshop on Geopolymers and Geopolymer Concrete), September, 28-29, 2005, Perth, Australia. CD ROM, qGGC2005.pdf

[22]"Toxicity Characterisation Leaching Procedure (TCLP)," EPA Method 1311, United States Environmental Protection Agency Publication SW-846, Cincinnati, OH, 1999.

[23]Standard Test Methods for Determining Chemical Durability of Nuclear, Hazardous, and Mixed Waste Glasses and Multiphase Glass Ceramics:The Product Consistency Test, ASTM Committee C26 (2000).

[24]"Measurement of Leachability of Solidified Low-Level Radioactive Wastes by a Short-Term Test Procedure," ANSI/ANS-16.1-2003. American Nuclear Society, Illinois, USA.

[25]E. H. Oelkers and J. Schott, Experimental Study of Anorthite Dissolution and the Relative Mechanism of Feldspar Hydrolysis, *Geochimica et Cosmochimica Acta*, **59**, 5039-5053, (1995).

[26]S. R. Gislason and E. H. Oelkers, Mechanism, Rates and Consequences of Basaltic Glass Dissolution:II. An Experimental Study of the Dissolution Rates of Basaltic Glass as a Function of pH and Temperature, *Geochimica et Cosmochimica Acta*, **67**, 3817-32, (2003).

[27]L. Ly, E R Vance, D S Perera, Z Aly and K Olufson, Leaching of geopolymers in deionised water, *Adv. in Mater. and Mater. Proc. J.*, **8** 236-247, (2006).

[28]E. R. Vance, D. S. Perera, Z. Aly, M. Blackford, Y. Zhang, M. Rowles, J. V. Hanna, K. J. Pike, J. Davis and O. Uchida, "Solid State Chemistry phenomena in Geopolymers with Si/Al ~ 2," Proceedings of the GGC 2005 (International Workshop on Geopolymers and Geopolymer Concrete), September, 28-29, 2005, Perth, Australia. CD ROM, nGGC2005.pdf.

[29]Y. Bao, M. W. Grutzeck and C. M. Jantzen, Preparation and Properties of Hydroceramic waste Forms Made with Simulated Hanford Low-Activity Waste, *J. Amer. Ceram. Soc.*, **88** 3287-3302, (2005).

[30]Department of Energy, Nuclear Waste Materials Handbook, DOE/TIC-11400 (1982).

[31]J. G. S. Van Jaarsveld, J. S. J/. Van Derventer and L. lorenzen, Factors Affecting the Immobilization of Metals in Geopolymerized Flyash, *Met. Mater. Trans. B.* **29B**, 283-9, (1988).

[32]A. Palomo and M. Palacios, Alkali-Activated Cementitious Materials: Alternative Matrices for the Stabilisation of Hazardous Wastes Part II. Stabilisation of Chromium and Lead, *Cement and Concrete*, **33**, 289-295, (2003).

[33]J. Z. Hu, Y. L. Zhou, Q. Chang and H. Q. Qu, Study on the Factors of Affecting the Immobilization of Heavy Metals in Fly Ash-based Geopolymers, *Mater. Let.*, **60**, 820-822, (2006).

[34]J. Deja, Immobilization of Cr^{6+}, Cd^{2+}, Zn^{2+} and Pb^{2+} in alkali-activated slag binders, *Cement and Concrete Res.*, **32**, 1971-1979, (2002).

[35]E. Zamorani and G. Serrini, "Stabilization of Hazardous Inorganic Waste in Cement-Based Matrices," Cement Industry Solutions to Waste Management: 1st International Symposium: Papers, 1992, pp. 373-381.

[36]A. Singhal, S. Prakash and V. K. Tewari, Trial on sludge of lime treated spent liquor of pickling unit for use in the cement concrete and its leaching characteristics, *Building and Environment*, **42**, 196-202, (2007).

[37]"Preliminary Examination of the Potential of Geopolymers for Use in Mine Tailing management," CANMET Final Report, DSS Contract No. 23440-6-9195/01SQ, 1998.

[38]G. Aylward and T. Findlay, SI Chemical Data, 4th Edition, John Wiley and Sons, Brisbane, 1998.

PRACTICAL APPLICATIONS OF GEOPOLYMERS

Chett Boxley, Balky Nair
Ceramatec, Inc.
2425 S. 900 W.
Salt Lake City, UT, 84119

P. Balaguru
Rutgers, The State University of New Jersey
Piscataway, NJ, 08901

ABSTRACT

At Ceramatec, geopolymer materials have been developed for a variety of applications including structural, fire resistance, and environmental protection. Specifically, the use of geopolymers as engine components could potentially raise the operating temperature of engines and turbines. In response to this, new geopolymer materials with enhanced thermal stability up to temperatures as high as 1350°C have been developed. Additional applications of high-temperature materials were evaluated for a number of applications that require fire-resistance, such as the interior of aircraft. Geopolymer coatings and laminates have been developed that can substantially increase the fire resistance of polymers and polymer composites.

Recent work at Ceramatec has focused on the use of geopolymerized fly ash as a potential Portland cement replacement for concrete applications. Various types of fly ashes have been made into practically strong parts through careful selection of the activator, the water amount, and the curing temperature. This process also produced strong bonding materials using fly ash containing excess mercury and activated carbon within the fly ash material. Consequently, this new cement composition is not environmentally hazardous, and it does not interfere with the functionality of other organic additives commonly used in concretes (e.g. air entraining agents). If this new material can be successfully developed and utilized to replace Portland cement, the CO_2 emission from Portland cement production will be significantly reduced.

INTRODUCTION

Ceramic composites are popular in applications where materials are expected to encounter high temperatures; such as engine components, exhaust systems, and fire barriers. The use of ceramic materials as engine components can potentially raise the operating temperature of engines and turbines. A change of materials systems from the currently used nickel-based super-alloy systems to "traditional ceramic materials", can result in an increase of ~200°C; thus, allowing the potential operating temperature to be as high as 1200°C. This step increase in the operating temperature is well beyond what is possible through marginal engineering improvements using currently used materials. However, brittleness and unreliability in traditional monolithic ceramics are significant problems that limit their use in such applications.[1] These problems have been partly addressed through the significant volume of research directed at development of ceramic matrix composites (CMCs) over the last two decades.[2,3]

Despite the progress that has been achieved in development of CMCs, significant problems still remain that need to be addressed and overcome before their application in high-temperature applications is practical. The primary issues are:

1. The achievement of sufficient mechanical properties in CMCs is currently possible only through an unacceptable increase in processing costs.[4]

2. The materials that have been developed with sufficient mechanical properties have poor corrosion resistance under hydrothermal oxidizing conditions typical of engines and turbines.[5]

To this end, Ceramatec and its University partner (Rutgers) have been developing geopolymer compositions for use as low-cost advanced ceramic composite materials that have sufficient mechanical properties, and may be corrosion resistant under hydrothermal oxidizing conditions.[6] Traditionally, geopolymers are low-cost, inorganic polymers derived from naturally occurring geological materials, namely silica and alumina, which can sustain temperatures of up to 800°C. Therefore, modifications were made to these basic geopolymer compositions in order to attain a higher operating temperature. Geopolymer materials are generally fabricated from a two-part system consisting of a silicate liquid and an aluminosilicate powder, and they can cure at a reasonably low temperature of 80°C. We have previously reported that geopolymers offer significant advantages for processing of low cost ceramic matrix composites (CMCs) and it was possible to develop functional geopolymer systems for improved hydrothermal corrosion resistance and/or mechanical properties.[6]

Other promising applications (including fire-resistance potentially for the interior of aircraft or for modular housing applications) for high-temperature geopolymer materials are beginning to be evaluated at both Ceramatec and Rutgers. The low density of ceramic composites, as compared to metals, makes them attractive in applications where weight is a critical design parameter. With regard to fire resistance in aircraft cabins, in conjunction with Rutgers, we begun to design and develop a composite that can sustain at least 1350°C, have a flexural strength of at least 75 MPa, be fabricated at low temperature, and use economical materials.[7] In order to increase the flexural strength, three types of alumina fibers were added to the base matrix. The first fiber type was an alumina paper, the second fiber type consisted of randomly distributed short fibers, and the third type consisted of discrete short fibers that were uniform in length and orientation.[7] The primary focus of that research was to evaluate the effects of fiber percentages and fabrication methods on the density and flexural properties of the coupons.

This paper summarizes the research to date conducted regarding the use of geopolymer materials as hydrothermal barrier for engine/turbine applications. We have demonstrated that a hydrothermally stable matrix phase was formed through a geopolymer processing approach. We also summarize the main results from the research performed to develop a geopolymer material that was lightweight for use as a fire-resistant barrier in aircraft. A unique geopolymer composition was designed to retain its strength at temperatures up to 1350°C for 30 minutes or more. These materials employed the use of alumina fiber additions to the geopolymer matrix to increase its density and flexural strength characteristics. As an extension of this research, we have also performed experiments regarding the use of geopolymer materials as fire-resistant coatings for modular building materials. We have successfully coated multiple sample substrates (polymers, metals, plastics, etc.) with geopolymer materials that we are believed to be fire-resistant.

HYDROTHERMAL CORROSION-RESISTANT MATERIALS

Our initial research task focused on developing several low silica activity aluminosilicates that were selected based on the relevant phase diagrams. Several compositions were chosen for study in this project, and many of the compositions utilized the same starting materials, however, they had different molar ratios in order to yield unique compositions. Two fabrication methods were employed during these studies: the first fabrication route was through traditional solid-state processing methods, and the second method was geopolymerization. For the geopolymerized compositions the precursor materials are reacted with an activator solution, which consists of KOH and potassium silicate (PS) to form the geopolymer slurry. The eight compositions selected were chosen based upon the phase diagrams, six samples were based on a $CaO-Al_2O_3-SiO_2$ system (compositions 1, 2, 3, 4, 7, and 8); one composition (5) was based on fluorokinoshitalite ($BaMg_3[Al_2Si_2]O_{10}F_2$); and finally one composition (6) was based on Celsian ($BaAl_2Si_2O_8$).

As mentioned, in order to confirm that the materials synthesized via the geopolymerization route were thermodynamically stable (or at least quasi-stable), each type of composition was fabricated by two methods, solid-state reaction route and geopolymerization route. The phases (as determined by XRD) of the final materials made through these two processing methods were compared against each other. The crystalline phases of the compositions synthesized by the solid-state and geopolymerization methods are shown in Table I. For most of the compositions, these two methods result in very similar phases; thus, we can conclude that the geopolymerization route provides a reliable way of material processing. Samples from both synthesis routes were tested for density and flexure strength, and the results are shown in Table II. The densities of the materials fabricated from the solid-state route tended to be slightly higher than those fabricated by geopolymerization. Furthermore, the flexure strength of the geopolymer materials was too low to be measured; whereas those produced by solid state reaction yielded average values that were consistent with our expectations.

Table I. Crystalline phases of synthesized geopolymer compositions after calcination

Sample #	Crystalline phases
1a.	Gehlenite + Parawollanstonite + Leucite
1b.	Gehlenite + Anorthite + Parawollanstonite+ Leucite
2a.	Gehlenite + Parawollanstonite + Leucite + $KAlSiO_4$ + $KAlSi_3O_8$
2b.	Gehlenite + Parawollanstonite + Anorthite + Leucite
3a.	Gehlenite + Parawollanstonite + Anorthite + Leucite + Alumina
3b.	Gehlenite + Parawollanstonite + Anorthite + Leucite
4a.	Gehlenite + Parawollanstonite + Anorthite + Leucite + Alumina
4b.	Anorthite + Gehlenite + Parawollanstonite + Leucite
5a.	$KMg_3(Si_3Al)O_{10}F_2$ + $BaMg_3Al_2Si_2O_{10}F_2$ + Mg_2SiO_4 + $KMg_2Al_2Si_3O_{11}F$
5b.	$BaMg_3[Al_2Si_2]O_{10}F_2$ + Celsian + $KAlSiO_4$ + $MgAl_2O_4$ + $KMg_3[AlSi_3]O_{10}F_2$
6a.	Celsian + $KAlSiO_4$ + $BaAl_2O_4$
6b.	Celsian + KAlSiO4 + Al_2O_3
7a.	N/A
7b.	Gehlenite + Parawollanstonite + Al_2O_3 + $KAlSiO_4$
8a.	N/A
8b.	Gehlenite + Parawollanstonite + Al_2O_3 + Anorthite + $KAlSiO_4$

Table II. Physical properties of synthesized compositions after 1165°C for 8 hours calcination

Sample #	Via solid-state reaction		Via geopolymerization	
	Density (g/cc)	Flexure strength (MPa)	Density (g/cc)	Flexure strength (MPa)
1	1.18±0.17	1.98±0.43	0.73±0.01	Too low to be measured.
2	1.56±0.02	11.68±1.16	0.69±0.01	-
3	1.31±0.02	20.15±3.92	0.80±0.01	-
4	1.22±0.01	9.17±0.66	0.83±0.01	-
5	1.10±0.02	0.82±0.35	Undetermined	-
6	1.54±0.00	11.23±0.62	Undetermined	-
7	N/A	N/A	1.37±0.01	-
8	N/A	N/A	1.35	-

One of the important properties related to the turbine/engine applications is the high temperature capability of the geopolymer materials. In order to test the temperature limit of these eight compositions, samples made via the geopolymerization route were heat treated at different temperatures for 8 hrs. The samples were heated in a kiln until they started to melt or deform at a certain temperature. The high temperature behaviors of geopolymer specimens synthesized in this study are shown in Table III. Here the temperature capability was judged by the physical integrity of the samples after high temperature exposure. If a sample melted or became severely deformed (note: shrinkage was not counted as deformation here) upon certain temperature exposure, this temperature was over the sample's temperature capability.

In Table III, the common geopolymer could not withstand 1165°C, but the geopolymer samples synthesized in this study showed a much improved temperature capability. Composition #6 showed the highest temperature capability (1350°C), prior to deformation. The next best composition was #8, which was also heated to 1350°C, but showed slight deformation. Figure 1 shows the density and open porosity data measured for five of the samples fired at 1300°C for 8 hours. Samples 1, 2, and 3 melted prior to achieving this temperature. As mentioned only samples #6 and #8 did not deform at this temperature. Figure 2 shows the compressive strength recorded for the same samples after calcination at 1300°C for 8 hours. These results are consistent with density and porosity values recorded for the samples. In other words one would expect the samples with the lowest density and greatest open porosity to have the lowest compressive strength, which is indeed the case for samples #6 and 8.

Samples of composition 4, 5, 6, 7, and 8 made via the geopolymerization route were calcined at 1300°C for 8 hours, and then they were shipped to Brown University for hydrothermal corrosion testing. For hydrothermal corrosion testing, bar shape specimens (2" in length x 0.25" in width) were put into a tube furnace reactor at 1200°C, partial pressure of H_2O = 0.15 atm, total pressure (P) = 1 atm, gas flow rate of 0.7 L/min, and with anneal times up to 100 hours. The water partial vapor pressure was controlled by a water bath, which was at a temperature that fixed the vapor pressure at the desired level for the experiment. Specimens were monitored every 25 hours for mass change.

Specimens with composition #4, #6, and #8 were tested under the hydrothermal corrosion conditions. Weight change was used as an indication of the stability of the samples under water vapor containing conditions, since it is sensitive enough to show any volatile phase forming, like $Si(OH)_4$. Figure 3 shows the weight change of those samples during the 100 hour testing. All compositions had some minor weight loss, and composition #4 and #8 seemed much more stable than #6.

However, one of possible reasons that caused larger weight loss of #6 compared to #4 and #8 is the sample has lower density and higher open porosity (see Fig. 2). More porosity provides more accessible surface for the water vapor. So the stability of composition #6 can be improved if its density is increased by some way, such as firing the sample at higher temperatures. Besides weight loss, phase and microstructure evolution during the testing are also important indications for the hydrothermal stability, which are currently under investigation.

Table III. High temperature capability of geopolymer samples

Sample	1165°C	1225°C	1300°C	1350°C
#1	Good[a]	Deformed	Melted	-
#2	Good	Deformed	Melted	-
#3	Good	Good	Melted	-
#4	Good	Good	Slightly melted	-
#5	Good	Good	Deformed	-
#6	Good	Good	Good	Good
#7	Good	Good	Slightly deformed	Deformed
#8	Good	Good	Good	Slightly deformed
Common Geopolymer[b]	Deformed	Melted	-	-

Note: a. "Good" means the sample was unchanged after high temperature exposure for 8 hours.
b. Common Geopolymer has a composition of $Si_{32}O_{81}H_{24}K_7Al$.

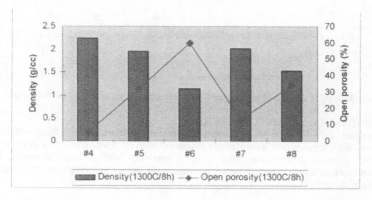

Figure 1. Density and open porosity of geopolymerized samples fired at 1300°C for 8 hours.

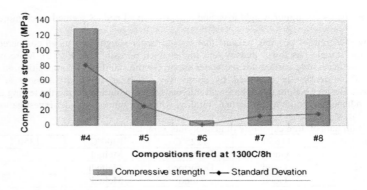

Figure 2. Compressive strength data of geopolymerized samples after an 8 hour calcination at 1300°C.

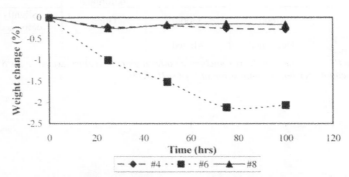

Figure 3. Weight change of geopolymer samples during hydrothermal corrosion test.

FIRE-RESISTANT COATINGS FOR AIRCRAFT

This part of the research program was performed in collaboration with Rutgers University to utilize the high temperature geopolymers discussed above.[7] Specifically this portion of the research was directed at obtaining geopolymer composites with different densities. Lower density compositions should insulate better, are inherently lighter in weight, and they may have lower flexural strength.

Matrix Composition and Fibers

A high-temperature geopolymer was used as the baseline sample matrix. We added three different types of fiber reinforcements to the baseline geopolymer matrix to increase the flexural strength of the samples. Reformatted alumina, milled alumina, and short ceramic fibers were evaluated. The reformatted alumina fiber is a paper type material that is available as rolls or cut sheets, and a thickness of 1.6mm or 6.35mm was used. The two thicknesses resulted in a fiber content of 9 or 13% respectively. The second type of reinforcement was milled alumina fibers. This material had cotton-like consistency, and the fiber content of the samples using these fibers was: 4.4, 4.5, 6.5 and 9.6%. The short ceramic fibers were 3 mm long and were very uniform. Fiber contents for these fibers were: 6.5, 8.5, 9.6 and 11.3%. The samples fabricated with these three fiber types had densities ranging from 1095 to 2901 kg/m^3. Three fabrication techniques were investigated for the sample preparation. These include hand lay-up, vibration, and a hot-press process.

For samples made with 1.6 and 6.35 mm alumina reformatted fiber, wet lay-up with hand impregnation technique was used. The geopolymer was prepared and poured onto alumina ply with an area of about 230 cm^2. Squeegees, brushes, and grooved rollers were used to impregnate the matrix into the paper. This process was also helpful for removing most of the entrapped air. The impregnated plies were stacked one on top of the other until the designated thickness was achieved. The laminate was then placed in a standard vacuum bagging system and placed in a heated pressed at a pressure of 48 MPa and a temperature of 150°C for a minimum of 3 hours. After the plate was cured, it was cut into 12 mm wide and 63.5 mm long coupons using a wet-saw with a diamond tipped blade.

For the samples with milled and short fibers, the matrix was mixed with fibers using a high shear mixer. The fiber reinforced mixture was placed on a plastic mold and vibrated using a table vibrator for 10 minutes. A 6 mm thick rubber plate was placed in between the mold and the vibrating table. A bungee was used to secure the mold to the table during vibration. This vibrating technique allows entrapped air to travel up through the matrix and escape from the top surface. The plates were left in the molds for 2 days to cure at room temperature. After 2 days the plates were removed from the molds and cured at 200°C for 24 hours. After the plates have cooled down, they were cut into coupons.

The last fabrication technique undertaken was a vacuum bagging system in conjunction with a heated-press process. Fibers were mixed with the matrix using high shear mixer and poured into a stainless-steel mold with approximate dimensions of 150 by 150 by 20 mm. The mold filled with the fiber-reinforced mixture was placed in the standard vacuum bagging system. Grooved rollers were used to gently distribute the mix within the mold. The vacuum pump was turned on and again the mix was rolled to ease the distribution. The entire system was then placed into a heated-press at a pressure of 48 MPa and temperature of 80°C. The plate would remain in the machine for 4 days while the vacuuming, pressure and heat were regulated during this time. Again, the plate was left to cool to room temperature and then cut to coupons. This process resulted in a uniform plate thickness with much lower imperfections as compared to plates made using the simple vibration technique.

Test Setup

The flexure tests were conducted over a simply supported span of 50 mm with a center point load in accordance with ASTM D790 (3). The span-to-depth ratios ranged from approximately 6:1 to 11:1, both of which fell within the acceptable limits of the standard flexure test. The tests were conducted on an MTS TestWorks® system under deflection control with a mid-span deflection rate of 0.25 mm/min. Load and deflection readings were taken using a computer for the entire test duration.

Results and Discussion

Flexural strength (modulus of rupture), strains at the maximum load, modulus of elasticity, and density for the various test parameters are presented in Table IV. The test parameters were: fiber type, fiber volume fraction, and fabrication technique. The primary response variables were: density, flexural strength, strain at failure, and modulus of elasticity. Stress-strain curve for the representative samples made using paper, milled fibers and short fibers are presented in Figures 4, 5 and 6 respectively. In Table IV, the flexural strength is presented as maximum stress, and the strain at failure is presented as maximum strain.

As expected both the strength and the stiffness values are strongly influenced by the density of the samples. All the samples can withstand 1350°C, as described in the previous section of this report. Samples heated to 400, 600, 800 and 1050°C were also tested in flexure, but these results are not presented in this paper because of space restriction. Most samples gained strength with exposure to high temperatures, but the increases were not substantial. For clarity, the discussion of the results is presented in three groups, wet lay-up group, vibration table group, and the vacuum bag group.

Table IV. Experimental Results for Fiber Infiltrated Geopolymer Samples

| Sample ID | Density (kg/m3) | Maximum | | E (GPa) | Fiber Type | Fiber Content (%) | Fabrication Series |
		Stress (MPa)	Strain (%)				
P1	1159	16	0.0918	18	Paper	9	Lay-up*
P2	1181	17	0.0937	17	Paper	9	Lay-up*
P3	1095	15	0.1046	12	Paper	9	Lay-up*
P4	1731	30	0.1216	23	Paper	13	Lay-up
P5	1772	32	0.1248	25	Paper	13	Lay-up
P6	1717	26	0.1037	26	Paper	13	Lay-up
M1	2777	62	0.1338	37	Milled	4.4	Vibration
M2	2688	61	0.1348	37	Milled	4.4	Vibration
M3	2748	65	0.1513	30	Milled	4.5	Vibration
M4	2703	60	0.1699	25	Milled	4.5	Vibration
M5	2754	55	0.1511	26	Milled	4.5	Vibration
M6	2654	63	0.1602	26	Milled	6.5	Vibration
M7	2675	64	0.1403	33	Milled	6.5	Vibration
M8	2901	78	0.1522	61	Short Ceramic	6.5	Vibration
M9	2695	91	0.1452	58	Short Ceramic	10.5	Vibration
V1	2573	51	0.1290	46	Milled	9.6	Vacuum
V2	2733	75	0.1757	42	Short Ceramic	9.6	Vacuum
V3	2735	72	0.1760	43	Short Ceramic	9.6	Vacuum
V4	2890	97	0.1889	55	Short Ceramic	11.3	Vacuum
V5	2770	93	0.1531	61	Short Ceramic	11.3	Vacuum
V6	2856	97	0.1937	52	Short Ceramic	11.3	Vacuum

*No pressure applied while curing

The following are the major observations that cover all the samples.

- It is feasible to fabricate samples using all three techniques.
- It will be easier to fabricate large samples such as plates, shells or pipes using the paper form of fibers or casting technique using vibration for compaction.
- Irrespective of fiber type and fiber content, strength and stiffness are influenced by the unit weight (density) of the samples.
- The stress-strain behavior is linear up to failure for almost all the samples.
- Flexural strength varies from 15 to 97 MPa where as the modulus of rupture varied from 12 to 61 GPa.
- Failure strain varied from 0.09 to 0.19 percent. Both higher modulus and increase in failure strain capacity contribute to increase in flexural strength. Note that toughness also increases with increase in modulus of elasticity and failure strain.
- As expected short ceramic fibers provided the best results because these fibers had much better mechanical properties, and they were easier to disperse. Fibers in paper form had the least strength, but they also had the lowest density. For applications that require better thermal insulation these formulations have an advantage. Note that these composites can be used as cores to fabricate sandwich beams or plates to satisfy the strength and stiffness requirements.

The stress-strain behavior of the samples made using alumina paper reinforcement is shown in Figure 4. There is an increase of about fifty percent in the strength when the fiber content was increased from 9% to 13%. This increase is also due to the change in fabrication method. As mentioned previously, the coupons with 9% fiber were not placed under pressure during curing. The stress-strain curves are not perfectly straight, due to some noise picked up by the computer. The authors were not able to identify the source of the noise.

For the samples with 13% fiber content, the density increased by about 60% and the strength increased by about 90% as compared to samples with 9% fibers. The strength increase was contributed by about 60% increase in modulus of elasticity and about 27% increase in failure strain capacity. The increases in modulus and strain capacity are consistent with increase in density.

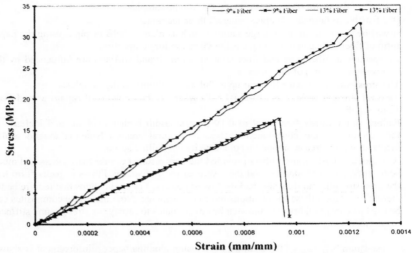

Strain (mm/mm)

Figure 4. Stress versus strain recorded for samples made using wet lay-up.

Samples of the second series type (vibration table) had two different fiber types designated as: milled alumina and short ceramic. From the stress-strain results presented in Figure 5, it can be seen that ceramic fibers provide a better performance. There is considerable increase in modulus of elasticity and strength. In addition there is a slight non-linearity near the peak load for the sample with ceramic fiber, indicating some fiber pull-out. This aspect can be utilized to develop ductile ceramic composites. On the other hand, a different fabrication technique can be used to increase strength but not the ductility. It should also be noted that the cost of ceramic fibers are much higher than the cost of milled alumina fibers.

At a stress level of about 8 MPa and a strain level of 0.00025, there is noticeable strain softening for samples made with milled fibers. A large number of samples show this behavior and hence this is not an experimental error. The authors believe that at these strains, the matrix develop micro cracks and the fibers were not able to compensate the loss of capacity created by this cracking. Absence of this behavior at larger fiber volume fractions supports this hypothesis. Ceramic fibers increases both the stress and strain at which the aforementioned cracking. Similar behavior was also reported by researchers working in fiber reinforced Portland cement composites. If there is better bonding due increase in fiber length or more fiber-force contribution due to increase in fiber content, discontinuity in load-deflection behavior at cracking of cement matrix vanishes.

There is also a significant increase in strength between these samples as compared to the lay-up samples, in spite of lower fiber volume fractions. The change in fabrication technique resulted in a much better compaction, higher densities and possible better anchoring of fibers. There was also considerable increase in modulus of elasticity and failure strain capacity.

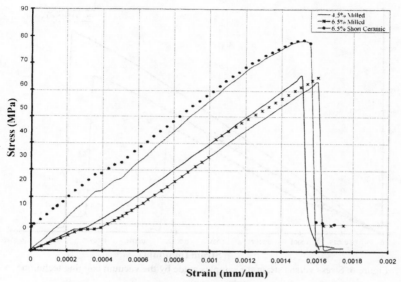

Figure 5. Stress versus strain for samples made using the vibrating table.

Stress-strain curves for the final series (vacuum bag samples) are presented in Figure 6. The vacuum bagging technique provided the sample with the highest flexural strength. But the performance was not consistently better as compared to samples made via vibration. Note that the fiber contents for this series were: 9.6% to 11.3% as compared to fiber contents of 4.4 to 8.5% for the vibration table samples. A careful review of the results from vibration table and the vacuum bag lead to the following observations.

- At fiber contents of 9.6 and 11.3% and vacuum bagging fabrication, the discontinuity at lower stress level disappears for the sample with ceramic fiber. However, the sample with milled fiber still experiences strain softening around a stress level of 11 MPa. Note that this stress level is higher than the level for samples with lower fiber contents, as in vibration table samples.
- Vacuum bagging combined with heated pressure typically provides a higher density but the fiber content plays a stronger role. For example samples with lower fiber contents of the second series have higher densities and higher strengths. As in the case of Portland cement fiber composites, higher fiber contents result in more entrapped air and reduced compaction and density. Vacuum bagging was not able to remove these air voids. However it might be possible to remove these air voids using higher compaction pressure. The authors chose not to increase the compaction pressure because it will be very difficult and expensive to increase compaction pressure for fabrication of large and complex shaped samples.
- Samples with ceramic fibers show a slight non-linearity at the maximum loads.

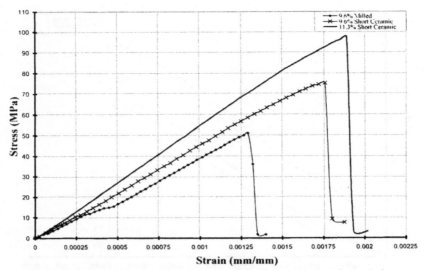

Figure 6. Stress versus strain for samples made by the vacuum bagging technique.

The density of the samples reported in this paper varies form 1095 to 2901 kg/m³. In general a three fold increase in density provides a six fold increase in strength. Corresponding increase in modulus of elasticity and failure strain are about 300 and 200% respectively. The density versus the maximum flexural strength is plotted in Figure 7. This plot includes all of the data represented in Table V as well as additional experimental data not presented in this paper. In general, increase in density provides exponential increase in strength. Although the density shows to have an affect on the flexural properties, the amount of fiber also plays a role.

- Even though the density has a strong influence on the mechanical properties, its influence is also affected by the type and volume fraction of fibers and fabrication technique.
- For the same density, ceramic fibers provide better mechanical properties. This should be expected because these fibers were more uniform in diameter and length and were easier to work with.
- For the same density, higher fiber volume fraction provides better performance.
- For different fiber contents, increase in strength is more consistent than increase in failure strains or modulus.
- It might be possible to obtain more than 100 MPa with 12 or 13% fiber content. Tests are in progress to verify this possibility.

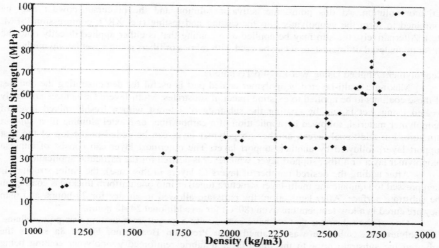

Figure 7. Flexural Strength versus Density.

GEOPOLYMERS AS FIRE-RESISTANT COATINGS FOR MODULAR HOUSING

We have concluded that functional coatings made from geopolymers for fire-resistance was a significant area of need for the Air Force. In response to this need, Ceramatec has been working on exploring the use of high-temperature geopolymer materials developed under this program as fire-resistant coatings for modular housing. Employing these geopolymer coatings for fire-resistance in modular housing, may allow human occupants enough time to vacate the area on fire. Therefore, we have developed a methodology to coat various material substrates with geopolymer compositions that are believed to be superior materials for fire-resistance. Below we outline the procedures for fabricating geopolymer-based coatings, as well as the procedure for applying them as coatings to various substrate materials.

Experimental Design

The processing steps for making the precursor powder are quite simple, and the composition was based upon composition 6b described in the first section of this paper. After combining the necessary ingredients in their proper proportions, milling media is added to the mixture along with methanol to make a slurry. The mixture is ball-milled for approximately 24 hours, whereupon the methanol is allowed to evaporate, and the dry powder is then sieved using a #60 sieve. The powder is then calcined at 1165°C for ~8 hours. Next, clean alumina media and methanol are again added to the dry powder and it is ball-milled again for 48 hours, after which the methanol is allowed to evaporate, and the media is separated from the powder by sieving.

After making the precursor powder, the alkali "*activator mixture*" is prepared prior to geopolymerization of the precursor powder (composition 6b). The activator mixture is formed by the mixing of potassium hydroxide (sodium hydroxide can also be used) with a potassium silicate solution. This mixture is paint-shaken for 10 minutes to ensure a uniform mixture of the activator solution. At this point, the activator solution and the precursor powder may be combined to form the geopolymer slip for casting and testing (i.e. XRD, compression, SEM, etc.). Alternatively, the slip may be applied as a coating that is either applied directly onto the substrate material of interest, or it can be used with fabrics or fibers using a wet lay-up method.

Geopolymers Applied Using Wet-Lay-up Technique

Making a multi-layered geopolymer coated part is useful for demonstrating the abilities of these coatings to be applied to existing material substrates. Multi-layered structures are easily created using wet lay-up by coating the substrate of interest with a freshly-mixed layer of geopolymer material, followed by application of a carbon fiber or Nextel alumina fiber support structure. For additional layers, another geopolymer layer can be added to the top of the first support layer, followed by another support layer. The outermost layer can consist of either a geopolymer layer or a fiber support layer, whichever is desired for the particular application.

After adding the desired number of layers (3 layers in this case), the entire structure is then pressed to laminate the multi-layer structure together into one uniform thickness coating on the substrate surface. After lamination, the parts are allowed to air dry for 24-hours, and then they are cured in a low-temperature oven (80°C) for an additional 24-48 hours.

One of the substrates that we have coated with a geopolymer coating was polyurethane. The polyurethane substrate was obtained from Precision Board Inc. Figure 8a shows the polyurethane substrate prior to the multi-layered fabric reinforced geopolymer coating being applied to the substrate. This sample was fabricated by applying a geopolymer coating directly to the substrate, using a paint roller, followed by a carbon fabric layer on top of the coating. Three layers of each were applied to the substrate, and the final outermost layer was a piece of carbon fabric. Figure 8b shows the final geopolymer coated part after pressing and curing the part. The coating exhibited excellent adherence to the substrate material. While we have not quantified the adherence of the coating to the substrate, we have been unable to physically separate the coating from the substrate using our hands.

Additional polyurethane substrates were coated with geopolymer films. However, these films were terminated using the geopolymer mixture (i.e. the outermost layer was geopolymer rather than carbon fabric). Termination using the geopolymer paste allowed us the ability to investigate whether it was possible to paint the outer surface of the coating, using traditional acrylic paint colors commonly used on walls in homes. Figure 9a shows a multi-layered carbon fabric/geopolymer coating on a polyurethane substrate prior to painting, and Figure 9b shows that same sample after it has been painted with green spray-on acrylic paint. The geopolymer surface was painted very easily, and has shown no signs of flaking or wear on the surface. Interestingly, the areas where the carbon fabric was not completely covered by the geopolymer paste did not paint very well. In other words the geopolymer was easily painted, but the carbon fabric was not. We have painted several other parts using different kinds of paint, and each part exhibited similar characteristics to the part shown in Figure 8.

Figure 8. a) uncoated polyurethane substrate, and b) multi-layered geopolymer coated carbon fabric.

Figure 9. a) unpainted geopolymer-terminated sample, and b) acrylic-painted multi-layered geopolymer carbon fabric.

Geopolymers Coatings Directly Applied to Substrate

We have also fabricated multiple demonstration parts that have the fire-resistant geopolymer coating that was applied directly to the substrate material. We first attempted to directly coat a very smooth polycarbonate piece, but the coating did not adhere very well to polycarbonate. As a consequence, the geopolymer coating was easily separated from the polycarbonate substrate by simply pulling it apart with our hands. After failing this initial rudimentary test, it was decided to intentionally roughen the surface with SiC sandpaper and then apply the geopolymer film. Figure 10 shows a roughened polycarbonate part after a single layer of geopolymer film was applied using a paint roller. Adhesion testing shows that roughening the surface significantly improved the adhesion between the substrate and the geopolymer film, and we were unable to separate the coating from the substrate using our hands.

We have performed some preliminary fire-resistance testing on some wood samples coated with geopolymers. The samples were heated using a propane torch for ~25 seconds. The sample did not ignite, burn, or release any heat or smoke even after exposure to high heat flux. More detailed experiments are required to prove-out this technology and demonstrate its true capabilities.

Figure 10. A roughened polycarbonate substrate coated once with a fire-resistant geopolymer material.

CONCLUSIONS

While more studies are required, we have demonstrated through preliminary testing the ability to use geopolymers for multiple applications including hydrothermal corrosion-resistant films, high-temperature light-weight coatings for aircraft, and fire-resistant films for modular housing.

ACKNOWLEDGEMENTS

The authors would like to gratefully acknowledge the contributions of Rutgers University, Mr. Gordon Roberts of Ceramatec, and the financial assistance provided by AFSOR and its Program Director Dr. Joan Fuller.

REFERENCES

[1] T. Koboyashi, Toughness Problems in Advanced Materials, *Int. J. Mater. Product. Tech.* **14**, 127-146, (1999).
[2] ,H. Ohnabe, S. Masaki, M. Onuzuka, K. Miyahara, and T. Sasa, Potential Application of Ceramic Matrix Composites to Aero-Engine Components, *Composites A-Appl. Sci. Manuf.*, **30**, 489-496, (1999).
[3] P. Spriet and G. Habarou, Applications of CMCs to Turbojet Engines: Overview of the SEP Experience, *Key Engineering Materials*, **127-132**, 1267-1276, (1997).
[4] N. Claussen, Processing of Advanced Ceramic Composites: Issues of Producibility, Affordability, Reliability, and Tailorability, *British Ceramic Trans.*, **98**, 256-257, (1999).
[5] P. Colomban, Corrosion of Ceramic-Matrix Composites, *Mat. Sci. Forum*, **251-254**, 833-844, (1997).
[6] B. Nair, Q. Zhao, T. Rahimian, R. Cooper, P. Balaguru, Matrix and Interphase Design of Geopolymer Composites, *Proceedings 107th Annual American Ceramic Society Meeting*, Baltimore, MD, (2005).
[7] C. DeFazio, M. Arafa, and P. Balaguru, Discrete Fiber Reinforced High Temperature Composites, *SAMPE*, Long Beach, CA, April 30th – May 4th, (2006).

GEOPOLYMER-JUTE COMPOSITE: A NOVEL ENVIRONMENTALLY FRIENDLY COMPOSITE WITH FIRE RESISTANT PROPERTIES

Amandio Teixeira-Pinto
Departamento de Engenharias
Universidade de Trás-os-Montes e
Alto Douro Quinta de Prados 5000-911 Vila
Real Portugal

Benjamin Varela, Kunal Shrotri, Raj S. Pai Panandiker, Joseph Lawson
Rochester Institute of Technology 76 Lomb Memorial Drive Rochester,
New York 14623-5604 United States of America

ABSTRACT

Geopolymers are a novel family of inorganic polymeric materials obtained by the alkali activation of a heat-treated aluminosilicate. Geopolymers can be considered green materials because they can be synthesized from natural resources and their chemistry is environmentally friendly without the production of toxic residues or CO_2 emissions.

Although several kinds of fibers have been used to reinforce a geopolymer matrix, little or non-work has been done in the use and characterization of natural fibers. Natural fibers don't require any kind of chemical treatment, they are inexpensive and biodegradable. For these reasons their use in the development of green composites is being studied.

The intention of this work was to develop a green composite material using geopolymers as the resin and woven jute fabric as reinforcement. The resulting plate was characterized using thermogravimetric analysis, dynamical mechanical analysis and tensile strength. In order to emphasize a practical application of this composite a flame exposure test was conducted and the results are discussed qualitatively.

INTRODUCTION

Numerous papers have been published on the overall performance of geopolymeric binders of several kinds, not only on the achievement of mortars and concretes for construction purposes, but also on other applications as waste processing or toxic metals confinement. Up to now, very few papers, if any, have presented results on the combination of geopolymeric resins with natural fibres that could act as a composite material to develop certain kinds of behaviours, such as flexibility, tensile strength and flammability resistance.

Geopolymers are a very promising kind of new materials that can be obtained at room temperature and normal atmospheric pressure by the alkali activation of alumino-silicates. These aluminosilicates can be obtained from natural sources like in the case of kaolin and volcanic ash or from industrial residuals like fly ash and blast furnace slags among others. These materials have in common a thermal history, caused by nature, as in volcanic ash or caused by heat treatment, which give them an amorphous structure with a great potential for chemical activity.

337

The development of composite materials using geopolymers as binders with fibres such as E-Glass, Carbon, Boron or Silicon Carbide as reinforcement has been studied in several projects [1,2,3], but to date, no reference has been made on the use of common fibres of organic type, like jute, linen or cotton. The advantage of using these natural fibres is that they do not involve any kind of chemical processing, they are inexpensive, easily available and biodegradable. Moreover, natural fibres such as jute have been studied recently with the purpose of developing green composite materials for structural applications due to their low cost and availability [4,5,6]. Jute is a long vegetable fiber obtained from the Corchorus plant in countries with tropical environments. Its production seconds only to cotton. Some of the limitations associated with these fibres are their fire resistance and compatibility with the matrix, which requires chemical treatment to improve both characteristics.

In this paper we study the feasibility of a geopolymer-jute fibre composite material and its characterization by flame exposure, tensile strength and dynamical mechanical analysis.

The composite material was prepared at the Universidade de Trás-os-Montes E Alto Douro and characterized at the Rochester Institute of Technology in the USA as part of a collaborative agreement between both institutions.

SAMPLE PREPARATION

The geopolymer resin was obtained from metakaolin, which was prepared in the laboratory by the calcination of commercial kaolin under standard conditions for 2 h at 750°C and then mixed with an alkali activator. The alkaline activation reaction was achieved by the use of an activator composed of 2 parts of sodium silicate solution (grade D, with 278 gr of Na_2SiO_3 and 84 gr of NaOH per kg) and 1 part of sodium hydroxide 15 molal of concentration. A ratio of both phases Liquid/Solid as high as 1 was used in order to improve the workability of the resin and to facilitate impregnation of the jute fabric. Metakaolin is a very fine powder, therefore some precaution had to be taken during the preparation of the resin in order to avoid clots. The liquid activator was the first ingredient to be placed in the mixing bowl followed by the dispersion of small amounts of metakaolin and mixed for a minimum of 10 minutes until a uniform slurry was obtained. Once the resin was prepared it was immediately placed in a refrigerator at a temperature in the range of -7 to -8°C in order to delay and control the hardening process.

Common jute fabric was used to prepare the composite material for all the experiments. Jute samples of 30 by 30 cm were cut and then impregnated with the geopolymer resin on both sides using a spatula and a cylindrical tool with several disks, which forced the resin to enter in the spaces of the jute fibre. The resin was applied in crossed layers, always trying to avoid excessive concentrations of paste until a final and homogeneous plate was obtained. The same process was repeated for a second jute fabric that was placed on top of the first one. This resulted in a thin plate of two layers with a thickness of approximately 1.5 mm. The plate was then covered with a plastic film and pressed for 60 minutes at 65°C under a pressure of 1 MPa. No ridges were observed on the plate's edges meaning that the distribution of the resin was uniform and without any excess. After this stage the plate was removed from the press and cooled to room temperature always wrapped in the same plastic material.

VISUAL INSPECTION

Figure 1 shows the appearance of the composite plate upon removal from the press. The surface looks very homogeneous with a semi-brilliant finish and no surface cracks even upon bending the plate, an indication of the flexibility obtained with the incorporation of the fibers.

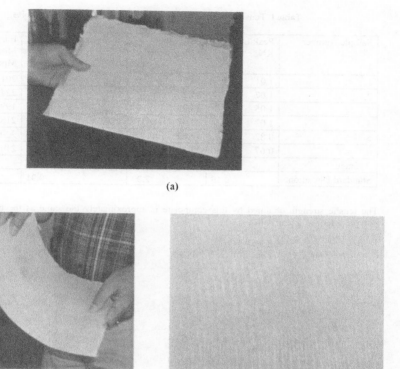

(a)

(b)   (c)

Figure. 1 The geopolymer-jute composite (a) after removal from the press, (b) bending of the plate and (c) no evidence of surface cracks after bending.

MECHANICAL TENSILE TESTING

Samples of the jute-geopolymer composite material of 25mm wide by 127 mm long and 1.5 mm thick were tested for tensile strength using an Instron MTS model 1125 testing machine at a strain rate of 2mm per minute. The results obtained from the tensile test are presented in Table I. From this table we can observe that the composite is very consistent in its mechanical strength, presenting an average maximum stress of 25.6 MPa, an average elastic modulus of 229.74 MPa and an average percent of elongation at fracture of 11.61%. There is some variation in the peak loads of samples but this can be expected since the composite was hand laid.

As it can be seen from figure 2, after breaking of the geopolymer matrix the fibers remain bonded and the failure mode resembles that of a ductile material.

Table I. Tensile test results for jute-geopolymer composite samples.

Sample Number	Peak Load KN	Peak Stress MPa	% Strain at Break	Elastic Modulus Mpa
1	1.01	23.76	11.8	201.21
2	1.07	25.28	11.7	224.21
3	1.05	24.71	13.1	199.55
4	1.02	24.02	11.6	218.23
5	0.82	31.62	10.3	320.98
6	0.67	24.22	11.2	214.28
Average	0.94	25.60	11.61	229.74
Standard Deviation	0.16	2.99	0.91	45.72

The tensile strength obtained by this composite is approximately the same as the one obtained with polypropylene as a matrix [4].

Figure 2. Failure mode of the geopolymer-jute composite under tension. The geopolymer matrix breaks but the fibres remain bonded to it.

THERMOGRAVIMETRIC ANALYSIS
Thermogravimetric analysis (TGA) tests were performed in order to examine their thermal behaviour over a large temperature range. The instrument used was TA Instruments Q500 model. Figure 4 presents the weight loss of the composite over a temperature range of $20°C$ to $900°C$ (curve A) and its derivative (curve B). It can be notice an initial weight loss at a lower temperature (~$50°C$) and another weight loss peak at about $100°C$. This weight loss can be attributed to the residual water content present in the system. Jute fibres are hydrophilic in nature and the geopolymer system is a water-based system so one can expect some water content held up by the jute fibers.

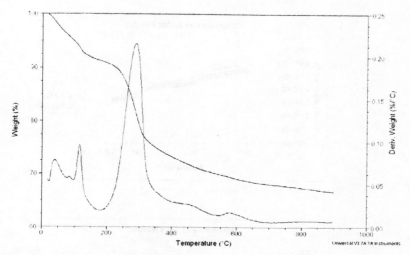

Figure 3. TGA curve of the jute composite

The major weight loss occurs between the temperature range of 225°C to 375°C. This weight loss is attributed to the decomposition of the jute fibers as it is reported in the literature [5]. We know that the decomposition of jute takes place at around the same temperature range. After 400°C we do not observe any major transition with the sample. At 900°C, more than 65% of the mass is retained. This gives a fair idea of the temperature stability that this composite exhibits and the potential application in fireproofing applications.

DYNAMIC MECHANICAL ANALYSIS

Dynamic mechanical analysis (DMA) is a very useful tool to characterize the thermomechanical behavior of the sample. To understand the dynamic mechanical behavior of the sample, tests were conducted in bending mode. The instrument used was Seiko DMS 110. Figure 5 presents the results obtained from the temperature sweep experiments from the DMA.

The storage modulus of the system (E') is closely related to the load bearing capacity of the material. Comparing the E' for pure jute (2090 MPa) with that of the composite (571 MPa), it is observed that the elastic modulus of the system is decreased approximately four fold after incorporation in the geopolymer matrix. But the viscous modulus, related to the impact resistance of the material, increased almost seven times for the composite (47.8 MPa) as compared to pure jute (6.98 MPa).

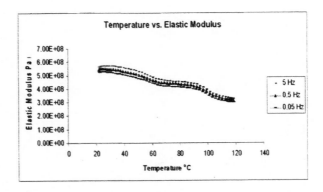

(a)

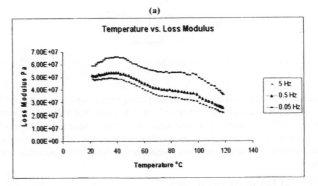

(b) **Figure 4.** DMA scans (a)-Variation in elastic/storage modulus with temperature; (b)-Variation in viscous/loss modulus with temperature.

From Figure 5a, it can be see that all the curves, at different frequencies, are close to each other. From this we can infer that the composite material is behaving as a homogenous material and that the interface between the jute and the geopolymer matrix is strong.

The variation of loss modulus against the temperature shows an opposite trend to what was observed with E'. The viscous modulus (E") value increases with a decrease in frequency. The low frequency measurements are more responsive to macroscopic motions in the sample. At a higher frequency the sample gets less time to respond to the applied deformation force and only small scale movement can be activated. At a lower frequency, the sample has enough time for a greater portion of its mass to respond to the applied force. We are able to pick up distinct transitions at a lower frequency suggesting that the sample is responding on a macroscopic level.

FLAMMABILITY TEST

Geopolymers have been studied for their potential application where high temperature and fire protection is required. Since the jute fabric is flammable, it is expected that the geopolymer matrix will improve its fire resistance. In order to qualitatively evaluate this effect a square sample of 55 by 55 mm of the composite was exposed to the flame of a Methyleacetylene-propadiene (MAPP) gas torch as presented in Figure 6.

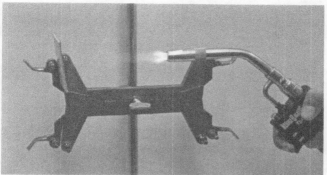

Figure 6. Experimental set up with MAPP gas torch.

The flame, which has a temperature of 2927 °C, was place approximately 20 cm away from the back face of the sample for 3 minutes and the temperature distribution on the front face was recorded using an Electrophysics Eztherm infrared camera. It was observed that at 10 seconds of exposure, a black spot formed on the opposite side of the plate followed by the emission of smoke caused by the combustion of the jute fabric. The sample remained stable for the rest of the test without igniting or releasing more smoke. Figure 7 presents some of the temperature recordings of the front face.

The temperature on the torch side couldn't be measured due to physical limitations. The infrared camera measures the temperature within a certain range which needs to be updated manually. The difference between the flame and the face is too large for the camera to record reasonable information. However, the plate was supported with Aluminum pins which has a melting temperature of 660°C, these pins melted. These is an indication that the temperature on this side was at least 660°C.

At the end of the test, the sample was removed and inspected for integrity. As it can be seen from Figure 8, the sample retained its original shape but not its original color in most part of it. There were spots were the original color was maintained. Its original flexible behaviour was lost and a brittle failure mode was observed.

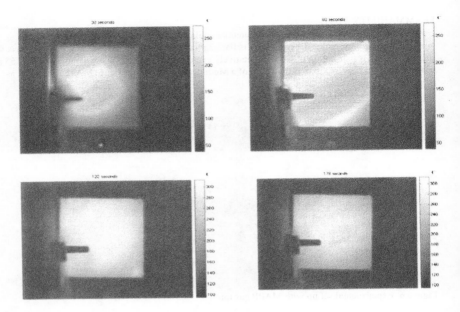

Figure 7. Recorded temperature distributions of front face at 30, 60, 122 and 178 seconds.

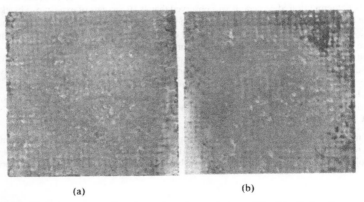

(a)   (b)

Figure 8. Physical appearance of the sample after flame exposure. The face facing the camera is shown in (a) and the one facing the torch is shown in (b).

CONCLUSIONS

The results obtained in this research indicate that raw jute fabric, without any chemical treatment, can be used to develop a green composite with a natural resin based on metakaolin material, able to present good mechanical and fire resistance properties. When exposed to a flame, the material doesn't ignite nor it decomposes in spite of being locally damaged by flame action.

In terms of mechanical behaviour, the values in several tests are consistent, pointing to a material with acceptable strength, adhesion between fibres and matrix and ductility performance. Moreover, the use of natural fibres like jute represents a friendly application of geopolymeric systems, avoiding the burden associated with other type of fibres.

REFERENCES

[1] R E. Lyon, P N. Balaguru P N, A. Foden, U. Sorathia, M. Davidovits, J. Davidovits." Fire Resistant Aluminosilicate Composites," *Fire and Materials* 2: 67-73 (1997). [2] R E. Lyon, U. Sorathia, P N. Balaguru, A. Foden, J. Davidovits, M. Davidovits," Fire Response of Geopolymer Structural Composites," *Proceedings of ICCI'96 conference, Tucson,* 972-981 (1996). [3] J A. Hammel, P N Balaguru, R. Lyon R."Influence of Reinforcement Type on the Flexural Properties of Geopolymer Composites, " *Proceedings of 43rd SAMPE International Symposium,* 1600-1608 (1998). [4] P.Wambua, J. Ivens, I. Verpoest, " Natural fibres: can they replace glass in fibre reinforced plastics?," *Composites Science and Technology* 63, 1259-1264 (2003). [5] D. Ray D, B K.Sarkar, A K Rana, N R Bos," The mechanical properties of vinylester resin matrix composites reinforced with alkali-treated jute fibres," *Composites Part A: Applied Science and Manufacturing,* 32,119-127 (2001). [6] B C. Mitra, Basak R K, M. Sarkar." Studies on jute-reinforced composites, its limitations, and some solutions through chemical modifications of fibres," *Journal of applied polymer science.* 67,1093-1100 (1998).

DESIGN, PROPERTIES, AND APPLICATIONS OF LOW-CALCIUM FLY ASH-BASED GEOPOLYMER CONCRETE

B. Vijaya Rangan
Emeritus Professor of Civil Engineering
Faculty of Engineering, Curtin University of Technology, Perth, Australia

ABSTRACT

The paper describes the studies conducted on heat-cured low-calcium fly ash-based geopolymer concrete, dry and steam-cured at 60 degrees C. Extensive test data are used to identify the effects of salient factors that influence the properties of the geopolymer concrete in the fresh and hardened states. These results are utilized to propose a simple method for the design of geopolymer concrete mixtures. Test data of various short-term and long-term properties of the geopolymer concrete are then presented. The last part of the paper describes the results of the tests conducted on large-scale reinforced geopolymer concrete members to illustrate the application of the geopolymer concrete in the construction industry. The economic merits of the geopolymer concrete are also briefly mentioned.

INTRODUCTION

Portland cement concrete is one of the most widely used construction materials. As the demand for concrete as a construction material increases, so also the demand for Portland cement. On the other hand, the cement industry is responsible for some of the CO_2 emissions, because the production of one ton of Portland cement emits approximately one ton of CO_2 into the atmosphere[1,2].

Several efforts are in progress to reduce the use of Portland cement in concrete in order to address the global warming issues. These include the utilization of supplementary cementing materials such as fly ash, silica fume, granulated blast furnace slag, rice-husk ash and metakaolin, and the development of alternative binders to Portland cement. In this respect, the geopolymer technology[3] shows considerable promise for application in the concrete industry. In terms of reducing the global warming, the geopolymer technology could reduce the CO_2 emission to the atmosphere caused by cement and aggregates industries by about 80% [1]. The paper presents the results of the studies conducted on heat-cured low-calcium fly ash-based geopolymer concrete. Here, heat-cured includes both dry and steam-cured as mentioned later.

MATERIALS IN GEOPOLYMER CONCRETE

Geopolymer concrete can be manufactured by using the low-calcium (ASTM Class F) fly ash obtained from coal-burning power stations. The chemical composition and the particle size distribution of the fly ash must be established prior to use. An X-Ray Fluorescence (XRF) analysis may be used to determine the chemical composition of the fly ash.

Low-calcium fly ash has been successfully used to manufacture geopolymer concrete when the silicon and aluminum oxides constituted about 80% by mass, with the Si-to-Al ratio of about 2. The content of the iron oxide usually ranged from 10 to 20% by mass, whereas the calcium oxide content was less than 3% by mass. The carbon content of the fly ash, as indicated by the loss on ignition by mass, was as low as less than 2%. The particle size distribution tests revealed that 80% of the fly ash particles were smaller than 50 μm [4,5,6].

Coarse and fine aggregates used by the concrete industry are suitable to manufacture geopolymer concrete. The grading curves currently used in concrete practice are applicable in the case of geopolymer concrete[4,5,6].

A combination of sodium silicate solution and sodium hydroxide (NaOH) solution can be used as the alkaline liquid. It is recommended that the alkaline liquid is prepared by mixing both the solutions together at least 24 hours prior to use[6].

The sodium silicate solution is commercially available in different grades. The sodium silicate solution A53 with SiO_2-to-Na_2O ratio by mass of approximately 2, i.e., $SiO_2 = 29.4\%$, $Na_2O = 14.7\%$, and water $= 55.9\%$ by mass, is recommended.

The sodium hydroxide with 97-98% purity, in flake or pellet form, is commercially available. The solids must be dissolved in water to make a solution with the required concentration. The concentration of sodium hydroxide solution can vary in the range between 8 Molar and 16 Molar. The mass of NaOH solids in a solution varies depending on the concentration of the solution. For instance, NaOH solution with a concentration of 8 Molar consists of 8x40 = 320 grams of NaOH solids per litre of the solution, where 40 is the molecular weight of NaOH. The mass of NaOH solids was measured as 262 grams per kg of NaOH solution with a concentration of 8 Molar. Similarly, the mass of NaOH solids per kg of the solution for other concentrations was measured as 10 Molar: 314 grams, 12 Molar: 361 grams, 14 Molar: 404 grams, and 16 Molar: 444 grams[6]. Note that the mass of NaOH solids is only a fraction of the mass of the NaOH solution, and water is the major component.

In order to improve the workability, a high range water reducer super plasticizer and extra water may be added to the mixture[6].

MIXTURE PROPORTIONS OF GEOPOLYMER CONCRETE

The primary difference between geopolymer concrete and Portland cement concrete is the binder. The silicon and aluminum oxides in the low-calcium fly ash reacts with the alkaline liquid to form the geopolymer paste that binds the loose coarse aggregates, fine aggregates, and other un-reacted materials together to form the geopolymer concrete.

As in the case of Portland cement concrete, the coarse and fine aggregates occupy about 75 to 80% of the mass of geopolymer concrete. This component of geopolymer concrete mixtures can be designed using the tools currently available for Portland cement concrete.

The compressive strength and the workability of geopolymer concrete are influenced by the proportions and properties of the constituent materials that make the geopolymer paste. Experimental results[6] have shown the following:

- Higher concentration (in terms of molar) of sodium hydroxide solution results in higher compressive strength of geopolymer concrete.
- Higher the ratio of sodium silicate solution-to-sodium hydroxide solution ratio by mass, higher is the compressive strength of geopolymer concrete.
- The addition of naphthalene sulphonate-based super plasticizer, up to approximately 4% of fly ash by mass, improves the workability of the fresh geopolymer concrete; however, there is a slight degradation in the compressive strength of hardened concrete when the super plasticizer dosage is greater than 2%.

- As the H_2O-to-Na_2O molar ratio increases, the compressive strength of geopolymer concrete decreases.

The effect of the Na_2O-to-Si_2O molar ratio on the compressive strength of geopolymer concrete is not significant.

As can be seen from the above, the interaction of various parameters on the compressive strength and the workability of geopolymer concrete is complex. In order to assist the design of low-calcium fly ash-based geopolymer concrete mixtures, a single parameter called 'water-to-geopolymer solids ratio' by mass was devised. In this parameter, the total mass of water is the sum of the mass of water contained in the sodium silicate solution, the mass of water in the sodium hydroxide solution, and the mass of extra water, if any, added to the mixture. The mass of geopolymer solids is the sum of the mass of fly ash, the mass of sodium hydroxide solids, and the mass of solids in the sodium silicate solution (i.e. the mass of Na_2O and SiO_2).

Tests were performed to establish the effect of water-to-geopolymer solids ratio by mass on the compressive strength and the workability of geopolymer concrete. The test specimens were 100 mm diameter x 200 mm cylinders, heat-cured in an oven at various temperatures for 24 hours. The results of these tests, plotted in Figure 1, show that the compressive strength of geopolymer concrete decreases as the water-to-geopolymer solids ratio by mass increases[6]. This test trend is analogous to the well-known effect of water-to-cement ratio on the compressive strength of Portland cement concrete. Obviously, as the water-to-geopolymer solids ratio increased, the workability increased as the mixtures contained more water.

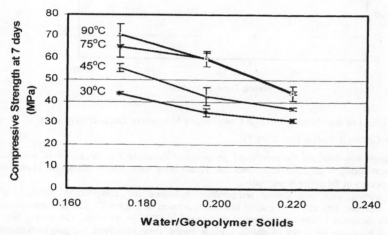

FIGURE 1 Effect of Water-to-Geopolymer Solids Ratio by Mass on Compressive Strength of Geopolymer Concrete[6] (See Text for Water/Geopolymer Solids Ratio)

MIXING, CASTING, AND COMPACTION OF GEOPOLYMER CONCRETE

Geopolymer concrete can be manufactured by adopting the conventional techniques used in the manufacture of Portland cement concrete. In the laboratory, the fly ash and the aggregates were first mixed together dry in 80-litre capacity pan mixer for about three minutes. The alkaline liquid was mixed with the super plasticiser and the extra water, if any. The liquid component of the mixture was then added to the dry materials and the mixing continued for another four minutes. The fresh concrete could be handled up to 120 minutes without any sign of setting and without any degradation in the compressive strength. The fresh concrete was cast and compacted using a vibrator by the usual methods used in the case of Portland cement concrete[6, 7, 8].

The compressive strength of geopolymer concrete is influenced by the wet-mixing time, as illustrated by the test data plotted in Figure 2. The test specimens were 100 mm diameter x 200 mm cylinders, steam-cured at 60°C for 24 hours and tested in compression at an age of 21 days. Figure 2 shows that the compressive strength significantly increased as the wet-mixing time increased. The slump values of fresh concrete were also measured. These results showed that the slump values decreased from 240 mm for two minutes of wet-mixing time to 210 mm when the wet-mixing time increased to sixteen minutes.

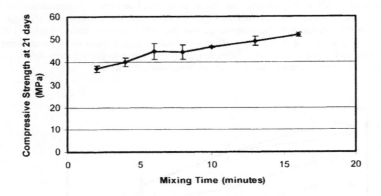

FIGURE 2 Effect of Wet-Mixing Time on Compressive Strength of Geopolymer Concrete[6]

CURING OF GEOPOLYMER CONCRETE

Although low-calcium fly ash-based geopolymer concrete can be cured in ambient conditions, heat-curing is generally recommended. Heat-curing substantially assists the chemical reaction that occurs in the geopolymer paste.

Both curing time and curing temperature influence the compressive strength of geopolymer concrete. The effect of curing time is illustrated in Figure 3[6]. The test specimens were 100 mm diameter x 200 mm cylinders heat-cured at 60°C in an oven. The curing time varied from 4 hours to 96 hours (4 days). Longer curing time improved the polymerization process resulting in higher compressive strength. The rate of increase in strength was rapid up to 24 hours of curing time; beyond 24 hours, the gain in strength is only moderate. Therefore, heat-curing time need not be more than 24 hours in practical applications.

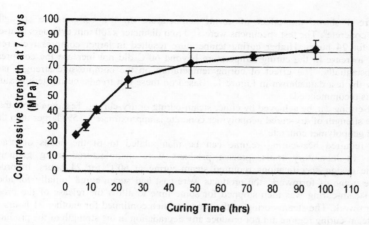

FIGURE 3   Effect of Curing Time on Compressive Strength of Geopolymer Concrete[6]

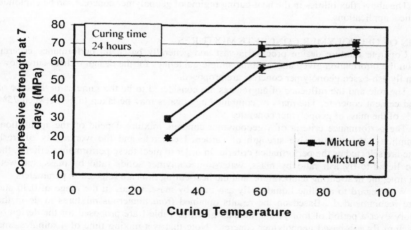

FIGURE 4 Effect of Curing Temperature on Compressive Strength of Geopolymer Concrete[6]

Figure 4 shows the effect of curing temperature on the compressive strength of geopolymer concrete[6]. The test specimens were 100 mm diameter x200 mm cylinders heat-cured in an oven for 24 hours. Higher curing temperature resulted in larger compressive strength, although an increase in the curing temperature beyond 60°C did not increase the compressive strength substantially. The effect of curing temperature on the compressive strength is also indicated by the test data shown in Figure 1. Based on these test trends, curing temperature of about 60°C is recommended.

Heat-curing can be achieved by either steam-curing or dry-curing. Test data show that the compressive strength of dry-cured geopolymer concrete is approximately 15% larger than that of steam-cured geopolymer concrete[6].

The required heat-curing regime can be manipulated to fit the needs of practical applications. In laboratory trials, precast products were manufactured using geopolymer concrete; the design specifications required steam-curing at 60°C for 24 hours. In order to optimize the usage of formwork, the products were cast and steam-cured initially for about 4 hours. The steam-curing was then stopped for some time to allow the release of the products from the formwork. The steam-curing of the products then continued for another 21 hours. This two-stage steam-curing regime did not produce any degradation in the strength of the products.

Also, the start of heat-curing of geopolymer concrete can be delayed for several days. Tests have shown that a delay in the start of heat-curing up to five days did not produce any degradation in the compressive strength. In fact, such a delay in the start of heat-curing substantially increased the compressive strength of geopolymer concrete[6].

The above flexibilities in the heat-curing regime of geopolymer concrete can be exploited in practical applications.

DESIGN OF GEOPOLYMER CONCRETE MIXTURES

Concrete mixture design process is vast and generally based on performance criteria. Based on the information given above, some simple guidelines for the design of heat-cured low-calcium fly ash-based geopolymer concrete are proposed.

The role and the influence of aggregates are considered to be the same as in the case of Portland cement concrete. The mass of combined aggregates may be taken to be between 75% and 80% of the mass of geopolymer concrete.

The performance criteria of a geopolymer concrete mixture depend on the application. For simplicity, the compressive strength of hardened concrete and the workability of fresh concrete are selected as the performance criteria. In order to meet these performance criteria, the alkaline liquid-to-fly ash ratio by mass, water-to-geopolymer solids ratio by mass, the wet-mixing time, the heat-curing temperature, and the heat-curing time are selected as parameters.

With regard to alkaline liquid-to-fly ash ratio by mass, values in the range of 0.30 and 0.45 are recommended. Based on the results obtained from numerous mixtures made in the laboratory over a period of four years, the data given in Table I are proposed for the design of low-calcium fly ash-based geopolymer concrete. Note that wet-mixing time of 4 minutes, and steam-curing at 60°C for 24 hours after casting are proposed. The data given in Figures 3 and 4 may be used as guides to choose other curing temperatures and curing times.

Sodium silicate solution is cheaper than sodium hydroxide solids. Commercially available sodium silicate solution A53 with SiO_2-to-Na_2O ratio by mass of approximately 2, i.e., $Na_2O = 14.7\%$, $SiO_2 = 29.4\%$, and water = 55.9% by mass, and sodium hydroxide solids (NaOH) with 97-98% purity are recommended. Laboratory experience suggests that the ratio of sodium silicate solution-to-sodium hydroxide solution by mass may be taken approximately as 2.5[6].

TABLE I: Data for Design of Low-Calcium Fly Ash-Based Geopolymer Concrete Mixtures

Water-to-geopolymer solids ratio, by mass	Workability	Design compressive strength (wet-mixing time of 4 minutes, steam curing at 60°C for 24 hours after casting), MPa
0.16	Very Stiff	60
0.18	Stiff	50
0.20	Moderate	40
0.22	High	35
0.24	High	30

Notes:

- The fineness modulus of combined aggregates is taken to be in the range of 4.5 and 5.0.

- When cured in dry-heat, the compressive strength may be about 15% larger than the above given values.

- When the wet-mixing time is increased from 4 minutes to 16 minutes, the above compressive strength values may increase by about 30%.

- Standard deviation of compressive strength is about 10% of the above given values.

The mixture design process is illustrated by the following Example. The terms 'fineness modulus', 'standard grading curves', and 'saturated-surface-dry' used here are explained in References 6 to 10:

Mixture proportion of heat-cured low-calcium fly ash-based geopolymer concrete with design compressive strength of 45 MPa is needed for precast concrete products.

Assume that normal-density aggregates are to be used and the unit-weight of concrete is 2400 kg/m³. Take the mass of combined aggregates as 77% of the mass of concrete, i.e. 0.77x2400= 1848 kg/m³. The combined aggregates may be selected to match the standard grading curves used in the design of Portland cement concrete mixtures. For instance, the aggregates may comprise 277 kg/m³ (15%) of 20mm aggregates, 370 kg/m³ (20%) of 14 mm aggregates, 647 kg/m³ (35%) of 7 mm aggregates, and 554 kg/m³ (30%) of fine sand to meet the requirements of standard grading curves. The fineness modulus of the combined aggregates is approximately 5.0.

The mass of low-calcium fly ash and the alkaline liquid = 2400 – 1848 = 552 kg/m³. Take the alkaline liquid-to-fly ash ratio by mass as 0.35; the mass of fly ash = 552/ (1+0.35) = 408 kg/m³ and the mass of alkaline liquid = 552 – 408 = 144 kg/m³. Take the ratio of sodium silicate solution-to-sodium hydroxide solution by mass as 2.5; the mass of sodium hydroxide solution = 144/ (1+2.5) = 41 kg/m³; the mass of sodium silicate solution = 144 – 41 =103 kg/m³.

Therefore, the trial mixture proportion is as follow: combined aggregates = 1848 kg/m^3, low-calcium fly ash = 408 kg/m^3, sodium silicate solution = 103 kg /m^3, and sodium hydroxide solution = 41 kg/m^3.

To manufacture the geopolymer concrete mixture, commercially available sodium silicate solution A53 with SiO_2-to-Na_2O ratio by mass of approximately 2, i.e., Na_2O = 14.7%, SiO_2 = 29.4%, and water = 55.9% by mass, is selected. The sodium hydroxide solids (NaOH) with 97-98% purity is purchased from commercial sources, and mixed with water to make a solution with a concentration of 8 Molar. This solution comprises 26.2% of NaOH solids and 68.6% water, by mass.

For the trial mixture, water-to-geopolymer solids ratio by mass is calculated as follows: In sodium silicate solution, water = 0.559x103 = 58 kg, and solids = 103 − 58 = 45 kg. In sodium hydroxide solution, solids = 0.262x41 = 11 kg, and water = 41 − 11 = 30 kg. Therefore, total mass of water = 58+30 = 88 kg, and the mass of geopolymer solids = 408 (i.e. mass of fly ash) +45+11 = 464 kg. Hence the water-to-geopolymer solids ratio by mass = 88/464 = 0.19.

Using the data given in Table I, for water-to-geopolymer solids ratio by mass of 0.19, the design compressive strength is approximately 45 MPa, as needed. The geopolymer concrete mixture proportion is therefore as follows:

20 mm aggregates = 277 kg/m^3, 14 mm aggregates = 370 kg/m^3, 7 mm aggregates = 647 kg/m^3, fine sand = 554 kg/m^3, low-calcium fly ash (ASTM Class F) = 408 kg/m^3, sodium silicate solution (Na_2O = 14.7%, SiO_2 = 29.4%, and water = 55.9% by mass) = 103 kg/m^3, and sodium hydroxide solution (8 Molar) = 41 kg/m^3(Note that the 8 Molar sodium hydroxide solution is made by mixing 11 kg of sodium hydroxide solids with 97-98% purity in 30 kg of water).

The aggregates are assumed to be in saturated-surface-dry condition. The geopolymer concrete must be wet-mixed at least for four minutes and steam-cured at 60°C for 24 hours after casting.

The workability of fresh geopolymer concrete is expected to be moderate. If needed, commercially available super plasticizer of about 1.5% of mass of fly ash, i.e. 408x (1.5/100) = 6 kg/m^3 may be added to the mixture to facilitate ease of placement of fresh concrete.

Numerous batches of the Example geopolymer concrete mixture have been manufactured and tested in the laboratory over a period of four years. These test results have shown that the mean 7th day compressive strength was 56 MPa with a standard deviation of 3 MPa (see Mixture-1 in Tables II and III). The mean slump of the fresh geopolymer concrete was about 100 mm.

SHORT-TERM PROPERTIES OF GEOPOLYMER CONCRETE
The behavior and failure mode of fly ash-based geopolymer concrete in compression is similar to that of Portland cement concrete. Test data have shown that the strain at peak stress is in the range of 0.0024 to 0.0026 and the modulus of elasticity increased as the compressive strength of geopolymer concrete increased. The measured Poisson's ratio of fly ash-based geopolymer concrete was in the range 0.12 and 0.16. These test values are similar to those of Portland cement concrete[6].

The tensile strength of fly ash-based geopolymer concrete was measured by performing the cylinder splitting test on 150 mm diameter x300 mm concrete cylinders. These test results have shown that the tensile splitting strength of geopolymer concrete is only a fraction of the compressive strength, as in the case of Portland cement concrete[6].

The unit-weight of concrete primarily depends on the unit mass of aggregates used in the mixture. Test results[6] show that the unit-weight of the low-calcium fly ash-based geopolymer concrete is similar to that of Portland cement concrete. When granite-type coarse aggregates were used, the unit-weight varied between 2330 and 2430 kg/m^3.

LONG-TERM PROPERTIES OF GEOPOLYMER CONCRETE
Compressive Strength

Geopolymer concrete mixture proportions used in laboratory studies are given in Table II[7]. In Mixture-2, additional water (designated as "Extra water" in Table II) was added to increase the workability of fresh concrete. Numerous batches of these mixtures were manufactured during a period of four years. For each batch of geopolymer concrete made, 100 mm diameter x200 mm cylinders specimens were prepared. At least three of these cylinders were tested for compressive strength at an age of seven days after casting. The unit-weight of specimens was also determined at the same time. For these numerous specimens made from Mixture-1 and Mixture-2 and heat-cured at 60°C for 24 hours after casting, the average results are presented in Table III[7].

TABLE II - Geopolymer Concrete Mixture Proportions

Materials		Mass (kg/m^3)	
		Mixture-1	Mixture-2
Coarse aggregates:	20 mm	277	277
	14 mm	370	370
	7 mm	647	647
Fine sand		554	554
Fly ash (low-calcium ASTM Class F)		408	408
Sodium silicate solution(SiO$_2$/Na$_2$O=2)		103	103
Sodium hydroxide solution		41(8Molar)	41(14Molar)
Super Plasticizer		6	6
Extra water		None	22.5

TABLE III - Mean Compressive Strength and Unit-weight of Geopolymer Concrete

Mixture	Curing type	7th Day compressive strength (heat-curing at 60°C for 24 hours), MPa		Unit-weight, kg/m³	
		Mean	Standard deviation	Mean	Standard deviation
Mixture-1	Dry curing (oven)	58	6	2379	17
	Steam curing	56	3	2388	15
Mixture-2	Dry curing (oven)	45	7	2302	52
	Steam curing	36	8	2302	49

In order to observe the effect of age on compressive strength of heat-cured geopolymer concrete, 100 mm diameter x200 mm cylinders were made from several batches of Mixture-1 given in Table II. The specimens were heat-cured in the oven for 24 hours at 60°C. These test data show that the compressive strength increased with age in the order of 10 to 20 percent when compared to the 7th day compressive strength[7].

The test data shown in Table III demonstrate the consistent quality, reproducibility, and long-term stability of low-calcium fly ash-based geopolymer concrete.

Creep and Drying Shrinkage

The creep and drying shrinkage behavior of heat-cured low-calcium fly ash-based geopolymer concrete was studied for a period of one year[7]. The geopolymer concrete mixture proportions used in that study were Mixture-1 and Mixture-2, as given in Table II. The test specimens were 150 mm diameter x300 mm cylinders, heat-cured at 60°C for 24 hours. The creep tests commenced on the 7th day after casting the test specimens and the sustained stress was 40% of the compressive strength on that day. The test results obtained for specimens made using Mixture-1 and heat-cured in an oven are shown in Figure 5[7]. The test trends were similar for both Mixture-1 and Mixture-2, heat-cured either dry-cured in an oven or steam-cured.

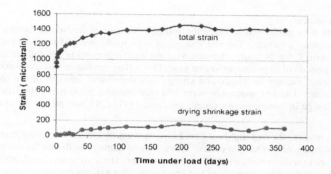

FIGURE 5 Total Strain and Drying Shrinkage Strain of Heat-Cured Geopolymer Concrete[7]

Test results show that heat-cured fly ash-based geopolymer concrete undergoes very little drying shrinkage in the order of about 100 micro strains after one year. This value is significantly smaller than the range of values of 500 to 800 micro strains experienced by Portland cement concrete.

The creep coefficient, defined as the ratio of creep strain-to-elastic strain, after one year of loading for heat-cured geopolymer concrete with compressive strength of 40, 47 and 57 MPa is between 0.6 and 0.7; the creep coefficient for geopolymer concrete with compressive strength of 67 MPa is between 0.4 and 0.5. The specific creep, defined as the creep strain per unit of sustained stress, data are shown in Figure 6[7](1CR, 2CR etc in this Figure, are specimen designations). These values are about 50% of the values recommended by the draft Australian Standard AS3600[9] for Portland cement concrete.

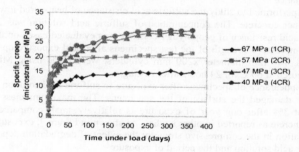

FIGURE 6 Effect of Compressive Strength on Creep of Heat-Cured Geopolymer Concrete[7]

Sulfate Resistance

Tests were performed to study the sulfate resistance of heat-cured low-calcium fly ash-based geopolymer concrete. The test specimens were made using Mixture-1 (see Table II) and heat-cured at 60°C for 24 hours after casting; they were immersed in 5% sodium sulfate solution for various periods of exposure up to one year. The sulfate resistance was evaluated based on the change in mass, change in length, and change in compressive strength of the specimens after sulfate exposure. The test specimens were 100 mm diameter x200 mm cylinders for change in mass and change in compressive strength tests and 75x75x285 mm prisms for change in length test.

Test results[7] showed that heat-cured low-calcium fly ash-based geopolymer concrete has an excellent resistance to sulfate attack. There was no damage to the surface of test specimens after exposure to sodium sulfate solution up to one year. There were no significant changes in the mass and the compressive strength of test specimens after various periods of exposure up to one year. The change in length was extremely small and less than 0.015%.

The deterioration of Portland cement concrete due to sulfate attack is attributed to the formation of expansive gypsum and ettringite which causes expansion, cracking, and spalling in the concrete. Low-calcium fly ash-based geopolymer concrete undergoes a different mechanism to that of Portland cement concrete and the geopolymerisation products are also different from hydration products. The main product of geopolymerisation is not susceptible to sulfate attack like the hydration products. Because there is generally no gypsum or ettringite formation in the main products of geopolymerisation, there is no mechanism of sulfate attack in heat-cured low-calcium fly ash-based geopolymer concrete. However, presence of high calcium either in the fly ash or in the aggregates could cause the formation of gypsum and ettringite in geopolymer concrete.

Sulfuric Acid Resistance

Tests[7] were performed to study the sulfuric acid resistance of heat-cured low-calcium fly ash-based geopolymer concrete. The concentration of sulfuric acid solution was 2%, 1% and 0.5%. The sulfuric acid resistance of geopolymer concrete was evaluated based on the mass loss and the residual compressive strength of the test specimens after acid exposure up to one year. The test specimens, 100 mm diameter x200 mm cylinders, were made using Mixture-1 (see Table II) and heat-cured at 60°C for 24 hours after casting.

The visual appearance of specimens after exposure to sulfuric acid solution showed that acid attack slightly damaged the surface of the specimens. The maximum mass loss of test specimens of about 3% after one year of exposure is relatively small compared to that for Portland cement concrete as reported in other studies. As shown in Figure 7, exposure to sulfuric acid caused degradation in the compressive strength; the extent of degradation depended on the concentration of the acid solution and the period of exposure[7].

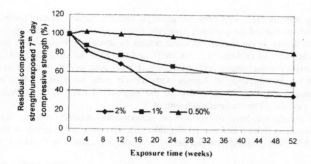

FIGURE 7   Sulfuric Acid Resistance of Heat-cured Geopolymer Concrete[7]

REINFORCED GEOPOLYMER CONCRETE BEAMS AND COLUMNS

In order to demonstrate the application of heat-cured low-calcium fly ash-based geopolymer concrete, twelve reinforced columns and twelve reinforced beams were manufactured and tested[8].

In the column test program, the primary parameters were longitudinal reinforcement ratio, load eccentricity, and compressive strength of geopolymer concrete. The longitudinal reinforcement ratio was 1.47% and 2.95% and the concrete compressive strength was 40 and 60 MPa. The column cross-section was 175 mm square. The columns were subjected to eccentric compression and bent in single curvature bending. The columns were pin-ended with an effective length of 1684 mm.

The beam cross-section was 200mm wide by 300mm deep, and 3300mm in length. The test parameters were concrete compressive strength and longitudinal tensile reinforcement ratio. The concrete compressive strength was 40, 50, and 70 MPa and longitudinal tensile reinforcement ratios were 0.64, 1.18, 1.84, and 2.69%. The beams were simply supported over a span of 3000mm, and subjected to two concentrated loads placed symmetrically on the span. The distance between the loads was 1000mm.

The behavior and failure modes of reinforced geopolymer concrete columns were similar to those observed in the case of reinforced Portland cement concrete columns. As expected, the load capacity of columns was influenced by the load-eccentricity, the concrete compressive strength, and the longitudinal reinforcement ratio. When the load eccentricity decreased, the load capacity of columns increased. The load capacity also increased when the compressive strength of concrete and the longitudinal reinforcement ratio increased.

The load-carrying capacity of reinforced geopolymer concrete columns was calculated using the moment-magnifier method incorporated in the daft Australian Standard for Concrete Structures AS 3600[9] and the American Concrete Institute Building Code ACI 318-02[10]. The calculated failure loads correlated well with the test values[8]. These results demonstrate that the methods of calculations used in the case of reinforced Portland cement concrete columns are applicable for reinforced geopolymer concrete columns.

The behavior and failure mode of reinforced geopolymer concrete beams were similar to those observed in the case of reinforced Portland cement concrete beams. The flexural capacity of beams was influenced by the concrete compressive strength and the tensile reinforcement ratio. The flexural strength of reinforced geopolymer concrete beams was calculated using the conventional flexural strength theory of reinforced concrete beams as described in standards and building codes such as the draft Australian Standard, AS 3600[9] and the ACI Building Code, ACI 318-02[10]. For beams with tensile reinforcement ratio of 1.18%, 1.84%, and 2.69%, the test and calculated values agreed well. In the case of beams with tensile steel ratio of 0.64%, as expected, the calculated values were conservative due to the neglect of the effect of strain hardening of tensile steel bars on the ultimate bending moment.

Mid-span deflection at service load of reinforced geopolymer concrete beams was calculated using the elastic bending theory and the serviceability design provisions given in the draft Australian Standard, AS 3600[9]. Good correlation of test and calculated deflections at service load was observed[8].

In all, these results demonstrated that reinforced heat-cured low-calcium (ASTM Class F) fly ash-based geopolymer concrete structural members can be designed using the design provisions currently used in the case of reinforced Portland cement concrete members.

ECONOMIC BENEFITS OF GEOPOLYMER CONCRETE

Heat-cured low-calcium fly ash-based geopolymer concrete offers several economic benefits over Portland cement concrete. The price of one ton of fly ash is only a small fraction of the price of one ton of Portland cement. Therefore, after allowing for the price of alkaline liquids, the price of fly ash-based geopolymer concrete is estimated to be about 10 to 30 percent cheaper than that of Portland cement concrete.

In addition, the appropriate usage of one ton of fly ash earns approximately one carbon-credit that has a redemption value of about 10 to 20 Euros. Based on laboratory test data[6,7,8], one ton low-calcium fly ash can be utilized to manufacture approximately 2.5 cubic meters of high quality fly ash-based geopolymer concrete, and hence earn monetary benefits through carbon-credit trade.

Furthermore, the very little drying shrinkage, the low creep, the excellent resistance to sulfate attack, and good sulfuric acid resistance offered by the heat-cured low-calcium fly ash-based geopolymer concrete may yield additional economic benefits when it is utilized in infrastructure applications.

CONCLUDING REMARKS

The paper presented information on heat-cured fly ash-based geopolymer concrete. Low-calcium fly ash (ASTM Class F) is used as the source material, instead of the Portland cement, to make concrete.

Low-calcium fly ash-based geopolymer concrete has excellent compressive strength and is suitable for structural applications. The salient factors that influence the properties of the fresh concrete and the hardened concrete have been identified. Data for the design of mixture proportion are included and illustrated by an Example.

The elastic properties of hardened geopolymer concrete and the behavior and strength of reinforced geopolymer concrete structural members are similar to those observed in the case of Portland cement concrete. Therefore, the design provisions contained in the current standards and codes can be used to design reinforced heat-cured low-calcium fly ash-based geopolymer concrete structural members.

Heat-cured low-calcium fly ash-based geopolymer concrete also shows excellent resistance to sulfate attack, good sulfuric acid resistance, undergoes low creep, and suffers very little drying shrinkage. The paper has identified several economic benefits of using geopolymer concrete.

REFERENCES

[1]Davidovits, J. (1994) "High-Alkali Cements for 21st Century Concretes. In Concrete Technology, Past, Present and Future." In proceedings of V. Mohan Malhotra Symposium. Editor: P. Kumar Metha, ACI SP- 144. pp. 383-397.

[2]McCaffrey, R. (2002). Climate Change and the Cement Industry. Global Cement and Lime Magazine (Environmental Special Issue), 15-19.

[3]Davidovits, J. (1988) "Soft Mineralogy and Geopolymers." In proceeding of Geopolymer 88 International Conference, the Université de Technologie, Compiègne, France.

[4]Gourley, J. T. (2003). Geopolymers; Opportunities for Environmentally Friendly Construction Materials. Paper presented at the Materials 2003 Conference: Adaptive Materials for a Modern Society, Sydney.

[5]Gourley, J. T., & Johnson, G. B. (2005). Developments in Geopolymer Precast Concrete. Paper presented at the International Workshop on Geopolymers and Geopolymer Concrete, Perth, Australia.

[6]Hardjito, D. and Rangan, B. V. (2005) Development and Properties of Low-Calcium Fly Ash-based Geopolymer Concrete, Research Report GC1, Faculty of Engineering, Curtin University of Technology, Perth, available at espace@curtin or www.geopolymer.org.

[7]Wallah, S.E. and Rangan, B.V. (2006), Low-Calcium Fly Ash-Based Geopolymer Concrete: Long-Term Properties, Research Report GC2, Faculty of Engineering, Curtin University of Technology, Perth, available at espace@curtin or www.geopolymer.org.

[8]Sumajouw, M.D.J. and Rangan, B.V. (2006), Low-Calcium Fly Ash-Based Geopolymer Concrete: Reinforced Beams and Columns, Research Report GC3, Faculty of Engineering, Curtin University of Technology, Perth, available at espace@curtin or www.geopolymer.org.

[9]Committee BD-002 Standards Australia (2005), Concrete Structures: Draft Australian Standard AS3600-200x, Standards Australia.

[10]ACI Committee 318 (2002), Building Code Requirements for Structural Concrete, American Concrete Institute, Farmington Hills, MI.

RATE OF SULPHURIC ACID PENETRATION IN A GEOPOLYMER CONCRETE

Xiujiang Song, Marton Marosszeky, Zhen-Tian Chang
School of Civil and Environmental Engineering, University of New South Wales
Sydney, NSW 2052, Australia

ABSTRACT

This paper presents the results of an experimental investigation to study the rate of sulphuric acid penetration in a Geopolymer concrete. Cubic samples were made with a fly ash based Geopolymer concrete, epoxy coated except for one surface and immersed in a 10% sulphuric acid. At each time after immersion in the acid for 7, 28, 56, 84, 140 and 180 days, two samples were split and tested with a Universal pH Indicator for pH variations across the sections. Three distinct layers, in the colour of pink, yellow and light blue, were observed on the sample sections with the pH value of approximate 3, 7 and 11 respectively. The yellowish transitional zone with the pH of approximate 7 was found to be rather narrow between the unaffected zone (light blue) and the acid attacked zone (pink). Electron Probe Microanalysis (EPMA) was conducted to study the distribution profiles of the element sodium (Na) and sulphur (S) in the selected sample areas. The results correlated well with the pH variations and the significant changes in the distribution profile of Na and S. Based on the experimental data up to 180 days, the acid affected depth including both the pink and yellow coloured zones were found to be proportional to the square root of immersion time. An empirical formula has been developed by regression analysis for evaluation of the rate of sulphuric acid penetration into the fly ash based Geopolymer concrete.

INTRODUCTION

Geopolymer concretes have been reported to have fairly high acid resistance (in terms of very low mass loss after acid exposure) in recent studies[1-4]. However, when exposed to an acid solution a Geopolymer concrete would reduce its alkalinity due to penetration of the acid into its pore system. The reduction of alkalinity at below surface depths in a Geopolymer concrete is a clear indication of acid attack. Reduction in alkalinity could also result in corrosion of steel reinforcement, if any, in a Geopolymer concrete. This investigation concerns the rate of sulphuric acid penetration in a Geopolymer concrete.

The alkalinity reduction phenomenon in a Geopolymer concrete exposed to sulphuric acid was previously studied using a chemical method[3]. A Universal pH Indicator was sprayed on split sample sections. Pink, yellow and light blue coloured layers were observed for the pH value of approximate 3, 7 and 11 respectively. The pH vale of 3 in the pink coloured zone indicates the ingress of sulphuric acid. Alternatively, the influence of sulphuric acid in the concrete matrix can be identified by an increase in the sulphur (S) content. It can be expected that relatively high sodium (Na) content should be detected in the unaffected Geopolymer concrete matrix because of the use of sodium hydroxide and sodium silicate in the mix design. This paper presents the results by the Electron Probe Microanalysis (EPMA) to study the distribution profiles of these two elements, sodium (Na) and sulphur (S), within the three coloured layers in a sample after immersion in 10% sulphuric acid. These quantitative results are then compared to the colours indicated by the Universal pH Indicator for an evaluation of the penetration rate of the sulphuric acid in the Geopolymer concrete.

MATERIALS AND METHODS

Materials and Geopolymer Concrete Mix

The Geopolymer binder used in this investigation is synthesised from two Class F fly ashes and a mixed alkaline activator of sodium hydroxide and sodium silicate. The median particle size is 20.4 micrometers for the fly ash No.1 and 3.14 micrometers for the fly ash No.2. The sodium hydroxide solution has the concentration of 14M (moles per litre). The sodium silicate solution is composed of 30.6% of SiO_2, 9.5% of Na_2O and 59.9% of H_2O by mass. Table 1 shows the mix design of the Geopolymer concrete used in this experiment. All raw materials were locally sourced and met the requirement of respective Australian standards.

Table 1: Mix Proportions of the Geopolymer concrete

Material types	Fly ash No.1	Fly ash No.2	Aqueous NaOH	Aqueous Na_2SiO_3	Silica sand	10mm Basalt
Mass ratio (%)	13.5	5.8	3.4	4.1	26.2	47.9

Geopolymer concrete samples were cast in 50mm steel cubic moulds and wrapped with plastic sheets. The samples were initially cured at 70°C for 12 hours in an oven followed by air curing in the laboratory at approximate 23°C. The resultant Geopolymer concrete had the 28-day compressive strength of 70MPa. At 21 days, twelve concrete cubes were coated with an acid resistant epoxy and only one surface of each sample was uncoated for exposure to acid. The samples were immersed in a 10% sulphuric acid at the concrete age of 28 days.

Alkalinity Examination by the Universal pH Indicator

At each time after immersion in the 10% sulphuric acid for 7, 28, 56, 84, 140 and 180 days, two concrete cubic samples were removed and rinsed briefly with tap water. A quarter of each concrete cube was cut off using a diamond saw (dry cut) perpendicularly to the uncoated surface. The saw cut concrete surfaces were then sprayed with the Universal pH Indicator to detect the alkalinity state by discolouration. Three distinct layers, in the colour of pink, yellow and light blue, were normally observed on the sample sections with the pH value of approximate 3, 7 and 11 respectively. The depth of the pink and yellow layers was measured from the acid exposed surface. The average of six readings taken from two samples each testing time was recorded as the "acid affected depth" at the testing time.

Determination of Element Distribution Profiles by EPMA

The Geopolymer concrete samples used for this study were immersed in a 10% sulphuric acid for 140 days. After a quarter of a cubic sample was cut away for the above alkalinity examination, the remaining three quarters of the sample was used for preparation of a polished sample (10×10×18mm) for element profile analysis by EPMA technique using a CAMECA SX50. Three representative areas on the sample for EPMA analysis were marked as "Edge",

"Centre" and "Front" with their physical locations shown in Figure 1. The "Front" area was further divided into A and B by the line (in yellow) for measurement of the "acid affected depth". Area A was in the unaffected zone (light blue colour) and the area B contained the yellow-coloured neutral boundary zone.

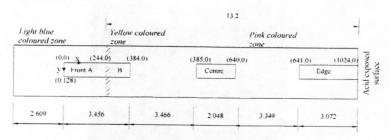

Length unit: mm, x and y coordinates: pixel

Figure 1: Areas for EPMA on a sample section after 140 days in 10% sulphuric acid

The location of each of the three scanned areas in Figure 1 is indicated in two ways. One is its distance to the acid exposed surface (right end), such as 0-3.072 mm for the "Edge" area. The other is expressed in pixels in the EPMA scanning analysis, for example, the "Edge" area is expressed with its x-pixels of 641-1024 and its y-pixels of 0-128. The EPMA results are plotted with the Matlab software into contour maps for two elements, sodium (Na) and sulphur (S). They were expressed by weight in percentage (wt%).

Microscopic Examination and Analysis by SEM and EDS

Separate samples were also cut from three layers in pink, yellow and light blue colour. A Hitachi 4500II Scanning Electronic Microscopy (SEM) coupled with an Oxford Energy-dispersive Spectrometer (EDS) was used to examine microscopic images and to undertake semi-quantitative analysis of the chemical composition of Geopolymer gels.

RESULTS

Alkalinity Reduction Pattern

When the Universal pH Indicator was sprayed on a saw cut concrete sample section, three distinct layers in the colour of pink, yellow and light blue were observed as shown in Figure 2. The outer layer near the uncoated surface directly exposed to acid solution was typically in pink colour with an approximate pH value of 3. Next to the pink layer was a narrow yellowish layer with a pH value of 7. Further inside the sample section showed a light blue colour with an approximate pH value of 11. The three coloured layers indicated the acid attacked zone (pink), the unaffected zone (light blue) and a narrow transitional zone (yellow) between them.

The depth of the pinkish layer was found to increase with the exposure time of samples immersed in the acid, while the yellowish layer remained to be a narrow layer. At each time after immersion in the acid for 7, 28, 56, 84, 140 and 180 days, the depth in a sample from the acid exposed surface to the innermost of the yellow layer was measured as the "acid affected depth".

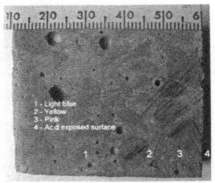

Figure 2: Three distinct colours on a saw cut sample section

Distribution of Element Na and S

Figure 3 presents the EPMA result as a two-dimensional distribution map of element Na in the three areas ("Front", "Centre" and "Edge") indicated in Figure 1 (note that the gaps between the three areas were not shown in Figure 3). The sample was analysed after 140 days immersion in a 10% sulphuric acid. According to the coloured scale bar, the Na content is much higher on the left side within pixels 0-288 range. However, it was noted that there were some blanks, such as those within pixels 0-128, due to the existence of aggregate particles, which contain very little Na.

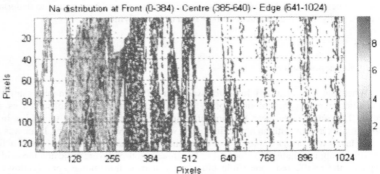

Figure 3: Two-dimensional distribution map of Na in three areas

Figure 4 shows the result in Figure 3 as a one-dimensional profile of the mean Na content along the X-direction. An abrupt drop in Na content occurred between pixels 256-288, which correlated with the narrow yellowish neutral zone on a sample section. Figure 4 clearly shows that there is very little Na left in the acid attacked pink zone (i.e. 288-1024 pixels). In the unaffected zone with light blue colour between pixels 0-120 in Figure 4, the relatively lower Na mean value is mainly due to the influence of aggregates.

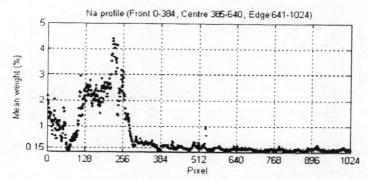

Figure 4: One-dimensional distribution profile of the mean Na in three areas

Figure 5 shows the one-dimensional profile of the mean sulphur (S) content. Higher S content was found in the acid attacked areas between pixels 288-1024.

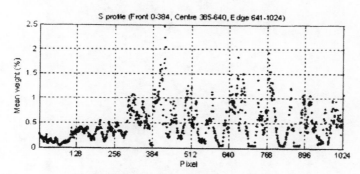

Figure 5: Distribution profile of mean S in three areas

To further study Na and S profiles across the "Front" area, a small area with very few aggregate particles was selected within x = 250-350 and y = 35-54 pixels in Figure 3. The EMPA results of the mean Na and S distributions in this selected area are presented in Figure 6. These profiles further verified significant changes in the Na and S content in the opposite manner across the transitional zone between the acid attacked and the unaffected zones; while the Na content dropped significantly in the area of x = 288 to 350 pixels, the S content increased in the same area.

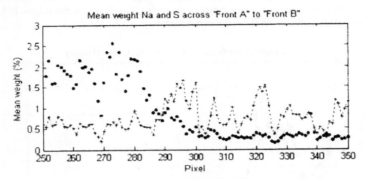

Figure 6: Distribution profiles of Na (•) with S (+) across transition zone

SEM images and EDS analyses

Typical SEM images in the three coloured layers on a sample section are presented in Figure 7 to Figure 9. The results of microscopic EDS analysis of the major elements of typical Geopolymer gel in each SEM image are presented in Table II.

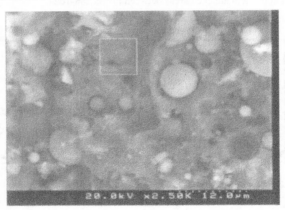

Figure 7: SEM image of Geopolymer gel in the light blue coloured zone

Figure 7 shows the solid and amorphous Geopolymer gel together with some un-reacted fly ash spheres in the light blue coloured zone, which has not been affected by sulphuric acid. By contrast, the matrix texture was very different in the acid attacked pink coloured zone. As shown in Figure 9, sulphate bearing deposits were observed at the bottom-right corner and, the chemical composition of the gel on the top-left corner included much higher S and much less Na (see Table II).

Figure 8: SEM image of Geopolymer gel in the yellow coloured zone

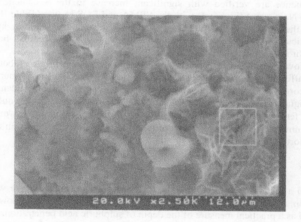

Figure 9: SEM image of Geopolymer gel in pink coloured zone

Table II: Major elements (atomic percentage) of Geopolymer gel from EDS analyses

Element	Na	Mg	Al	Si	S	Ca	Fe
Figure 7	6.3	0.2	10.2	22.2	0.9	0.7	3.3
Figure 8	5.1	0.8	7.7	21.4	1.6	0.3	8.6
Figure 9	0.4	0.2	5.8	26.1	6.4	0.1	4.2

In the yellowish transitional zone with a neutral pH value, the SEM image in Figure 8 was similar to that in Figure 7. The Na and S contents in the typical gel were between those in the acid attacked and unaffected zones but more closer to that in the unaffected zone. It is noted in Table II that a higher Fe content was found in the yellowish neutral transitional zone.

DISCUSSION

In the alkaline examination three distinct layers with pink, yellow and light blue colours were observed on the sample sections with the pH value of approximate 3, 7 and 11 respectively. These were classified as the acid attacked zone, transitional zone and unaffected zone.

In the previous research work with this fly ash based Geopolymer concrete[5], alkali Na appear to exist in two forms. While most sodium cations balance the negative charges of bridging oxygen created in formation of Al-O-Si bonds, there are some residual sodium hydroxides left from the initially alkaline reactions. Accordingly, the original Geopolymer concrete is alkaline with a certain level of sodium content. This alkaline environment (pH value of about 11) in the unaffected zone has been confirmed in both the alkaline examination test (light blue colure in Figure 2) and with the relatively high Na content in the EPMA analysis (Figures 3 and 4).

In the pink coloured zone (Figure 2) with a low pH of as low as 3, the sulphuric acid attack and influence are verified with significant increase in the S content (Figure 5) and reduction in the Na content (Figure 4) with the EPMA and EDS results (Table II). The very low Na content in the pink zone indicated that not only the residual sodium hydroxides were neutralised by acids but also most Na cations in the S-O-Al structures had been replaced by hydrogen ions through ion- exchanged reactions[6].

The yellowish transitional zone (Figure 2) with a neutral pH value (about 7) was found between the unaffected zone and the sulphuric acid attacked zone. This rather narrow transitional zone was well correlated with the dramatic drop in the Na content and significant increase in the S content in Figure 6. The yellow colour shown in the transitional zone could be related to precipitation of ferric salts[7-8], which are not soluble in a neutral solution but would be dissolved in acidic environments. The higher Fe content found in the yellowish transition zone in Table II provided a support to this judgement.

APPLICATION

The acid affected depth D (in mm) in a Geopolymer concrete sample was measured in this investigation. It includes both the acid attacked low pH zone and the narrow transition zone and is then used as a practical measure of the depth of sulphuric acid penetration.

The depth D in this experimental investigation was found to be proportional with the square root of the exposure time (t in day) in a 10% sulphuric acid when measured after the immersion for 7, 28, 56, 84, 140 and 180 days. The following equation (1) has been derived from the regression analysis of the experimental data in Figure 10. The introduction of the square root formula is based on and correlated well with the Fick's law of diffusion.

$$D = 1.14\sqrt{t} \tag{1}$$

Such a linear relationship in Figure 10 is similar to that of the neutralisation depth of Portland cement concrete samples immersed in the acidic water containing carbon dioxide[9] and, to the typical empirical equations of carbonation depth of normal concrete in atmospheric environments containing carbon dioxide gas[10].

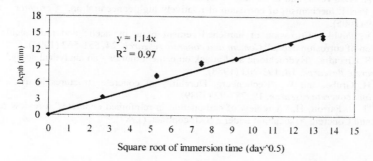

Figure 10: Acid penetration depth against square root of time in sulphuric acid

CONCLUSIONS

On a fly ash based Geopolymer concrete sample after sulphuric acid exposure, three distinctly coloured zones were identified on sample sections with the Universal pH Indicator. A narrow transitional zone (in yellow) with a neutral pH value was found between the unaffected zone (in light blue) with a high pH of 11 and the acid attacked zone (in pink) with a low pH value of 3. The results of the element analysis by EPMA technique correlated well with the chemical analysis method. The highest sodium (Na) appeared in the unaffected zone and dramatically reduced across the transitional zone. By contrast, a significant increase in the sulphur (S) content appeared in the acid attacked zone.

The acid affected depth, including both the acid attacked low pH zone and the neutral transitional zone, is used as a practical measure of the depth of sulphuric acid penetration. Based on the experimental data up to 180 days, the acid affected depth is found to be proportional to the square root of time in 10% sulphuric acid. A formula of $D = 1.14\sqrt{t}$ has been derived by the regression analysis to evaluate the rate of sulphuric acid penetration in the Geopolymer concrete.

REFERENCES

[1]B. Bakharev, "Resistance of Geopolymer materials to acid attack," *Cement and concrete research*, **35**, 658-70 (2005).

[2]S. E. Wallah and B. V. Rangan, "Low-calcium fly ash based Geopolymer concrete: long-term properties," *Research report GC2, Faculty of Engineering, Curtin university of technology*, (2006).

[3]X. J. Song, M. Marosszeky, M. Brungs, and Z. T. Chang, "Response of Geopolymer concrete to sulphuric acid attack," *The 4th World Congress Geopolymer*, 157-60 (2005).

[4]J. Davidovits, D. C. Comrie, J. H. Paterson, and D. J. Ritcey, "Geopolmeric concretes for environmental protection," *ACI Concrete International*, **12**, 30-40 (1990).

[5]X. J. Song, "Development and performance of class F fly ash based Geopolymer concretes against sulphuric acid attack," PhD thesis, March 2006.

[6]A. Allahverdi, and F. Skvara, "Sulphuric acid attack on hardened paste of Geopolymer cements Part I: mechanism of corrosion at relatively high concentration," *Ceramics – silikaty*, **49**, 225-9 (2005).

[7]V. Pavlik, "Corrosion of hardened cement paste by acetic and nitric acids part I: calculation of corrosion depth," *Cement and concrete research*, **24**, 551-562 (1994).

[8]S. Chandra, "Hydrochloric acid attack on cement mortar - an analytical study," *Cement and Concrete Research*, **18**, 193-203 (1988).

[9]H. Grube, and W. "Rechenberg, Durability of concrete structures in acidic water," *Cement and concrete research*, **19**, 783-92 (1989).

[10]L. J. Parrott. H. "A review of carbonation in reinforced concrete," a review report by Cement and Concrete Association under a BRE contract, (1987).

CORROSION PROTECTION ASSESSMENT OF CONCRETE REINFORCING BARS WITH A GEOPOLYMER COATING

W. M. Kriven and M. Gordon
Department of Materials Science and Engineering
University of Illinois at Urbana-Champaign
1304 East Green Street
Urbana, IL, 61801

B. L. Ervin and H. Reis
Department of Industrial and Enterprise Systems Engineering
University of Illinois at Urbana-Champaign
104 South Mathews Avenue
Urbana, IL, 61801

ABSTRACT
Steel reinforcing bars (rebars), with and without a geopolymer coating, were embedded in mortar and subjected to accelerated general corrosion using impressed current. The percent corrosion (mass loss) of the bar was estimated using Faraday's law *in situ* and confirmed by direct measurement of mass loss after extraction.

Results are very encouraging and indicate that the geopolymer coating provides the means to reduce activity of the corrosion process. This reduction in corrosion activity would result in a decrease in maintenance and rehabilitation efforts and in an increase in service life of steel reinforced concrete structures with respect to corrosion. For example, after nine days of accelerated corrosion, the mortar specimen reinforced with a geopolymer-coated bar had about 5% corrosion, while the mortar specimen reinforced with a bar without a geopolymer coating had accumulated over 20% corrosion. Both reinforced mortar specimens were monitored using guided mechanical waves to nondestructively assess the interface conditions between the steel and concrete, with results provided.

INTRODUCTION
Structures made of steel reinforced concrete (RC) are aging and deteriorating at a rapid pace. One of the more destructive effects is the corrosion of reinforcing steel that give these structures their strength.[1,2] This is particularly true in areas with marine environment and road salt deposits. With an increase in the number of RC bridge collapses caused by corrosion within the past decade, many of which were well below their anticipated design lives, the problem is becoming an epidemic. The direct costs associated with corrosion of RC bridges in the U.S. alone are estimated at \$8.33 billion per year[3]. Indirect costs (e.g., productivity loss, user inconvenience) are estimated at nearly ten times the direct costs[3]. There is a clear risk to public safety, with substantial financial incentive to develop advanced approaches to eliminating corrosion attack to RC structures. To aid in this development, a geopolymer was investigated for use as a rebar coating prior to being embedded in concrete.

Geopolymers are cementitious materials made at ambient temperatures from aluminosilicate materials (e.g. metakaolin, fly ash) and highly caustic, aqueous solutions of sodium or potassium silicate[4]. The macro/microstructures of geopolymers are highly heterogeneous, consisting of the reacted geopolymer binder, pore space, unreacted aluminosilicate material, and weakly and strongly bound water.[5] The geopolymer binder[4,6] is a rigid, inorganic, aluminosilicate, hydrated gel consisting of cross-linked AlO_4^- and SiO_4 tetrahedra, where charge neutrality is maintained by Group I cations such as Na^+, K^+ or Cs^+. The geopolymer binder phase is X-ray amorphous and consists exclusively of Al-O-Si and Si-O-Si bonds.[4,6] Geopolymers are chemically compatible with cements and concretes.

Regardless of the exact structure, once hardened, geopolymers have a number of attractive properties such as high compressive strength relative to standard Portland cement (60-70 MPa), fire-resistance, acid-resistance, and rapid-set time[7-26]. The high degree of chemical durability combined with the low cost of fabrication has made geopolymers attractive for waste encapsulation applications[27-29]. Geopolymers may also represent a "green", low CO_2 alternative to conventional cements. Davidovits has estimated that the manufacture of geopolymer cements produces approximately six times less CO_2 than the production of traditional Portland cement[30].

To explore the feasibility of using geopolymer coatings on reinforcing bars for corrosion protection, mortar specimens reinforced with a steel bar with and without a geopolymer coating were corroded in a uniform manner using impressed current. While both types of specimen were undergoing accelerated general corrosion, guided mechanical waves nondestructively monitored the interface conditions between the steel and concrete.

EXPERIMENTAL PROCEDURE

The same mortar mix and rebar type was used for both specimens. The mortar mix had the following weight proportions of sand, Type I cement, and water: 2.11:1:0.45. The reinforcing steel bars used were ½ inch (12.7 mm) in diameter and 42 inches (1.07 m) in length. The mortar specimens had dimensions of 4 x 4 inches (102 x 102 mm) in cross-section and 36 inches (91 cm) of continuous mortar. The rebar was exposed 3 inches (76 mm) on each end.

The geopolymer used to coat the rebar was prepared by mixing MetaMax metakaolin powder (Engelhard Corporation, Iselin, NJ) with a potassium silicate solution ($K_2O \cdot 2SiO_2 \cdot 10 H_2O$) using a high-speed dispersion mixer for 15 minutes. It was assumed that the metakaolin had a stoichiometric composition of $Al_2O_3 \cdot 2 SiO_2$, giving the final geopolymer a nominal composition of $K_2O \cdot Al_2O_3 \cdot 4 SiO_2$. The rebar was positioned horizontally in wooden forms and the geopolymer was applied using Nitrile gloves leading to an approximately 0.039 inch (1 mm) thick coating. The rebar was coated with geopolymers before the mortar mix was poured into the form.

The reinforcement was weighed prior to being embedded in the mortar specimen. The experimental setup for accelerated corrosion testing (commonly referred to as impressed current or anodic polarization) is shown in Figure 1. While others researchers[31-38] have used slight variations, the basic premise is to partially submerge the reinforced mortar specimen in a bath of salt water (5% NaCl). This allows for oxygen, water, and chlorides to reach the surface of the rebar. For a faster corrosion rate, a positive current was supplied to the rebar (anode) to draw the negatively charged chlorides. A negative current was supplied to the surrounding copper mesh (cathode) to complete the circuit and the specimens were driven potentiostatically by applying 30 volts. To induce a more uniform corrosion, the specimens were wrapped in a thin cloth centered on the specimen for a length of 30 inches (76.2 cm), and a small pump was used to pour salt water onto the top of the mortar specimens. Please notice that for the accelerated corrosion tests, the specimens were completely saturated with water and then partially submerged in a salt bath of 5% NaCl.

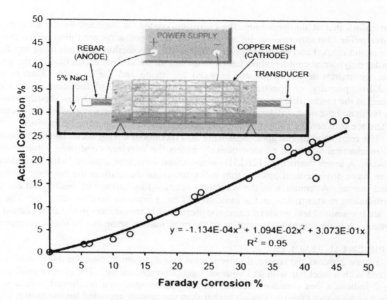

Figure 1: Measured corrosion after bar extraction as a function of predicted corrosion using Faraday's law for accelerated corrosion testing of the reinforced mortar specimens. The accelerated corrosion test setup is shown along with the transducer arrangement used to transmit and receive guided mechanical waves.

A constant potential was applied with the current level continuously monitored using an ammeter. Faraday's law provides a relationship between the time of a sustained current level and the amount of steel weight loss. The weight of metal lost from corrosion is calculated theoretically by:

$$\Delta w = \frac{AIt}{ZF} \tag{1}$$

where A is the atomic weight of Iron (56 grams), I is the corrosion current (amp), t is the time elapsed (seconds), Z is the valency of the reacting electrode (2 for iron), and F is Faraday's constant (96,500 amp-sec). The original bar weight w is measured on a scale before embedding in the mortar. The amount of corrosion is expressed as the following:

$$\text{Corrosion} \% = 100 * \frac{\Delta w}{w} \tag{2}$$

After each specimen has undergone accelerated corrosion, the rebar was extracted and the actual mass loss was measured. Figure 1 shows the relationship between the actual percentage of corrosion sustained to the rebar and the predicted percentage using Faraday's law. The third order polynomial trendline provides a coefficient of determination of 0.95 and is used for the estimation of corrosion percentage during accelerated corrosion testing. The non-linearity of the relation shows that at the beginning. the corrosion process is impeded by the mortar and the protective oxide film surrounding the rebar. However. as pathways are found within the mortar (via pores and cracks) and the oxide layer is destroyed, the reaction proceeds more rapidly. The third order polynomial captures this change in the rate of the corrosion reaction. In the corrosion process. the mortar acts as a resistor between the anode and the cathode. Factors such as permeability, porosity. compaction of the mortar matrix, purity of the steel rebar, electrical properties of the mortar fine aggregates, salt deposits, and the overall electrical path between the copper mesh and the mortar determine to a large extent how much resistance will be present, and the difference between the Faraday's estimation of corrosion and the actual corrosion[16].

The reinforced mortar specimens were also monitored during accelerated corrosion using guided mechanical waves to nondestructively assess the interface conditions between the steel and mortar. A lower frequency (128 kHz) longitudinal wave was invoked in the embedded rebar to create wave displacement profiles that will interact substantially at the interface between the steel and mortar. Attenuation of the waveform is caused by leakage of mechanical energy into the surrounding mortar matrix and is indicative of bond between the two materials. The guided waves are transmitted and received using the piezoelectric transducers in the arrangement shown in Figure 1. A more complete description of the experimental setup can be found in reference[23].

EXPERIMENTAL RESULTS

Figure 2 shows the recorded current level as a function of time for the mortar specimen reinforced with a steel bar with and without a geopolymer coating. The current level shown in Figure 2 indicates that corrosion activity in the mortar specimen reinforced with a steel bar without a geopolymer coating is much higher than the activity recorded for the mortar specimen reinforced with a steel bar coated with geopolymers.

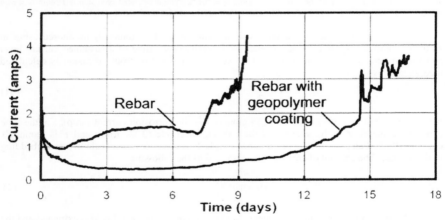

Figure 2: Impressed current as a function of time for the mortar specimen reinforced with steel bars with and without a geopolymer coating. Both specimens were completely saturated in water and then partially submerged in a salt bath of 5% NaCl. Copper mesh wrapped around the mortar specimens was used as the cathode. Both specimens were driven potentiostatically (30 volts).

Figure 3 shows the percent corrosion as a function of time for the mortar specimen reinforced with a bar with and without a geopolymer coating. For a mortar specimen reinforced with a bar without a geopolymer coating, it requires nine days to reach approximately 25% corrosion, while for the mortar specimen reinforced with a geopolymer coated bar, it takes approximately 17 days to reach the same corrosion level. The results show a delay in corrosion of almost twice as long caused by the geopolymer coating. Figure 3 also shows that after nine days of accelerated corrosion the mortar specimen reinforced with a geopolymer coated bar had about 5% corrosion while the mortar specimen reinforced with a bar without a geopolymer coating had over 20% corrosion.

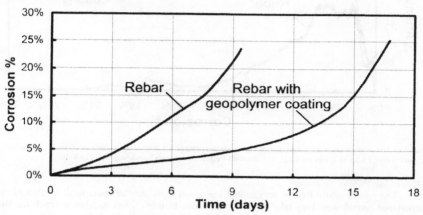

Figure 3: Corrosion percentage as a function of time for mortar specimens reinforced with steel bars with and without a geopolymer coating. The potential was kept constant at 30 volts.

A discussion of guided mechanical waves and their use towards a corrosion monitoring system can be found in references[39-41]. Figure 4 shows the wave attenuation of an input frequency of 128 kHz as a function of corrosion percentage for mortar specimens reinforced with steel bars with and without a geopolymer coating. There is wave attenuation for both reinforced mortar specimens when corrosion product (rust) first starts to accumulate between the steel and mortar interface. Wave energy leaks out into the mortar matrix as the pressure from the corrosion product builds and the bond between the steel and mortar increases. Eventually, the mortar specimen reinforced with a steel bar without a geopolymer coating cracks, the pressure drops, and the amount of wave energy leaking into the mortar decreases. The mortar specimen reinforced with a geopolymer coated bar did not crack from corrosion product accumulation or have a decrease in waveform attenuation. Because the specimens did not crack from the corrosion products accumulation, the monitoring effort was only possible up to about six percent mass loss under the current setup. The results suggest that the interface conditions between the steel and mortar were substantially affected by application of a geopolymer coating.

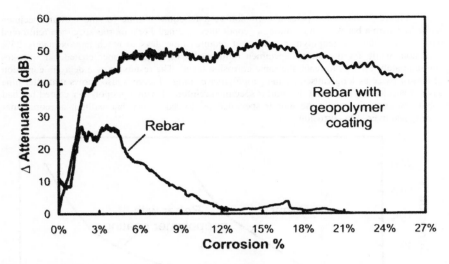

Figure 4: Wave attenuation as a function of corrosion percentage for mortar specimens reinforced with a steel bar with and without a geopolymer coating. The invoked frequency for both specimens was 128 kHz.

CONCLUSIONS

The results show that in accelerated corrosion tests. mortar specimens reinforced with geopolymer coated steel bars take approximately two to three times as long to reach the same level of percent corrosion when compared with mortar specimens reinforced with steel bars without geopolymer coatings. While the tentative results indicate that from the corrosion point of view it may be beneficial to coat the rebar with geopolymers, its effects upon the bond between the rebar and the surrounding concrete/mortar is not fully understood yet. Because the level of bond will affect the mechanics of load transfer between the rebar and the surrounding concrete/mortar, caution needs to be exercised. Also. the corrosion induced by impressed current does not mimic the corrosion (either general or pitting) found in real structures.

It should also be noted that the results presented in this paper are for a particular mortar mix, rebar size, temperature, under no applied external loading. no wet/dry cycling, mainly uniform accelerated corrosion. and one particular guided wave setup. Extending this type of procedure for changes in these factors has not yet been explored.

ACKNOWLEDGEMENTS

The work of Kriven and Gordon was supported by the Air Force Office of Scientific Research Contract number FA9550 04-C-0063. The work of Reis and Ervin was partially supported by the National Science Foundation under Grant No. CMS 02-01305.

REFERENCES

[1] *Uhlig's Corrosion Handbook*, 2nd Edition, Ed. R.W. Revie, John Wiley, N. Y. (2000).

[2] A. Aktan, and K. Grimmelsman, "The Role of NDE in Bridge Health-Monitoring," *Proc. SPIE*, **3587**, 2-15 (1999).

[3] G.H. Koch, M.P.H. Brongers, N.G. Thompson, Y.P. Virmani, and J.H. Payer, "Corrosion Cost and Preventive Strategies in the United States," FHWA-RD-01-156, Federal Highway Administration (2001).

[4] J. Davidovits, "Geopolymers - Inorganic Polymeric New Materials," *J. Therm. Anal.*, **37**, 1633-1656 (1991).

[5] I. Lecomte, M. Liegeois, A. Rulmont, R. Cloots, and F. Maseri, "Synthesis and Characterization of New Inorganic Polymeric Composites Based on Kaolin or White Clay and on Ground-granulated Blast Furnace Slag," *J. Mater. Res.*, **18**, 2571-2579 (2003).

[6] V. F. F. Barbosa, K. J. D. MacKenzie, and C. Thaumaturgo, "Synthesis and Characterisation of Materials Based on Inorganic Polymers of Alumina and Silica: Sodium Polysialate Polymers," *Int. J. Inorg. Mater.*, **2**, 309-317 (2000).

[7] W. M. Kriven, J. L. Bell and M.Gordon, "Microstructure and Microchemistry of Fully-Reacted Geopolymers and Geopolymer Matrix Composites," *Ceramic Transactions*, **153**, 227-250 (2003).

[8] D. C. Comrie and W. M. Kriven, "Composite Cold Ceramic Geopolymer in a Refractory Application," *Ceramic Transactions*, **153**, 211-225 (2003).

[9] W. M. Kriven and J. L. Bell, "Effect of Alkali Choice on Geopolymer Properties," *Cer. Eng. and Sci. Proc.*, **25**, 99-104 (2004).

[10] W. M. Kriven, J. L. Bell and M. Gordon, "Geopolymer Refractories for the Glass Manufacturing Industry," *Cer. Eng. and Sci. Proc.*, **25**, 57-79 (2004).

[11] J. L. Bell and W. M. Kriven, "Nanoporosity in Geopolymeric Cements," Microscopy and Microanalysis '04, (Proc. 62nd Annual Meeting of Microscopy Society of America), **10**, 590-591 (2004).

[12] M. Gordon, J. Bell and W. M. Kriven, "Comparison of Naturally and Synthetically-Derived, Potassium-Based Geopolymers," *Ceramic Transactions*, **165**. Advances in Ceramic Matrix Composites X, edited by J. P. Singh, N. P. Bansal and W. M. Kriven 95-106 (2005).

[13] P. Duxson, G. C. Lukey, J. S. J. van Deventer, S. W. Mallicoat and W. M. Kriven, "Microstructural Characterization of Metakaolin-based Geopolymers," *Ceramic Transactions*, **165**, Advances in Ceramic Matrix Composites X, edited by J. P. Singh, N. P. Bansal and W. M. Kriven 71-85 (2005).

[14] P. Duxson, J. L. Provis, G. C. Lukey, S. W. Mallicoat, W. M. Kriven and J. S. J. van Deventer, "Understanding the Relationship between Geopolymer Composition, Microstructure and Mechanical Properties," *Colloids and Surfaces A – Physicochemical and Engineering Aspects*, **269**, 47-58 (2005).

[15] J. L. Bell, M. Gordon and W. M. Kriven, "Use of Geopolymeric Cements as a Refractory Adhesive for Metal and Ceramic Joins," *Cer. Eng. Sci. Proc.*, Edited by D.-M. Zhu, K. Plucknett and W. M. Kriven, **26**, 407-413 (2005).

[16] S. Mallicoat, P. Sarin and W. M. Kriven, "Novel Alkali-Bonded Ceramic Filtration Membranes," *Cer. Eng. Sci. Proc.*, Edited by M. E. Brito, P. Filip, C. Lewinsohn, A. Sayir, M. Opeka, W. M. Mullins, D.-M. Zhu and W. M. Kriven, **26**, 37-44 (2005).

[17] M. Gordon, J. Bell and W. M. Kriven, "Thermal Conversion and Microstructural Evaluation of Geopolymers or "Alkali Bonded Ceramics" (ABCs)," *Ceramic Transactions*, **175**. Advances in

Ceramic Matrix Composites XI. Edited by N.P. Bansal, J.P. Singh and W. M. Kriven, 215- 224 (2005).

[18]J. L. Provis, P. Duxson, G. C. Lukey, F. Separovic, W. M. Kriven and J. S. J. van Deventer, "Modeling Speciation in Highly Concentrated Alkaline Silicate Solutions," *Industrial and Engineering Chemistry Research*, **44**, 8899-8908 (2005)

[19]W. M. Kriven, J. Bell, M. Gordon and Gianguo Wen, "Microstructure of Geopolymers and Geopolymer-based Materials," pp 179-183 in Geopolymer, Green Chemistry and Sustainable Development Solutions, edited by Joseph Davidovits. Proc. World Congress Geopolymer, 2005, St. Quentin, France. Published by the Geopolymer Institute, St. Quentin, France (2005).

[20]P. Duxson, S. W. Mallicoat G. C. Lukey, W. M. Kriven and J. S. J. van Deventer, "Effect of Alkali and Si/Al Ratio on the Development of Mechanical Properties of Metakaolin-based Geopolymers," *Colloids and Surfaces A-Physicochemical and Engineering Aspects*, in press (2006).

[21]J. L. Bell, P. Sarin and W. M. Kriven, "Formation of Nanocrystalline Zeolites in Geopolymer Gels," Microscopy and Microanalysis '06, (Proc. **64**[th] Annual Meeting of Microscopy Society of America), **12** (2006).

[22]J. L. Bell, M. Gordon and W. M. Kriven, "Nanoporosity and Microporosity in Geopolymer Gels," Microscopy and Microanalysis '06, (Proc. 64th Annual Meeting of Microscopy Society of America), 12 (2006).

[23]W. M. Kriven, J. L. Bell and M. Gordon. "Microstructure and Nanoporosity in As-set Geopolymers," Mechanical Properties and Performance of Engineering Ceramics II, *Cer. Eng. Sci. Proc.*, Cocoa Beach, **27**, Issue 2, pp 313-324 (2006).

[24]A. Allahverdi and F. Skvara, "Nitric Acid Attack on Hardened Paste of Geopolymeric Cements - Part 1," *Ceramics-Silikaty*, **45**, 81-88 (2001).

[25]A. Allahverdi and F. Skvara, "Nitric Acid Attack on Hardened Paste of Geopolymeric Cements - Part 2," *Ceramics-Silikaty*, **45**, 143-149 (2001).

[26]T. Bakharev, "Resistance of Geopolymer Materials to Acid Attack," *Cem. Con. Res.*, **35**, 658-670 (2005).

[27]J. W. Phair, J. S. J. van Deventer, and J. D. Smith, "Effect of Al Source and Alkali Activation on Pb and Cu Immobilisation in Fly-ash Based "Geopolymers," *Appl. Geochem.*, **19**, 423-434 (2004).

[28]J. G. S. Van Jaarsveld, J. S. J. Van Deventer, and A. Schwartzman, "The Potential Use of Geopolymeric Materials to Immobilise Toxic Metals: Part II. Material and Leaching Characteristics," *Minerals Engineering*, **12**, 75-91 (1999).

[29]J. G. S. VanJaarsveld, J. S. J. VanDeventer, and L. Lorenzen, "The Potential Use of Geopolymeric Materials to Immobilise Toxic Metals .1. Theory and Applications," *Miner. Eng.*, **10**, 659-669 (1997).

[30]J. Davidovits, "Environmentally Driven Geopolymer Cement Applications."; in Geopolymers 2002: Turning Potential into Profit. CD publication, edited by J. Davidovits (2002).

[31]G. Al-Sulaimani, M. Kaleemullah, I. Busanbul, and F. Rasheeduzzafar, "Influence of Corrosion and Cracking on Bond Behavior and Strength of Reinforced Concrete Members," *ACI Structural J.*, **87**, 220-31 (1990).

[32]J. Cabrera, "Deterioration of Concrete Due to Reinforcement Steel Corrosion," *Cement and Composites*, **18**, 47-59 (1996).

[33]P. Mangat and M. Elgarf, "Flexural Strength of Concrete Beams with Corroding Reinforcement," *ACI Structural J.*, **96**, 149-58 (1999).

[34]L. Amleh and S. Mirza, "Corrosion Influence on Bond between Steel and Concrete," *ACI Structural J.*, **96**, 415-23 (1999).

[35]K. Stanish, R. Hooton and S. Pantazopoulou, "Corrosion Effects on Bond Strength in Reinforced Concrete," *ACI Structural J.*, **96**, 915-21 (1999).

[36]Y. Auyeung, P. Balaguru and L. Chung, "Bond Behavior of Corroded Reinforcement Bars," *ACI Materials J.*, **97**, 214-20 (2000).

[37]C. Lee, J. Bonacci, M. Thomas, M. Maalej, S. Khajehpour, N. Hearn, S. Pantazopoulou and S. Sheikh, "Accelerated Corrosion and Repair of Reinforced Concrete Columns using Carbon Fibre Reinforced Polymer Sheets," *Can. J. Civ. Eng.*, **27**, 941-48 (2000).

[38]K. Soudki and T. Sherwood, "Behaviour of Reinforced Concrete Beams Strengthened with Carbon Fibre Reinforced Polymer Laminates Subjected to Corrosion Damage," *Can. J. Civ. Eng.*, **27**, 1005-10 (2000).

[39]B. Ervin, D. Kuchma, and J. Bernhard, and H. Reis, "Estimation of General Corrosion Damage to Steel Reinforced Mortar using Frequency Sweeps of Guided Mechanical Waves," *INSIGHT-The Journal of the British Institute for Non-Destructive Testing.* **48**, 682-692 (2007).

[40]H. Reis, B. Ervin, D. Kuchma, and J. Bernhard, "Estimation of Corrosion Damage in Steel Reinforced Mortar Using Guided Waves," *ASME J. of Pressure Vessel Technology*, **127**, 255-61 (2005).

[41]K. Hietpas, B. Ervin, J. Banasiak, D. Pointer, D. Kuchma, H. Reis, and J. Bernhard, "Ultrasonics and Electromagnetics for a Wireless Corrosion Sensing System Embedded in Structural Concrete," *Smart Structures and Systems*, **1**, 267-82 (2005).

Author Index

Alexopoulos, S., 95
Araki, R., 159

Baba, H., 77
Balaguru, P., 321
Baumann, M. J., 197
Bell, J. L., 273
Bernardo, E., 3
Biasetto, L., 3
Boddapati, S. R., 57
Bordia, R. K., 57
Boxley, C., 321
Breitbach, G., 95
Buchwald, A., 257

Caruso, F., 273
Case, E. D., 197
Chang, Z.-T., 363
Cheng, Y., 209
Colombo, P., 3
Costacurta, S., 3
Courtney, L., 209

Davis, J., 309
Dombrowski, K., 257

Ervin, B. L., 373

Fletcher, R. A., 249
Fransaer, J., 13

Garcia-Luna, A., 283
Garrido, L., 41, 49
Genova, L., 41, 49
Gordon, M., 373

Heard, D., 41
Hinoki, T., 89
Hoffschmidt, B., 95
Huang, W., 171

Ikeda, K., 293
Itoh, S., 159

Jain, H., 183
Janney, M. A., 105
Jia, D., 67
Johnson, A. P. R., 273

Kawashita, M., 159
Keyte, L. M., 235
Kim, D., 67
Kiyama, S., 309
Kohyama, A., 89
Komphanchai, S., 249
Kriven, W. M., 67, 273, 373
Krivenko, P., 221
Krivenko, P. V., 235

Lawson, J. L., 337
Li, C. G., 27

Lloyd, R. R., 235
Lorenzen, L., 235

MacKenzie, K. J. D., 249
Marosszeky, M., 363
Miyamoto, K., 159
Miyazaki, T., 151
Moawad, H. M. M., 183
Muntingh, Y., 235

Nair, B., 321
Neirinck, B., 13

Ohtsuki, C., 151

Panandiker, R. S. P., 337
Park, J.-S., 89
Park, Y.-H., 89
Peña-Alonso, R., 3
Perera, D. S., 309
Pippel, E., 3
Plucknett, K. P., 41, 49
Provis, J. L., 235
Pyzik, A. J., 27

Quinlan, M., 41, 49

Rangan, B. V., 347
Reis, H., 373

Sermon, P. A., 209

Shrotri, 337
Smith, I. O., 197
Soare, L. C., 283
Song, X., 363
Sorarù, G. D., 3
Stafford, R. J., 113
Stobbe, P., 95
Sugino, A., 151
Suzuki, A., 77

Takaoka, G. H., 159
Tas, A. C., 135
Teixeira-Pinto, A., 337

Unuma, H., 127

Vakifahmetoglu, C., 3
Vance, E. R., 309
Van der Biest, O., 13
van Deventer, J. S. J., 235
Varela, B., 337
Vleugels, J., 13

Wang, R., 113
Weil, M., 257
Woltersdorf, J., 3

Xu, H., 235

Yee, P., 309

Printed and bound by CPI Group (UK) Ltd, Croydon, CR0 4YY

16/04/2025

14658450-0005